Grundlehren der mathematischen Wissenschaften 235

A Series of Comprehensive Studies in Mathematics

E.B. Dynkin
A.A. Yushkevich

Controlled
Markov Processes

With 36 Figures

Springer-Verlag
Berlin   Heidelberg   New York

E. B. Dynkin

Department of Mathematics, White Hall, Cornell University.
Ithaca, New York, USA

A. A. Yushkevich

Department of Applied Mathematics, MIIT, ul. Obrazscova 15,
Moscow, USSR

Translators:

J. M. Danskin, C. Holland

Originally published as *Upravliaemye markovskie protsessy i ikh prilozheniia*
by Nauka, Moscow, 1975.

AMS Subject Classifications: 49E99, 93E XX, 60JOS, 60J20, 62M05

ISBN 0-387-90387-9   Springer-Verlag   New York   Heidelberg   Berlin
ISBN 3-540-90387-9   Springer-Verlag   Berlin   Heidelberg   New York

Library of Congress Cataloging in Publication Data
Dynkin, Evgenii Borisovich, 1924–
   Markov control processes and their applications.
   (Grundlehren der mathematischen Wissenschaften; 235)
   Translation of Upravliaemye markovskie protsessy i ikh prilozheniia.
   Bibliography: p.
   Includes index.
   1. Mathematical optimization.   2. Control theory.   3. Markov processes.   I. IUshkevich, Aleksandr Adol'fovich, 1930–
joint author.   II. Title.   III. Series: Die Grundlehren der mathematischen Wissenschaften in Einzeldarstellungen; 235.
QA402.5.D9413   519.2′33   79–1465

Printed in the United States of America.

9 8 7 6 5 4 3 2 1

Preface

This book is devoted to the systematic exposition of the contemporary theory of controlled Markov processes with discrete time parameter or in another terminology multistage Markovian decision processes. We discuss the applications of this theory to various concrete problems. Particular attention is paid to mathematical models of economic planning, taking account of stochastic factors.

The authors strove to construct the exposition in such a way that a reader interested in the applications can get through the book with a minimal mathematical apparatus. On the other hand, a mathematician will find, in the appropriate chapters, a rigorous theory of general control models, based on advanced measure theory, analytic set theory, measurable selection theorems, and so forth. We have abstained from the manner of presentation of many mathematical monographs, in which one presents immediately the most general situation and only then discusses simpler special cases and examples. Wishing to separate out difficulties, we introduce new concepts and ideas in the simplest setting, where they already begin to work. Thus, before considering control problems on an infinite time interval, we investigate in detail the case of the finite interval. Here we first study in detail models with finite state and action spaces—a case not requiring a departure from the realm of elementary mathematics, and at the same time illustrating the most important principles of the theory. The results obtained for finite models are then carried over to a wider class of semicontinuous models, which is distinguished by means of hypotheses of topological character. Here we use the simplest facts about continuous functions and metric spaces, and about the Lebesgue integral. Finally, we study the most general case—the Borel models—which imposes considerably more requirements on the mathematical preparation of the reader. Some of the difficulties we encounter arise already for countable models. We consider them first. More serious complications connected with measurability problems are overcome by using the apparatus of analytic sets and isomorphism theorems for Borel spaces; the necessary results are proved in the Appendix. Such a system of exposition requires some repetition. As a rule a proof is presented in detail for the simplest class of models to which it is applicable. For wider classes we indicate only the necessary additions. General statements of problems are preceded by concrete examples from various domains of application. A number of such examples are described in the Introduction. We shall return to them in the course of the whole book, to the extent that the general theory provides the means for their solution.

The theory of multistage Markovian decision processes was first laid out in the pioneering works [1,2] of Wald on sequential analysis and statistical decision functions. It was developed, under the name of dynamic programming, by Bellman in the early 1950's. In this theory one takes account of the double rôle of the control at each step, the immediate payoff and the influence on the future evolution of the system. Already in Bellman's first monograph [1], the stochastic case was considered along with the deterministic case; in the stochastic case the control does not exactly determine the future states of the system but rather influences their probability distribution. Further essential contributions were made by Howard [1] and by Dubins and Savage [1]. The fundamental results relative to controlled Markov processes with arbitrary state and action spaces were obtained by Blackwell [4,5] and Strauch [1]. Their papers form the basis for the theory of Borel models which we present here. Another exposition of the results of Blackwell-Strauch is contained in the monograph [1] of Hinderer.

In the first two parts of the book we prove the existence of optimal and ε-optimal strategies for various classes of controlled Markov processes; we investigate the properties of these strategies and indicate various methods for finding them. Along with the general theory we consider concrete examples: the allocation of resources between production and consumption and among various branches of production, equipment replacement, stabilization of linear systems under the influence of random perturbations, the allocation of stakes, and so forth. We first consider processes which are nonhomogeneous in time, and then specific problems connected with the homogeneous case: the existence of stationary optimal strategies and the maximization of average reward per unit time. In the third part we consider models with incomplete information. We succeed in making them into models with complete information which were studied in the first part, by introducing spaces whose points are probability distributions. The last chapter is devoted to new results on concave models and models of economic development, taking account of random factors.

The other chapters also contain some novelties, such as canonical strategies, summable models and an investigation of general models with incomplete information. The proofs of several well known results are considerably revised.

Nowhere in the book do we touch on control processes in continuous time; this is a theme requiring a special monograph.* Nor have we set as our goal that of including all results on processes in discrete time. Information on a number of these subjects is contained in the historical-bibliographical notes at the end of the book.

The material of the book may be separated into three levels, depending on the demands on the reader. The first supposes only the knowledge of the elements of the theory of probability and analysis. The corresponding sections are accessible to specialists with engineering and economic backgrounds. They are the Introduction, Chapters 1 and 4, §§1–6 of Chapter 6, §§1–7 and §11 of Chapter 7. To this

---

* Controlled diffusion processes are treated in recent monographs by Fleming and Rishel [1] and Krylov [3]. For an exposition of controlled jump Markov models treated from the same point of view as in this book see Yushkevich [3]–[5], Yushkevich and Fainberg [1].

type of reader we recommend also §§7–11 of Chapter 2, §§9–12 of Chapter 6 and §12 of Chapter 7, where we consider the applications of the general methods to concrete problems; here it will be necessary to look over formulations from other sections to which references are made. The next level is oriented to people working in areas of applied mathematics. In the corresponding sections we make use only of the most elementary information from measure theory and metric space theory, all the necessary formulations being presented in the text. It is for this level that we intend Chapter 2, §§8–10 of Chapter 7, §§1–3, 5 of Chapter 8, and Chapter 9, the last section of that chapter requiring a knowledge of the elements of functional analysis. The remaining portions of the book—Chapters 3 and 5, §§7,8 of Chapter 6, §4 of Chapter 8, and Appendices 1–5, are mainly interesting for mathematicians, although the beginning sections of Chapters 3 and 5, and §7 of Chapter 6, where the fundamental results are formulated, may be found useful by other readers. For a complete understanding of these portions of the book it suffices to have had the usual undergraduate course in mathematics. Information lying outside the scope of these courses is presented in the appendices, and, in part, in the basic text. The apparatus presented, with detailed proofs, in the appendices, is of wide application in many realms of contemporary mathematics.

We enumerate formulas throughout each section. The following examples help one to understand the system of references adopted in the book: formula (3.2.7) is formula (7) of section 2 of Chapter 3; formula (2.7) is formula (7) of section 2 of the same chapter; formula (0.5) is formula (5) of the Introduction.

Contents

Introduction

Controlled stochastic processes arise in many diverse fields.

Consider, for example, economic planning. This could be the planning of the work of an individual enterprise, of the entire national economy or of a sector of it. At the beginning of each period, starting from the state which has been achieved, one chooses the plan for the following period. The development of the system may be described mathematically as a controlled deterministic process, *if* one is able to assume that the state of the system at the end of each period is uniquely determined by the state at the beginning of the period and the plan for that period. However it is not always possible to disregard the influence of such factors as meteorological conditions, demographic shifts, fluctuations in demand, delays in the coordination of complex production processes, scientific discoveries, and inventions. These factors are better taken into account by stochastic models, in which, knowing the state at the beginning of the period, and the plan, one may calculate only the probability distribution for the state at the end of the period. Thus we arrive at a controlled stochastic process.

We shall illustrate what has been said on the simplest model of the allocation of resources between consumption and production.

One can already make interesting qualitative deductions from an extremely simplified model with one resource, which is used both for production and for consumption. We suppose that during one period, by using y units of the resource in the production sphere, one obtains $F(y)$. If we denote by y_t the quantity directed into production in the period t or in the time interval $(t-1, t)$, and by c_t the quantity of the product consumed in that period, then we have the obvious relation

$$y_t + c_t = F(y_{t-1}). \qquad (1)$$

The influence of random factors is introduced by assuming that F depends not only on y but on a random parameter s_t, so that

$$y_t + c_t = F(y_{t-1}, s_t). \qquad (2)$$

A more elaborate economic model may be reduced to the one just described. It is the following. Suppose that the output X_t over the period $(t-1, t)$ depends on two inputs, the existing capital stock K_{t-1} and the existing supply of labor L_{t-1}:

$$X_t = \Phi(K_{t-1},L_{t-1}).$$

The function Φ is called the *production function*. The output X_t is divided into consumption C_t and investment $X_t - C_t$. The obvious relation $K_t = K_{t-1} + X_t - C_t$ may be rewritten in the form

$$K_t = K_{t-1} + \Phi(K_{t-1},L_{t-1}) - C_t. \tag{3}$$

Often it is assumed that the production function Φ satisfies the condition

$$\Phi(\lambda K,\lambda L) = \lambda\Phi(K,L) \qquad \text{for} \quad \lambda > 0.$$

Dividing equation (3) by L_t and putting

$$c_t = \frac{C_t}{L_t}, \qquad y_t = \frac{K_t}{L_t}, \qquad l_t = \frac{L_t}{L_{t-1}},$$

we obtain the equation

$$y_t = l_t^{-1}[y_{t-1} + \Phi(y_{t-1},1)] - c_t. \tag{4}$$

We will suppose that $l_t = l$ does not depend on the time t, so that the labor resources change according to an exponential law. Then equation (4) takes on the form (1) if one puts

$$F(y) = l^{-1}[y + \Phi(y,1)].$$

We note that the quantities c_t and y_t have a simple economic meaning, expressing respectively the consumption and the capital per worker.

If we wish to take account of random factors, then we need to introduce the random parameter s into the production function Φ. Then F will depend on s as well, and we obtain equation (2).

A more adequate model of economic planning must include not just one resource, but rather a set of resources. One such model is due to David Gale, and is a generalization of an earlier model of von Neumann. It is based on the representation of a production process as a pair of nonnegative m-dimensional vectors (ξ,η), the ith coordinates denoting respectively the input and the output of the ith resource over the time interval $(t-1, t)$. For each period $(t-1, t)$ the production process (ξ_t,η_t) has to be chosen from a given set $\mathfrak{T}_t$ of processes which are technically realizable in that period. In addition the input on each step must not exceed the output from the preceeding step, so that $\xi_t \leq \eta_{t-1}$, η_0 denoting a given vector of initial resources*.

* The notation $\xi \leq \eta$ for m-dimensional vectors means that no coordinate of ξ exceeds the corresponding coordinate of η.

The set $\mathfrak{X}_t$ may depend on a random parameter s_t, describing for example the state of scientific-technical knowledge or of the exterior environment. Then the choice of the process (ξ_t, η_t) must depend on s_t; it may depend also on the preceding values of the random parameter as well, but not on the future ones, these being as yet unknown.

Another example is the problem of the regulation of stocks of water. Water is stored in a reservoir and later expended for irrigation in periods of drought. Suppose that s_t is the yearly quantity of water available to replenish the reservoir. At the beginning of the period $(t-1, t)$, having a stock x_{t-1} of water, we plan the quantity a_t of water to be used during that period for irrigation. If the volume of the reservoir were infinite, then we would have the equation $x_t = x_{t-1} - a_t + s_t$. However if that volume is finite and equal to M, then we get instead of this the relation

$$x_t = (x_{t-1} - a_t + s_t) \wedge M, \tag{5}$$

where $a \wedge b$ denotes the smaller of the numbers a and b. The value of s_t depends on the quantity of rainfall, the character of flooding, the thawing of glaciers and so forth, and it is natural to consider it as a random variable.

The next example may be interpreted as a problem of allocating stakes between two variants of a game. With the stake x the gain in the first game is equal to σx, and in the second τx, where σ and τ are random variables with different probability distributions. The game is repeated several times. Denote by x_{t-1} the total sum of which the player disposes at the instant $t-1$. Suppose he stakes $\alpha_t x_{t-1}$ on the first game and $\beta_{t-1} x_{t-1}$ on the second, where $\alpha_t + \beta_t = 1$. Then

$$x_t = (\alpha_t \sigma_t + \beta_t \tau_t) x_{t-1}. \tag{6}$$

Instead of two games one may consider two ways of investing money, for instance putting it into the savings bank or purchasing lottery tickets, or putting it into two productive sectors with different output coefficients. In the latter case it is natural to replace equation (6) by

$$x_t = (\alpha_t \sigma_t + \beta_t \tau_t)(x_{t-1} - c_t), \tag{7}$$

taking account of the fact that the output at the time $t-1$ is not entirely distributed between the productive sectors, but goes partially into consumption.

In every control problem there arises the question as to the objective of the control. In the problem of distribution of resources between consumption and production it is natural to evaluate the plan according to the consumption program $c_1, c_2, \ldots, c_t, \ldots$. The most simple and complete theory is obtained if one assumes that the value of this program is equal to the sum of values of the quantities c_t, which leads to the expression

$$q_1(c_1) + q_2(c_2) + \cdots + q_t(c_t) + \cdots, \tag{8}$$

(it is natural to assume that the value of the consumption c_t depends on the time t). In mathematical economics it is usually assumed that the function q_t is concave*. The sum (8) defines the so-called *objective function*, and the aim of the planner is to maximize it.

In the multisector model of von Neumann-Gale, they took as the objective function

$$q_1(\xi_1, \eta_1) + q_2(\xi_2, \eta_2) + \cdots + q_t(\xi_t, \eta_t),$$

where $q_t(\xi, \eta)$ measures the utility of the production process (ξ, η).

The control by expenditures of water seeks to achieve the largest possible harvest. One may suppose that the average harvest over the period t is a function $f(a)$ of the amount a of water released for irrigation. Estimating the harvest z in the year by the function $g_t(z)$, we arrive at the objective function

$$q_1(a_1) + q_2(a_2) + \cdots + q_t(a_t) + \cdots, \tag{9}$$

where $q_t = g_t[f(a)]$. If the plan is set up for n periods, then what we need to consider in the sums (8) and (9) are the first n terms.

If the controlled process is stochastic, then the objective function (8) or (9) is a random variable. Of two random variables it is natural to prefer that with the larger mathematical expectation. Therefore, in the stochastic variant of the problem of control the sums (8) and (9) are replaced by their mathematical expectations.

Let us return to the problem of the allocation of stakes between two games. Here it is natural to seek the maximum possible final payoff x_n, where the value of x_n might for example be given by a mathematical expectation of $r(x_n)$, where r is an arbitrary nondecreasing function. We note that the optimal behavior depends on the form of the function r. Generally speaking, it is more important to receive one dollar more when the total amount is small than when it is large, and therefore one often assumes the function r to be concave. However it may happen that we need a specific sum h, and that the objective is to win this sum with the maximum probability. In this case we have to put

$$r(x) = \begin{cases} 1 & \text{for} \quad x \geq h, \\ 0 & \text{for} \quad x < h. \end{cases}$$

A more general statement of the problem is as follows: given any two probability distributions of the gain, the player prefers one of them or regards them as indifferent. It follows from a general theorem of von Neumann and Morgenstern that under wide assumptions such an ordering of distributions is defined by the magnitude of the mathematical expectation $r(x_n)$, where the function r is uniquely defined up to a positive multiplier and a constant additive term[†].

---

* The graph of a concave function lies above any of its chords.
† See Arrow [1], Part 2, and Dynkin and Ovsevič [1].

In the problem of the allocation of resources between two sectors it is reasonable to consider the objective function (7).

Now we present two further problems of the optimal control of stochastic processes.

The first of these is the replacement problem. Suppose that some device (mechanism, equipment) has a random lifetime. At the beginning of each period we have to take one of two decisions: replace the device by a new one, or continue to make use of the old one. The probability of a breakdown of the device, and the profit gained by using it, depend on the time it has already served. On replacement we bear the costs for new equipment, and on breakdown we suffer certain additional losses. The objective of the controller is to obtain the largest possible profits (measured, because they are stochastic, in terms of mathematical expectation).

The second problem is the maintenance of a stationary régime of the working of a mechanism undergoing random perturbations. The simplest description of a process of this type is given by the equations

$$y_t = x_{t-1} - a_t,$$
$$x_t = y_t + s_t,$$
(10)

where $s_1, s_2, \ldots, s_t, \ldots$ are random perturbations, $a_1, a_2, \ldots, a_t, \ldots$ are corrective actions (a_t being chosen knowing x_{t-1}). We suffer losses from the deviation of y_t from 0 and we have to bear expenses depending on the magnitude of the a_t. One may, for example, wish to minimize the mathematical expection of the sum

$$b \sum_1^n y_t^2 + \sum_1^n a_t^2.$$
(11)

Another possibility is to minimize the average expenses per unit time interval as $n \to \infty$.

To this point we have been supposing that complete information about the controlled process is available. But in most real applications obtaining complete information is impossible or else it is excessively complicated and expensive.

For example, the value of x_t in the problem of maintenance of a stationary régime of work might be observed with some (unknown) error. In the problem of choosing between two sectors the probability distributions for the random coefficients σ and τ are usually unknown, although one may well have partial information on them based on previous experience and calculations. Under these circumstances each successive step not only yields a material payoff, but also leads to additional knowledge. These two sides of the problem pervade almost every sphere of human endeavor, with as a rule one of them dominant. For a productive enterprise the fundamental objective is the material product, while the accumulation of production experience is an important auxiliary result. For scientific institutions the situation is quite the contrary.

Part I

Control on a Finite Time Interval

Chapter 1

Finite and Denumerable Models

§1. Deterministic Controlled Processes

In order to introduce the reader to the ideas on which the solution of problems of optimal control is based, we consider the following simple schema.

In figure 1.1 we depict a system of 4 points and 8 directed segments ("arrows") joining these points. Along each arrow we give a number—the *value* for this arrow. One is allowed to move in the direction of the arrows in any way, the value for the path being the sum of the values of all the arrows included in it*. Among the paths issuing from the point x and consisting of four arrows it is required to find the path with the maximal value; this one will be said to be *optimal*. The answer, as we shall see further on, is the path distinguished in figure 1.1 by boldface arrows.

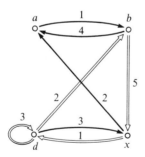

Figure 1.1

In the choice of the optimal path, at each step we need to take account not only of the point where we find ourselves, but also of how many steps there remain to be done. Therefore it is appropriate to replace figure 1.1 by the schema depicted in figure 1.2. In that schema the 5 columns depict the 4 points of figure 1.1 at the instants $t = 0, 1, 2, 3, 4$ of time. The arrows leading from the X_{t-1} column to the X_t column show the transitions possible at the tth step. If one chooses any path of the four components in figure 1.1, then the corresponding chain of arrows in figure 1.2 depicts the graph of the motion.

---

* The value of a path can be considered as a payoff or a reward received for passing through this path. It serves as a definite criterion for the quality of the decision adopted.

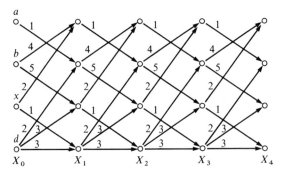

Figure 1.2

of the four components in figure 1.1, then the corresponding chain of arrows in figure 1.2 depicts the graph of the motion.

In distinction from figure 1.1, schemas similar to figure 1.2 make it possible to set up systems changing in time (see figure 1.3). If one crosses out in the figure 1.3 several first or last columns, then once again one obtains a schema of analogous type, except the index of the leftmost column becomes different from zero.

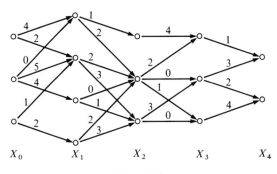

Figure 1.3

In the general case (see figure 1.4) we have some finite sets of points X_m, $X_{m+1}, \ldots, X_n$ (columns) and finite sets of arrows $A_{m+1}, \ldots, A_n$, the arrows of the set A_t leading from X_{t-1} to X_t. The points of the set X_n will be called *terminal*. There is at least one arrow leading out of each nonterminal point. A succession of arrows forms a *path* if the initial point of each of them, other than the first, coincides with the endpoint of the preceding arrow, and the last arrow ends at X_n. We suppose a function q is given on the set of all arrows. The sum of the values of this function on all the arrows of the path is called the *value* of that path. Among the paths issuing from a given point x, it is required to choose the path with the largest value; this is called the *optimal path*.

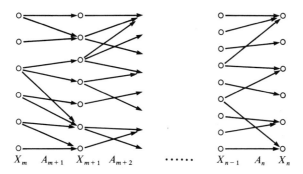

$$X_m \quad A_{m+1} \quad X_{m+1} \quad A_{m+2} \qquad \cdots\cdots \qquad X_{n-1} \quad A_n \quad X_n$$

Figure 1.4

The value of an optimal path issuing from the point x will be called the *value of the point* x and will be denoted by $v(x)$. So that the function $v(x)$ will be defined on the set X of all points, we put it equal to 0 on X_n.

The problem will of course be solved if we look through all the paths issuing from x and compare their values. But this method is rarely applicable, because of the enormous number of possibilities present even in the simplest schemas. Suppose however that we already know the function $v(x)$. Then the problem is easily solved using the following criterion: for optimality of the path l it is necessary and sufficient that for each arrow a, belonging to l,

$$q(a) = v(x) - v(y), \tag{1}$$

where x is the initial point and y is the endpoint of the arrow a. Indeed, adding the equations (1) along all the arrows of the path l, we find that the value of l is equal to the value $v(x)$ at the initial point x of that path. On the other hand, for any arrow a, if its initial and endpoint are x and y respectively,

$$v(x) \geq q(a) + v(y), \tag{2}$$

since the right hand side is equal to the value of the path consisting of the arrow a and the optimal path issuing from y. If (2) holds with strict inequality for any arrow of the path l, then we find that the value of l is less than the value $v(x)$ at the initial point x, and thus the path l is not optimal.

We note further that at a nonterminal point x

$$v(x) = \operatorname*{Max}_{a \in A(x)} [q(a) + v(y)], \tag{3}$$

where y is the endpoint of the arrow a and $A(x)$ is the bundle of arrows issuing from x. Indeed, for any arrow a of $A(x)$ inequality (2) is satisfied, and for the arrow a which is the first arrow in the optimal path from the point x, that inequality becomes an equality, because of (1). Formula (3) expresses the values of v on X_{t-1} in terms of its values on X_t and makes it possible to calculate its values recursively moving from right to left.

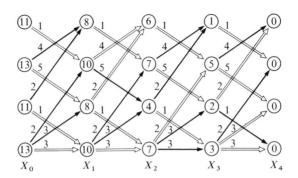

Figure 1.5

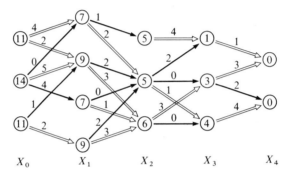

Figure 1.6

In figures 1.5 and 1.6 we depict such a calculation of the values $v(x)$ for figures 1.2 and 1.3. From each nonterminal point x we have drawn in boldface the arrow satisfying the criterion (1). Those, and only those paths consisting entirely of boldface arrows, are optimal.

$$* \quad * \quad *$$

We have taken the value of a path to be the sum of the values of the arrows entering into it. One may consider a more general problem in which in evaluating the path one enters not only the arrows but the points of the path as well. Suppose that $l = x_m a_{m+1} x_{m+1} \cdots a_n x_n$ is a path consisting of the initial point x_m, and then, for each t with $m + 1 \leq t \leq n$, a passage from the point $x_{t-1} \in X_{t-1}$ along the arrow $a_t \in A_t$ to the point $x_t \in X_t$. We define the value $I(l)$ of the path l by the formula

$$I(l) = r(x_m) + q(a_{m+1}) + r(x_{m+1}) + \cdots + q(a_n) + r(x_n), \qquad (4)$$

where r is a function on points and q a function on arrows. If one puts

$$\tilde{q}(a) = r(x) + q(a),$$

where x is the initial point of the arrow a, then the value (4) can be written in the form

$$\sum_{t=m+1}^{n} \tilde{q}(a_t) + r(x_n).$$

Therefore we do not lose anything in generality by supposing that r is equal to 0 at nonterminal points. The value of the path l will then be the sum

$$I(l) = \sum_{t=m+1}^{n} q(a_t) + r(x_n).$$

We will call the functions q and r the *running* and *terminal* rewards respectively.

The criterion for optimality of a path, and formula (3), remain in force in the presence of a terminal reward as well. The only difference consists in that the values $v(x_n)$ on X_n are no longer zero but rather are given by $r(x_n)$.

$$* \; * \; *$$

It is convenient to introduce an operator V, transforming functions on arrows into functions on nonterminal points according to the formula*

$$Vg(x) = \sup_{a \in A(x)} g(a), \qquad (5)$$

and an operator U, transforming a function on points into a function on arrows according to the formula

$$Uf(a) = q(a) + f(y), \qquad (6)$$

where y is the endpoint of the arrow a. In particular $Uv(a)$ is the maximal value of the path starting with the arrow a. We denote that quantity by $u(a)$ and call it the *value of the arrow* a. In view of (3)

$$v(x) = Vu(x), \qquad x \in X \backslash X_n;$$
$$u(a) = Uv(a), \qquad a \in A, \qquad (7)$$

and

$$v(x) = r(x), \qquad x \in X_n. \qquad (8)$$

Equations (7), along with the boundary condition (8), define u and v.

The value $Vg(x)$ for x in X_{t-1} is defined by the values of g on A_t, and the value $Uf(a)$ for a in A_t by the values of f on X_t. The function g may be considered as a collection $\{g_{m+1}, \ldots, g_n\}$, where g_t is the restriction of g to A_t, and the function f as a collection $\{f_m, f_{m+1}, \ldots, f_n\}$, where f_t is the restriction of f to X_t. The operators V and U may also be replaced by collections $\{V_{m+1}, \ldots, V_n\}$ and $\{U_{m+1}, \ldots, U_n\}$, where V_t transforms functions on A_t into functions on X_{t-1}

---

* Since we are for the present dealing with finite sets $A(x)$, the supremum may of course be replaced by a maximum.

and U_t carries functions on X_t into functions on A_t. In these notations formulas (5)–(8) may be rewritten as follows:

$$V_t g_t(x) = \sup_{a \in A(x)} g_t(a), \qquad x \in X_{t-1},$$

$$U_t f_t(a) = q_t(a) + f_t(x), \qquad a \in A_t \text{ (where } x \text{ is the endpoint of } a);$$

$$v_{t-1} = V_t u_t, \qquad u_t = U_t v_t, \qquad m + 1 \le t \le n, \tag{9}$$

$$v_n = r. \tag{10}$$

§2. Controlled Markov Processes and Models

Now we suppose that the choice of the arrow at the point x no longer determines the state into which we go, but rather a probability distribution for that state. We depict an example of such a schema in figure 1.7. In the A_1-column we indicate

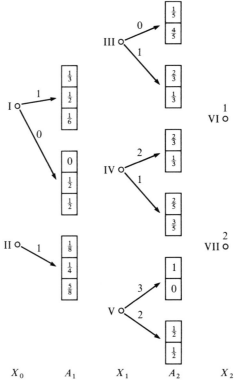

Figure 1.7

three probability distributions on the set X_1, corresponding to the three arrows issuing from X_0. In the A_2-column we indicate six probability distributions on the set X_2, corresponding to arrows starting in X_1. As before, we endeavor to go along a path with the maximal value, i.e. the maximum sum of the numbers over the arrows of the path and those circled by the terminal states. However now the path which we go over depends not only on our choice but also on chance, and we wish to maximize the mathematical expectation of the value.

It is natural to reason as follows. In state III the mathematical expectation of the value is equal to

$$0 + \tfrac{1}{5} \cdot 1 + \tfrac{4}{5} \cdot 2 = 0 + \tfrac{9}{5} = \tfrac{9}{5}$$

if we choose the first arrow, and equal to

$$1 + \tfrac{2}{3} \cdot 1 + \tfrac{1}{3} \cdot 2 = 1 + \tfrac{4}{3} = \tfrac{7}{3}$$

if we choose the second arrow. The value of the state III is equal to the maximum of these two numbers, i.e. $\tfrac{7}{3}$, and it is clear that one ought to choose the second arrow at III. We have depicted the optimal arrow in boldface in figure 1.8. Analogously,

$$v(IV) = \max(2 + \tfrac{2}{3} \cdot 1 + \tfrac{1}{3} \cdot 2, 1 + \tfrac{2}{5} \cdot 1 + \tfrac{3}{5} \cdot 2)$$
$$= \max(2 + \tfrac{4}{3}, 1 + \tfrac{8}{5}) = \tfrac{10}{3},$$

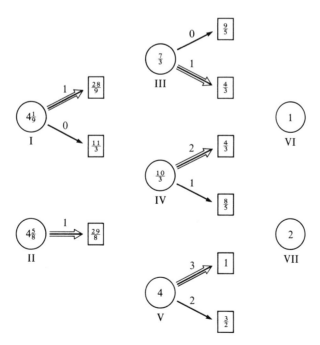

Figure 1.8

and in state IV the first arrow is preferable. We have

$$v(V) = \max(3 + 1 \cdot 1 + 0 \cdot 2, 2 + \tfrac{1}{2} \cdot 1 + \tfrac{1}{2} \cdot 2)$$
$$= \max(3 + 1, 2 + \tfrac{2}{3}) = 4,$$

so that in state V the advantage is with the first arrow. Further, choosing in state I the first arrow, and then proceeding optimally, we obtain the value

$$1 + \tfrac{1}{3} \cdot \tfrac{7}{3} + \tfrac{1}{2} \cdot \tfrac{10}{3} + \tfrac{1}{6} \cdot 4 = 1 + \tfrac{28}{9} = 4\tfrac{1}{9},$$

and if we choose the second arrow we get the value

$$0 + 0 \cdot \tfrac{7}{3} + \tfrac{1}{2} \cdot \tfrac{10}{3} + \tfrac{1}{2} \cdot 4 = 0 + \tfrac{11}{3} = 3\tfrac{2}{3}.$$

The larger of these two numbers is $v(I)$. In state I we need to choose that arrow which will lead to the value $v(I)$, i.e. the first arrow. The selection, in each non-terminal state, of the arrow indicated in boldface in figure 1.8, describes an optimal method of behavior.

In order to justify this conclusion, we must first state the problem precisely.

Suppose that X_t $(t = m, m + 1, \ldots, n)$ and A_t $(t = m + 1, \ldots, n)$ are arbitrary finite sets. To each a of A_t we associate a probability distribution $p(\cdot|a)$ on X_t.* The function p defining the law of transition from A_t into X_t, will be called the *transition function*. It is natural to suppose that the point of the set X_m from which the path starts is also random, and that its *initial probability distribution* μ is given.

The transition from $x \in X_{t-1}$ into A_t is determined by our choice. Here we choose a, not from all of A_t, but rather from its subset $A(x)$, depending on the state x. The elements of the set $A(x)$ will be called *actions* (controls) at the point x. The set $A(x)$ is defined, and nonempty, for all nonterminal states x. We will suppose that the $A(x)$ are pairwise disjoint, and that their union over all x of X_{t-1} is equal to A_t. In other words, each control a may be used in one and only one state. We denote that state by $j(a)$, so that the notation $x = j(a)$ is equivalent to the notation $a \in A(x)$. The mapping j will be called a *projection*, and the set $A(x) = j^{-1}(x)$ a *fibre*.

On the set of all actions there is given a running reward function $q(a)$, and on the set of terminal states a terminal reward function $r(x)$.

Thus we have arrived at the concept of a *controlled Markov process* on the time interval $[m,n]$. This process is given by the following elements:

a) The sets $X_m, X_{m+1}, \ldots, X_n$, the *state spaces*;
b) The sets $A_{m+1}, \ldots, A_n$, the *action spaces*;
c) The mapping j of the action set $A = \bigcup_{t=m+1}^n A_t$ into the set $X = \bigcup_{t=m}^n X_t$ of states, such that $j(A_t) = X_{t-1}$—this is called the *projection*;

---

* To give a probability distribution p on a finite (or countable) set E means that one assigns to each $x \in E$ a nonnegative number $p(x)$ such that the sum of these numbers will be equal to 1. For each subset $\Gamma \subseteq E$, $p(\Gamma)$ denotes the sum of $p(x)$ over all x of Γ. If for some subset Γ of E we have $p(\Gamma) = 1$, then we will say that the distribution p is *concentrated* on Γ. We write $p(\cdot|a)$ instead of $p(x|a)$ in order to distinguish the law giving the probability distribution from the number $p(x|a)$ corresponding to a concrete x.

d) The probability distributions $p(\cdot|a)$ on X_t, depending on $a \in A_t$ $(t = m + 1, \ldots, n)$*, called the *transition function*;

e) The function q on the action set A, called the *running reward function*;

f) The function r on the set X_n of terminal states, called the *terminal reward function*;

g) The probability distribution μ on X_m, called the *initial distribution*.

In the study of control processes it is useful to contract the interval $[m,n]$ to a smaller segment $[m_1,n]$. In the contracted process the elements enumerated in a)–f) above are uniquely determined by the original process. However, the distribution at the instant m_1 depends on the method of control on the interval of time $[m,m_1]$. It is therefore natural to give a special name to the object defined by the elements a)–f). We will call this object a *model*[†]. If we are given a model Z, then to each distribution μ on X_m there corresponds a uniquely defined controlled process Z_μ, for which μ is the initial distribution. In the case when μ is concentrated at the point x, we will write Z_x in place of Z_μ.

Our aim is to find a control procedure under which the mathematical expectation of the value

$$I(l) = \sum_{t=m+1}^{n} q(a_t) + r(x_n) \tag{1}$$

of the path

$$l = x_m a_{m+1} x_{m+1} \cdots a_n x_n, \tag{2}$$

where $x_t \in X_t$, $m \le t \le n$; $a_t \in A(x_{t-1}) \subset A_t$, $m + 1 \le t \le n$, is as large as possible. We have however to make clear what is to be understood by a method of control. This will be done in the following section.

* * *

In preparation we shall make a number of observations and consider some examples.

We have been supposing that the set of actions which are possible at the various states do not intersect. This is convenient in the general theory, but is not always convenient in the analysis of concrete examples. If the sets $A(x)$ intersect, then it is natural to suppose that the transition function and the running reward at the instant t depend not only on a_t but also on x_{t-1}. However this more general case immediately reduces to the case we have been considering, if we understand by "action" the pair $x_{t-1}a_t$; here the mapping j assigns to each such pair its first component.

Further, the choice of an action may fail to yield a deterministic value for the running reward, but rather yield only a probability distribution for that value. We may however once again turn to the consideration of a scheme in which such a random payoff is replaced by its mathematical expectation[‡]. We deal with such

---

* Often it is convenient to regard $p(\cdot|a)$ as a probability distribution on X concentrated on X_t for $a \in A_t$.

† It might have been better to call it a "Markov model". But we have left the word "Markov" off for brevity.

‡ See the end of the following section for details.

a situation for example when the reward on the step t depends on x_{t-1}, a_t, and x_t. Such a reward may be replaced by

$$\bar{q}(x_{t-1}a_t) = \sum_{x_t \in X_t} q(x_{t-1}a_t x_t)p(x_t | x_{t-1}a_t).$$

(3)

Finally, the definition of models is greatly simplified when their elements do not change in time, as in the example with which we started out in §1. Here we speak of *homogeneous models*. In order to define such a model we need to define a projection j of the action space A onto the state space X, a probability distribution p on X, depending on $a \in A$ (a transition function), and a running reward function q on A.

If we are given a homogeneous model Y, then to each integer n there corresponds an n-step model Z in our previous sense (nonhomogeneous model), which is constructed as follows. Consider $n + 1$ copies $X_0, X_1, \ldots, X_n$ of the space X and n copies $A_1, \ldots, A_n$ of the space A, and suppose that j maps A_t into X_{t-1}, and that the distributions $p(\cdot | a)$ for $a \in A_t$ is concentrated on $X_t{}^*$. We have already used this construction at the very beginning of §1. The terminal reward can be set equal to zero, or can be defined in some other way.

It is more natural to consider homogeneous models on the time interval $[0, \infty)$ (see Chapter 6).

* * *

Now we will discuss how the concrete problems which we considered in the introduction reduce to the above general schema.

We begin with the problem of the replacement of machinery. To this problem there corresponds the following homogeneous model. By the state we understand the total working time that the operating equipment has worked. We suppose that this time is described by a nonnegative integer x (see figure 1.9). At each state x two actions are possible: c—keep the old machine, and d—replace it. Under the action d the system goes into the state 0. Under the action c the transition $x \to x + 1$ takes place, if the machine does not break down. If such a breakdown does take place, then the equipment *has* to be replaced, and the transition $x \to 0$ takes place. The probability of breakdown depends of course on the service time x. We denote this by q_x and put $p_x = 1 - q_x$. It is natural to assume that q_x does not decrease as x increases. In order to restrict the problem to a finite state space, we suppose that for some $x = K$ this breakdown probability becomes equal to 1; then x will take on only the values $0, 1, 2, \ldots, K$. The transition function for this model is defined by the formulas

$$p(x + 1 | xc) = p_x, \qquad p(0 | xc) = q_x, \qquad p(0 | xd) = 1,$$

(4)

$x = 0, 1, 2, \ldots, K$, the probability of other transitions being equal to 0.

---

* We may define X_t formally as the collection of pairs (t,x), $x \in X$, and A as the collection of pairs (t,a), $a \in A$, and put

$$j(t,a) = (t - 1, j(a)), \qquad p(t,x | t,a) = p(x,a), \qquad q(t,a) = q(a).$$

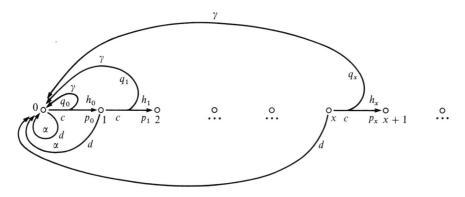

Figure 1.9

The running reward on the step t depends on the length of time the equipment has been in service, on our decision, and on whether or not a breakdown occurs on this step. Suppose that h_x is the income under the transition $x \xrightarrow{c} x + 1$ (i.e. assuming the continued health of the equipment, already having survived through a time x). We assume that h_x does not increase with increasing x*. Further, let us denote by α the income over a period, when the machine is being replaced (the transition $x \xrightarrow{d} 0$). We will suppose that α does not depend on x and that $\alpha < h_x$ for any x. Finally, suppose that γ is the income under the transition $x \xrightarrow{c} 0$. Inasmuch as the replacement of a machine in the case of a breakdown now turns out to be more expensive, then $\gamma < \alpha$. The running reward which we have defined depends generally speaking on the whole triple (x_{t-1}, a_t, x_t). In accordance with the remarks which we made prior to Equation (3), this payoff may be replaced by its mathematical expectation (3) for fixed x_{t-1} and a_t. Then we will have

$$q(xc) = p_x h_x + q_x \gamma, \qquad q(xd) = \alpha, \tag{5}$$

$x = 0, 1, \ldots, K$. The terminal reward r_x, $x = 0, 1, \ldots, K$, may be put equal to zero or to any nonincreasing function of x; this last may be interpreted as an estimate of the residual value of the machine at the end of the control interval.

The remaining problems considered in the Introduction may be included in the following schema. The evolution of the system is described by the equation

$$x_t = F_t(x_{t-1}, a_t, s_t), \tag{6}$$

showing where it goes from x_{t-1} under the action a_t in the situation described by the random parameter s_t. In order to obtain equation (0.2), describing a one-sector model, it suffices to put

$$x_t = y_t + c_t, \qquad a_t = y_{t-1}, \qquad F_t(x, a, s) = F(a, s); \tag{7}$$

here the action a_t may be chosen from the interval $[0, x_{t-1}]$.

---

* This income may be random, and then we must understand by h_x its mathematical expectation. The same must be said about the quantities α and γ introduced further on.

Equation (0.5) in the water regulation problem is obtained with

$$F_t(x,a,s) = (x - a + s) \wedge M.$$

One arrives at equation (0.6) on the problem of the distribution of stakes, if one takes the pair (σ_t, τ_t) as s_t and puts

$$a_t = \alpha_t, \qquad F_t(x,a,s) = [a\sigma + (1 - a)\tau]x. \tag{9}$$

In the case of the distribution of resources between two productive sectors and consumption (equation (0.7)), the control a_t is given by two numbers: the amount $i_t = x_{t-1} - c_t$ invested in production, and the share γ_t of this investment directed into the first sector. Equation (0.7) is obtained from (6) with

$$a_t = (i_t, \gamma_t), \qquad s_t = (\sigma_t, \tau_t),$$
$$F_t(x,a,s) = i[\gamma\sigma + (1 - \gamma)\tau]. \tag{10}$$

Finally, to equations (0.10) there corresponds the function

$$F_t(x,a,s) = x - a + s. \tag{11}$$

In all of these examples it is more natural to describe the states and rewards by parameters with values in numerical intervals rather than finite sets. More general models including these cases will be discussed in the following chapters. Here however we shall write down the transition function of a system controlled by equation (6) only under the assumption that x, a, and s take on a finite number of values. Moreover, we will suppose that the values of the parameter s_t at different instants of time are mutually independent. Then

$$p(y|x,a) = \Pi_t\{F_t(x,a,s) = y\}, \qquad y \in X_t, x \in X_{t-1}, a \in A_t, \tag{12}$$

where Π_t is the probability distribution for the parameter s_t.

To the utility (0.8) in the one-sector model there corresponds the running reward function

$$q_t(x_{t-1} - a_t),$$

where the q_t are concave functions. The utility (0.9) in the problem of the regulation of water takes the form

$$q(a_t) = g_t(f(a_t)).$$

In the stake distribution problem the running reward is equal to 0 and the winnings are obtained only as a terminal reward.

In the problem of the distribution of resources between two sectors, the running reward is equal to

$$q(x_{t-1}a_t) = q_t(x_{t-1} - i_t).$$

In the problem of a stationary work régime

$$q(x_{t-1}a_t) = -b(x_{t-1} - a_t)^2 - ca_t^2.$$

§3. Strategies

We return to the example discussed at the beginning of §2. In this example, we defined an optimal way of behavior, by associating an optimal arrow at each nonterminal state. We proceeded analogously in §1. How does one describe this in general terms?

The mapping $x \to A(x)$ is an example of a set-valued function or a correspondence. In general a *correspondence* Φ of E into E' assigns to each point x of E a nonempty set $\Phi(x)$ in the other space E'. A single-valued function φ on E with values in E' will be called a *selector* of Φ if $\varphi(x)$ belongs to the set $\Phi(x)$ for all x of E. Thus, in the cases we have discussed we have given some selector of the correspondence $A(x)$ of $X \backslash X_n$ into A. We will agree to call all the selectors of this correspondence *simple strategies*.

Applying a simple strategy φ, we obtain a path $l = x_m a_{m+1} x_{m+1} \cdots a_n x_n$, where x_m is a random point with the distribution μ, $a_t = \varphi(x_{t-1})$, and x_t is a random point with the distribution $p(\cdot|a_t)(m + 1 \leq t \leq n)$.

More general methods of control are also possible. We may select each time, rather than a definite action, a probability distribution for them. A further generalization consists in taking account, when one chooses a_t, not only the state x_{t-1}, but the entire preceding *history*

$$h = x_m a_{m+1} x_{m+1} \cdots a_{t-1} x_{t-1}, \tag{1}$$

$x_s \in X_s$, $m \leq s \leq t - 1$; $a_s \in A(x_{s-1}) \subset A_s$, $m < s < t$; $m < t \leq n$. We thus arrive at the following definition. A *strategy* π is a function assigning to each history (1) a probability distribution $\pi(\cdot|h)$ on the set A_t of actions, concentrated on $A(x_{t-1})$.

It is natural to call a strategy *Markovian* if the distribution $\pi(\cdot|h)$ depends only on the last state x of the history h (i.e. with the present known, the strategy does not depend on the past).

The Markov strategy $\sigma(\cdot|x)$ defines the transition from a state to an action in quite the same way as the transition function $p(\cdot|a)$ defines the transition from an action to a state. The difference between them consists only in that p is given and σ has to be chosen. In the case of a non-Markovian strategy an additional distinction consists in that the transition mechanism depends on the past*.

---

* One could regard the mechanism of transition from an action to a state as depending upon the past, i.e. consider transition functions of the type $p(\cdot|x_m a_{m+1} \cdots x_{t-1}a_t)$. But this case reduces to ours if one introduces new action spaces

$$A_t' = A_m \times A_{m+1} \times \cdots \times A_t.$$

Denote by L the set of all paths (2.2). If we are given a transition function p and a strategy π, then to each initial distribution μ there corresponds a probability distribution P in the space L, defined by the formula*

$$P(x_m a_{m+1} x_{m+1} \cdots a_n x_n) = \mu(x_m)\pi(a_{m+1}|x_m)p(x_{m+1}|a_{m+1})$$
$$\cdots \pi(a_n|x_m a_{m+1} x_{m+1} \cdots x_{n-1})p(x_n|a_n). \qquad (2)$$

For each function ξ on the space L we put

$$P\xi = \sum_{l \in L} \xi(l)P(l). \qquad (3)$$

This is the *mathematical expectation of the random variable* ξ[†]. An example of such a function is the value (2.1) of the path l. We will denote its mathematical expectation by w:

$$w = PI(l) = P\left[\sum_{t=m+1}^{n} q(a_t) + r(x_n)\right]. \qquad (4)$$

For a controlled process Z_μ with a given initial distribution μ the quantity w is a function $w(\pi)$ of the strategy π. We will call it the *value* of the *strategy* π. The aim of the control is then the maximization of the function $w(\pi)$.

The least upper bound v of the function w over all π is called the *value of the process Z_μ* or the *value of the initial distribution* μ. The strategy π will be said to be *optimal for the process Z_μ* if $w(\pi) = v$.

We will be dealing simultaneously with the class of all control processes Z_μ corresponding to some model Z. Reflecting the dependence of the values w and v on μ, we will write $w(\mu,\pi)$ and $v(\mu)$. If μ is concentrated at the point x then we will write, instead of this, $w(x,\pi)$ and $v(x)$. It is natural to call $v(x)$ the *value of the state* x.

A strategy π will be said to be *optimal for the model Z*, or *uniformly optimal*, if π is optimal for the process Z_μ for any initial distribution μ.

In §§4–6 we will prove the existence of a simple strategy φ which is optimal for the model Z, and describe a method of computing v and constructing φ.

* * *

In our definition of a strategy, at each step we have the right to mix in any way the actions which are admissible at the instant in question, i.e. to choose the control randomly, with an arbitrary probability distribution. Do we not extend our possibilities if we decide to mix the strategies themselves?

Suppose that $\{\pi_k\}$ is a finite or countable set of strategies, and that $\{\gamma_k\}$ are nonnegative numbers adding to 1. If, given any initial distribution μ, we use the

---

* After the measure P has been defined, the path (2.2) may be considered as a stochastic process. If the strategy π is Markovian, then that process is Markovian.

[†] Usually the mathematical expectation is denoted by a letter E. We use for it the same letter as for the corresponding probability distribution. This is convenient because we will be dealing with several distributions. Sometimes, however, it is not necessary to introduce a special notation for the probability distribution, and in such cases we will use the letter E.

strategy π_k with the probability γ_k, then we obtain a probability distribution P on the space L of paths, defined by the formula

$$P = \sum_k \gamma_k P_k \tag{5}$$

where the probability distribution P_k corresponds to the strategy π_k. It turns out that one can obtain the distribution P by using some individual strategy π.
   In fact, put

$$D = \sum_k \gamma_k \pi_k(a_{m+1}|x_m) \cdots \pi_k(a_t|x_m a_{m+1} \cdots x_{t-1}).$$

Then put

$\pi(a_{t+1}|x_m a_{m+1} \cdots x_t)$

$$= \begin{cases} \dfrac{1}{D} \sum_k \gamma_k \pi_k(a_{m+1}|x_m) \cdots \pi_k(a_t|x_m a_{m+1} \cdots x_{t-1}) \pi_k(a_{t+1}|x_m a_{m+1} \cdots x_t) \\ \qquad\qquad\qquad\qquad\qquad\qquad\qquad\qquad \text{if } D > 0; \\ \pi_1(a_{t+1}|x_m a_{m+1} \cdots x_t) \qquad\qquad\qquad\qquad \text{if } D = 0. \end{cases} \tag{6}$$

Here $m \le t < n$, $x_m a_{m+1} \cdots x_t$ is any history, a_{t+1} is any control from A_{t+1}. If $t = m$ the denominator is put equal to 1. [One obtains the expression in the right side if, starting out from the measure (2), one writes out the conditional distribution for a_{t+1} given the history $x_m a_{m+1} \cdots x_t$]. From the fact that $\pi_k(\cdot|x_m a_{m+1} \cdots x_t)$ is a probability distribution concentrated on $A(x_t)$, and from the conditions $\sum \gamma_k = 1$, $\gamma_k \ge 0$, it follows that $\pi(\cdot|x_m a_{m+1} \cdots x_t)$ is also a probability distribution concentrated on $A(x_t)$. This means that formula (6) defines a strategy. From (6) it follows that for any path $l = x_m a_{m+1} \cdots x_n$

$$\pi(a_{m+1}|x_m)\pi(a_{m+2}|x_m a_{m+1} x_{m+1}) \cdots \pi(a_n|x_m a_{m+1} \cdots x_{n-1})$$
$$= \sum_k \gamma_k \pi_k(a_{m+1}|x_m)\pi_k(a_{m+2}|x_m a_{m+1} x_{m+1}) \cdots \pi_k(a_n|x_m a_{m+1} \cdots x_{n-1})$$

Multiplying both sides by $\mu(x_m)p(x_{m+1}|a_{m+1}) \cdots p(x_n|a_n)$, and taking account of formula (2), we find that to the strategy π there corresponds the measure

$$\sum_k \gamma_k P_k = P.$$

Thus the answer to the question posed above is negative.

* * *

   Having now a definition of a strategy, we may make more precise the remarks on page 11 about the possibility of replacing the random variable q by its expectation: the value $w(\mu,\pi)$ of any strategy π does not change on replacing the payoff $q(x_{t-1}a_t x_t)$ by its expectation $\bar{q}(x_{t-1}a_t)$ defined by formula (2.3). One may verify this directly by using formulas (2)–(4), taking account of the fact that the role of the action a_t is played by the pair $(x_{t-1}a_t)$.

§4. Existence of a Uniformly Optimal Strategy. Combination of Strategies

The strategy π is described by a finite collection of nonnegative numbers $\pi(a|h)$. The collections giving the strategies form a closed bounded set Π in a finite dimensional space. The function $w(\pi)$ is continuous, since it is expressed in terms of $\pi(a|h)$ by addition and multiplication operations. A continuous function defined on Π attains a maximum. The strategy which yields the maximum is optimal for the process Z_μ. In particular, for each $x \in X_m$ there exists a strategy π_x which is optimal for the process Z_x. Starting from the strategies π_x, we wish to construct a single strategy $\bar{\pi}$ which is optimal for the model Z.

The following procedure is the natural one: use the strategy π_x if the path starts at the point x. Formally we put

$$\bar{\pi}(\cdot|h) = \pi_{x(h)}(\cdot|h), \tag{1}$$

where $x(h)$ is the initial state of the history h. It is clear that this formula defines some strategy $\bar{\pi}$, and that $w(x,\bar{\pi}) = w(x,\pi_x) = v(x)$ for all $x \in X_m$.

It remains for us to show that *every strategy $\bar{\pi}$ for which*

$$w(x,\bar{\pi}) = v(x), \qquad x \in X_m,$$

is uniformly optimal, i.e. for any μ

$$\sup_\pi w(\mu,\pi) = w(\mu,\bar{\pi}).$$

It follows from (3.2)–(3.4) that for any strategy π

$$w(\mu,\pi) = \sum_{X_m} \mu(x)w(x,\pi). \tag{2}$$

In particular,

$$w(\mu,\bar{\pi}) = \sum_{X_m} \mu(x)w(x,\bar{\pi}).$$

But $w(x,\pi) \le w(x,\bar{\pi})$ for all $x \in X_m$, which means that $w(\mu,\pi) \le w(\mu,\bar{\pi})$.

We have proved that there exists a uniformly optimal strategy. (However we do not yet know whether it may be taken to be a simple strategy.)

For the uniformly optimal strategy $\bar{\pi}$ and any initial distribution μ

$$v(\mu) = w(\mu,\bar{\pi}) = \sum_{X_m} \mu(x)w(x,\bar{\pi}) = \sum_{X_m} \mu(x)v(x) = \mu v.$$

Therefore *the value of any initial distribution μ is expressed in terms of the values of the initial states by the formula*

$$v(\mu) = \mu v. \tag{3}$$

Formulas (2) and (3) make it possible to reduce the study of the control processes Z_μ for arbitrary μ to the study of the processes Z_x. The function $v(x)$, defined for $x \in X_m$, will be called the *value of the model Z*.

The strategy $\bar\pi$ which we constructed above starting from the collection $\{\pi_x\}_{x \in X_m}$ has the following property.

For any initial state $x \in X_m$, the distributions in the space L of paths corresponding to the strategies $\bar\pi$ and π_x under formula (3.2) coincide.

If this property is satisfied, then we will say that the strategy $\bar\pi$ is a *combination* of the strategies π_x. Here $w(x,\bar\pi) = w(x,\pi_x)$ for all $x \in X_m$. However formula (1) may in general fail to be satisfied. We will have to deal in what follows with a combination of strategies in which $\bar\pi$ is not constructed by formula (1), but rather in another way.

§5. The Derived Model. The Fundamental Equation

It is natural to represent a controlled process as a succession of steps. The first step consists of the choice of a probability distribution on A_{m+1}, depending on the initial state. If this choice has been made, then to each initial distribution μ on X_m there corresponds a probability distribution μ' on X'_{m+1}. Taking μ' to be an initial distribution at the instant $m + 1$, we split the maximization problem into two problems:

1) For any initial distribution on X_{m+1}, to choose the optimal behavior at the succeeding instants;

2) To choose the first step in such a way that the sum of the rewards across that step and the value of the optimal behavior at the succeeding instants, under the initial distribution μ', should be maximized.

The model obtained from Z by deleting the first columns X_m and A_{m+1} will be called *the derived model*, and it will be denoted by Z'.

An important rôle in control theory is played by the equation

$$w(x,\pi) = \sum_{A(x)} \pi(a|x)[q(a) + w'(p_a,\pi_a)], \qquad x \in X_m, \tag{1}$$

expressing the value w of any strategy π in the model Z in terms of the values w' of certain strategies in the model Z'. This is the *fundamental equation*. In this equation the initial distribution p_a and the strategy π_a for the model Z' are defined by the formulas

$$p_a = p(\cdot|a), \qquad \pi_a(\cdot|h') = \pi(\cdot|yah'), \tag{2}$$

where $a \in A_{m+1}$, $y = j(a)$, and h' is a history in the model Z'. (π_a prescribes the use of the strategy π, each history h' being prefaced by the prehistory $j(a)a$.

We note that in view of (4.2)

$$w'(p_a,\pi_a) = \sum_{X_{m+1}} p(y|a)w'(y,\pi_a). \tag{3}$$

In order to derive equation (1), we consider the spaces L and L' of paths in the models Z and Z'. Suppose that P is the distribution in L corresponding to the state x and the strategy π, and P_a the distribution in L' corresponding to the initial distribution p_a and the strategy π_a. One sees from formulas (2.1) and (3.2) that for any path l' of L'

$$I(xal') = q(a) + I(l'), \qquad P(xal') = \pi(a|x)P_a(l') \tag{4}$$

for all $a \in A(x)$. In view of (3.3) and (3.4)

$$w(x,\pi) = \sum_L P(l)I(l), \tag{5}$$

$$w'(p_a,\pi_a) = \sum_{L'} P_a(l')I(l'). \tag{6}$$

But $P(l)$ is different from zero only on paths beginning at x, i.e. for paths of the form xal'. Therefore, substituting the values of $I(l)$ and $P(l)$ from (4) into (5) and taking account of (6), we get (1).

In the case of a one-step model, when $m + 1 = n$, the derived model Z' degenerates into a single column X_n and does not contain any actions. Therefore the strategy π_a has no meaning. It is then clear directly from formulas (3.3)–(3.4) and (3.2), defining w and P, that in this case

$$w(x,\pi) = \sum_{A(x)} \pi(a|x)[q(a) + r(p_a)], \tag{7}$$

where

$$r(p_a) = \sum_{X_n} p(y|a)r(y), \tag{8}$$

r being the terminal reward function. In order for formulas (7)–(8) to be considered as a special case of formulas (1)–(3), we agree in the case of a degenerate model Z' to understand by $w'(x,\cdot)$ simply $r(x)$. This is completely consistent with the meaning of the value w. It is then clear that $v' = r$ as well.

§6. Reduction of the Problem of Optimal Control to the Analogous Problem for the Derived Model

Now we are able to justify the splitting of the problem of optimal control described at the beginning of the preceding section.

It follows from the fundamental equation (5.1) that for any $x \in X_m$ and any strategy π

$$w(x,\pi) \le \sup_{A(x)} [q(a) + w'(p_a,\pi_a)] \le \sup_{A(x)} [q(a) + v'(p_a)], \tag{1}$$

v' being the value of the model Z'.

Write

$$u(a) = q(a) + v'(p_a), \qquad a \in A_{m+1}. \tag{2}$$

It is natural to call this quantity the *value of the action a*. We note that in view of (4.3)

$$u = Uv',$$

where the operator U on functions of the states is defined by the formula*

$$Uf(a) = q(a) + \sum_y p(y|a)f(y), \qquad a \in A. \tag{3}$$

Making use of the function u, we may rewrite inequality (1) in the form

$$w(x,\pi) \le Vu(x), \tag{4}$$

where V is the supremum operator relative to the fibre $A(x)$ defined by (1.5).

It follows from (4) that $v \le Vu$. We will prove that $v = Vu$, by constructing a strategy for which equality holds in (4).

We begin with a new concept. Suppose that π' is any strategy in the model Z', and suppose that to each $x \in X_m$ there has been assigned some probability distribution $\gamma(\cdot|x)$ on A_{m+1}, concentrated on the fibre $A(x)$. Choosing at the first step an action a with the distribution γ, and using the strategy π' at the succeeding steps, we obtain a strategy π in the model Z, which is described by the formulas

$$\pi(\cdot|h) = \begin{cases} \gamma(\cdot|x) & \text{if } h = x \in X_m; \\ \pi'(\cdot|h') & \text{if } h = xah'. \end{cases}$$

We shall call this strategy the *product of γ and π'*, and denote it by $\gamma\pi'$.

Obviously if $\pi = \gamma\pi'$, then the strategy π_a described in §5 coincides with π' for any $a \in A_{m+1}$. Therefore, for the product $\gamma\pi'$, the fundamental equation (5.1) takes the form

$$w(x,\gamma\pi') = \sum_{A(x)} \gamma(a|x)[q(a) + w'(p_a,\pi')]. \tag{5}$$

If π' is a strategy optimal for Z' (the existence of such a strategy was established in §4) then $w'(p_a,\pi') = v'(p_a)$, and in view of (2) equation (5) takes the form

$$w(x,\gamma\pi') = \sum_{A(x)} \gamma(a|x)u(a).$$

If for each x the distribution $\gamma(\cdot|x)$ is concentrated on that subset $\bar{A}(x)$ of the fibre $A(x)$ where the function $u(a)$ takes on its maximum $Vu(x)$, then this last equation takes the form

$$w(x,\gamma\pi') = Vu(x), \qquad x \in X_m. \tag{6}$$

---

* Formulas (2)–(3) generalize the definitions of §1 (see (1.6)).

Thus we have proved that

$$v = Vu. \tag{7}$$

It is clear from (6) and (7) that the strategy $\gamma\pi'$ is optimal for the model Z.

In particular we may choose as $\gamma(\cdot\,|\,x)$ a distribution concentrated at any single point $\psi(x)$ of the set $\bar{A}(x)$. The condition $\psi(x) \in \bar{A}(x)$ is equivalent to the equation $u(\psi(x)) = Vu(x)$ or the equation $u(\psi(x)) = v(x)$.

Thus, we have the following results:

a) *The value v of the model Z may be expressed in terms of the value v' of the model Z' by the formulas*

$$v = Vu, \qquad u = Uv'. \tag{8}$$

where the operators V and U are given by the formulas (1.5) *and* (3);

b) *There exists a selector ψ of the correspondence $A(x)$ of X_m into A_{m+1} such that*

$$u(\psi(x)) = v(x); \tag{9}$$

c) *If π is an optimal strategy for the model Z' and the selector ψ is as in b) above, then the strategy $\psi\pi'$ is optimal for the model Z^*.*

* * *

Sometimes it is convenient to use the results a)–c) in a somewhat different form. The selector ψ may be considered as a strategy in a one-step model with the spaces X_m, A_{m+1}, X_{m+1} (and the correspondingly restricted projection j, transition function p, and running payoff q). In view of (5.6)–(5.7) the value of this strategy with the terminal reward f is given by the formula

$$T_\psi f(x) = q[\psi(x)] + \sum_{X_{m+1}} p(y\,|\,\psi(x)) f(y). \tag{10}$$

The operator T_ψ transforms functions on X_{m+1} into functions on X_m. In view of (2) and (10)

$$T_\psi f(x) = Uf[\psi(x)].$$

Putting $T = VU$ or, more explicitly,

$$Tf(x) = \sup_{a \in A(x)} \left[q(a) + \sum_{y \in X_{m+1}} f(y) p(y\,|\,a) \right], \tag{11}$$

one may replace formulas (8) and (9) by

$$v = Tv' \tag{12}$$

---

* As we already noted in §4, in a degenerate model Z' consisting of one column, $v' = r$. In this case we need to replace the formulation c) by the following: any of the functions ψ described under b) defines an optimal strategy for the model Z.

and

$$T_\psi v' = v. \tag{13}$$

In view of (10) and (12), the operator T may be defined in another way by the formula

$$Tf = \sup_\psi T_\psi f, \tag{14}$$

(since the selector ψ may at the point x be equal to any action a of the corresponding fibre).

* * *

If $\pi = \psi\pi'$, where ψ is a selector of the correspondence $x \to A(x)$ for $x \in X_m$ and π' is any strategy in the model Z', then the fundamental equation for the strategy π may, in view of (5.1), (5.3), and (10), be rewritten in the form

$$w(x, \psi\pi') = T_\psi w'(x, \pi'), \tag{15}$$

(the operator being applied to $w'(y, \pi')$ as a function of the argument $y \in X_{m+1}$).

§7. The Optimality Equations. Construction of Simple Optimal Strategies

Without loss of generality we may suppose that $m = 0$ in the original model Z. Consider models $Z_0, Z_1, \ldots, Z_n$, where $Z_0 = Z$ and Z_t is the derived model of $Z_{t-1}, t = 1, \ldots, n$. We will denote the values v and u for the model Z_t by v_t and $u_{t+1}, t = 0, 1, \ldots, n-1$ (v_t and u_t are defined on X_t and A_t respectively). The restrictions of the reward q and of the transition function p to A_t will be denoted by q_t and p_t. From the results of the preceding section, v_t and u_t are connected by the recurrence relations

$$v_{t-1} = V_t u_t, \qquad u_t = U_t v_t, \qquad t = 1, \ldots, n, \tag{1}$$

where

$$U_t f(a) = q_t(a) + \sum_{X_t} p_t(y|a)f(y), \qquad a \in A_t,$$

$$V_t g(x) = \sup_{A(x)} g(a), \qquad x \in X_{t-1},$$

while

$$v_n = r. \tag{2}$$

We will call equations (1) the *optimality equations*. Putting $T_t = V_t U_t$ (see formula (6.12)), we may rewrite the optimality equations in the form

$$v_{t-1} = T_t v_t. \tag{1'}$$

Equations (1) or (1'), along with the boundary condition (2), make it possible to calculate $v_n, v_{n-1}, \ldots, v_0$ in backward succession.

Further, for each $t = 1, 2, \ldots, n$, it is possible to choose the selector ψ_t of the correspondence $A(x)$ of X_{t-1} into A_t such that

$$u_t(\psi_t) = v_{t-1}. \tag{3}$$

In view of the result 6.c the simple strategy $\varphi = \psi_1 \psi_2 \cdots \psi_n$ is optimal for the model Z. Equations (3) may be rewritten in the form

$$T_{\psi_t} v_t = v_{t-1}, \tag{3'}$$

where the operator T_{ψ_t} carries functions on X_t into functions on X_{t-1} according to the formula

$$T_{\psi_t} f(x) = q_t[\psi_t(x)] + \sum_{X_t} p(y \mid \psi_t(x)) f(y) \tag{4}$$

(see (6.10) and (6.14)).

Now suppose that π is any strategy in the derived model Z_k $(k = 1, 2, \ldots, n)$ and that ψ_t are arbitrary selectors of the correspondence $A(x)$ of X_{t-1} into A_t $(t = 1, \ldots, k)$. From equation (6.15), we find by induction that

$$w_0(x, \psi_1 \psi_2 \cdots \psi_k \pi) = T_{\psi_1} T_{\psi_2} \cdots T_{\psi_k} w_k(x, \pi), \tag{5}$$

where w_k denotes the value w in the model Z_k. Formula (5) has a simple intuitive meaning: the result given by the strategy $\psi_1 \psi_2 \cdots \psi_k \pi$ does not change if one interrupts the action at the instant k, taking the terminal reward to be the value of the strategy π.

One may drop the subscripts in equations (1)–(3) and rewrite them in the form

$$v = Vu \quad \text{on } X \backslash X_n,$$
$$u = Uv \quad \text{on } A, \tag{6}$$

$$v = r \quad \text{on } X_n, \tag{7}$$

$$u(\varphi) = v \quad \text{on } X \backslash X_n, \tag{8}$$

or

$$v = Tv \quad \text{on } X \backslash X_n, \tag{9}$$

$$v = r \quad \text{on } X_n, \tag{10}$$

$$T_\varphi v = v \quad \text{on } X \backslash X_n. \tag{11}$$

Here

$$Uf(a) = q(a) + \sum_y p(y \mid a) f(y), \qquad a \in A,$$

$$Vg(x) = \sup_{A(x)} g(a), \qquad x \in X \backslash X_n,$$

$$T = VU,$$

$$T_\varphi f(x) = q[\varphi(x)] + \sum_y f(y) p(y \mid \varphi(x)), \qquad x \in X \backslash X_n.$$

Equations (1)–(2) (or (6)–(7)) generalize the equations (1.9)–(1.10) (resp. (1.7)–(1.8)) deduced earlier for the simplest control process, and formula (3) (or (8)) generalizes formula (1.1)*.

In summary, we have proved the following.

　　1) *The value v of the model satisfies the optimality equations, making it possible to calculate v;*
　　2) *There exists a simple uniformly optimal strategy;*
　　3) *Such a strategy can be found at each stage from equation* (3), *independently from its values at other stages.*

§8. The Markov Property

Suppose that $0 < k < n$. Suppose that on the segment $[0,k]$ we make use of the strategy ρ, and on the segment $[k,n]$ the strategy π (in more precise terms, π is a strategy in the derived model of order k). In analogy with §6 it is natural to say that we are using the strategy $\rho\pi$.

Consider the space L_0 of paths on the segment $[0,n]$ and the space L_k of paths on the segment $[k,n]$. Any function $\xi = \xi(x_k a_{k+1} \cdots x_n)$ in the space L_k may be treated as a function in the space L_0, not depending on $x_0 a_1 \cdots a_k$. Introduce into L_0 a probability distribution $P_x^{\rho\pi}$, corresponding to the initial state x and the strategy $\rho\pi$. Analogously we define a distribution P_y^{π} in the space L_k. It is clear from formula (3.2) that $P_x^{\rho\pi}\xi$ is equal to the expectation, relative to the measure P_x^{ρ}, of the random variable $F(x_k)$, where $F(y) = P_y^{\pi}\xi$.

Indeed, it follows immediately from (3.2) that for any path $y_0 b_1 \cdots b_k y_k b_{k+1} \cdots y_n$

$$P_x^{\rho\pi}(y_0 b_1 \cdots y_n) = P_x^{\rho}(cy_k)P_{y_k}^{\pi}(y_k d),$$

where $c = y_0 b_1 \cdots b_k$ and $d = b_{k+1} \cdots y_n$. Multiplying both sides by $\xi(y_k d)$ and summing over all paths, we get

$$P_x^{\rho\pi}\xi = \sum_{cy_k} P_x^{\rho}(cy_k) \sum_d P_{y_k}^{\pi}(y_k d)\xi(y_k d). \tag{1}$$

Since $P_{y_k}^{\pi}(yd) = 0$ if $y \neq y_k$, then

$$\sum_d P_{y_k}^{\pi}(y_k d)\xi(y_k d) = \sum_{yd} P_{y_k}^{\pi}(yd)\xi(yd) = F(y_k). \tag{2}$$

It remains to substitute (2) into (1) and to observe that

$$\sum_{y_k} P_x^{\rho}(cy_k)F(y_k) = P_x^{\rho}F(x_k).$$

This may be written out as follows:

$$P_x^{\rho\pi}\xi = P_x^{\rho}P_{x_k}^{\pi}\xi. \tag{3}$$

---

* In the analysis of the example at the beginning of §2 we in fact made use of the optimality equations and of equation (8) for an optimal strategy.

It follows from (3) that for any initial distribution μ

$$P_\mu^{\rho\pi}\xi = P_\mu^\rho P_{x_k}^\pi \xi. \tag{4}$$

Put

$$v(y) = P_\mu^\rho\{x_k = y\}, \qquad y \in X_k. \tag{5}$$

The right side of (4) is equal to

$$\sum_{y \in X_k} v(y)P_y^\pi \xi = P_v^\pi \xi,$$

so that

$$P_\mu^{\rho\pi}\xi(x_k a_{k+1} \cdots x_n) = P_v^\pi \xi(x_k a_{k+1} \cdots x_n). \tag{6}$$

Formula (6) shows that if the distribution of the state x_k is known, the probability for the portion of the trajectory on the segment $[k,n]$ does not depend on the distribution μ and the strategy ρ. Put into common terms, the forecast for the "future" (ξ) with a given "present" (v) does not depend on the "past" (μ,ρ). This is the *Markov property*. We emphasize that it is valid only for strategies of the special type $\rho\pi$, i.e. for strategies such that the choice of actions on the segment $[k,n]$ does not depend on the preceding history $x_0 a_1 \cdots a_k$. As a rule, for a general strategy the Markov property does not hold.

Making use of the Markov property, we may estimate the contributions of the intervals $[0,k]$ and $[k,n]$ to the value of the strategy $\rho\pi$. Applying formula (6) to the function $\xi = q(a_{k+1}) + \cdots + q(a_n) + r(x_n)$, we get

$$w(\mu,\rho\pi) = \sum_1^k P_\mu^{\rho\pi} q(a_t) + w(v,\pi). \tag{7}$$

Evidently

$$P_\mu^{\rho\pi} q(a_t) = P_\mu^\rho q(a_t)$$

for $t \leq k$ (one may deduce this formally from (3.2)). Therefore the sum in formula (7) expresses the value $w(\mu,\rho)$ of the strategy ρ for a zero terminal reward, and we may write

$$w(\mu,\rho\pi) = w(\mu,\rho) + w(v,\pi). \tag{8}$$

It is possible to give another interpretation to formula (7). By (4.2) and (5)

$$w(v,\pi) = \sum_y v(y)w(y,\pi) = P_\mu^\rho w(x_k,\pi).$$

Therefore (7) may be rewritten in the form

$$w(\mu,\rho\pi) = P_\mu^\rho\left[\sum_1^k q(a_t) + w(x_k,\pi)\right]. \tag{9}$$

Thus the value of the strategy $\rho\pi$ is equal to the value of the strategy ρ with a terminal reward at the instant k equal to $w(\cdot,\pi)$. A special case of this result, when $\rho = \psi_1 \cdots \psi_k$ is a simple strategy, was presented in §7 (see (7.5)).

§9. The Dynamic Programming Principle

The optimality equations of §7 are special cases of a more general relation which establishes the contributions of various intervals of time to the total value of the model.

Suppose that Z is a model on the segment $[0,n]$, and suppose that $0 \le s < t \le n$. We will denote by $Z_s^t[f]$ the model which is obtained from Z by restricting the time interval to $[s,t]$ and taking the terminal reward at the time t to be f. In particular, if $s = 1$, $t = n$, and $f = r$, we have the derived model Z'. We denote the value of the model $Z_s^t[f]$ corresponding to the terminal reward f by $v_s^t[f]$.

Clearly

$$v_s^t[r] = (VU)^{t-s} r = T^{t-s} r \quad \text{on } X.$$

Hence it follows that for any t on the interval $[0,n]$ one has the equation

$$v_0^n[r] = v_0^t[v_t^n[r]] \quad \text{on } X_0, \tag{1}$$

r being given on X_n.

Equation (1), which is equivalent to the optimality equations (7.6) with the boundary condition (7.7), expresses the *principle of dynamic programming*, according to which, in order to optimize the control on the interval $[0,n]$ with the terminal reward r, one must first optimize the control on the interval $[t,n]$ with the same terminal reward, and then optimize the control on the interval $[0,t]$ with the terminal reward $v_t^n[r]$. It follows in particular from equation (1) that if π'' is an optimal strategy for Z_t^n with the terminal reward r and π' is an optimal strategy for Z_0^t with the terminal reward $v_t^n[r]$, then the strategy $\pi = \pi'\pi''$ has the value $v_0^n[r]$, so that it is optimal for Z_0^n with the terminal reward r.

§10. The Bus, Streetcar, or Walk Problem

We will show by a simple example how one applies the general theory to concrete evaluations.

Consider the situation of an inhabitant of a large city, wishing to arrive at a certain place, and having various means of transport at his disposal. The choice can be made on considerations of cost or convenience, but the decisive factor is the time. Here it is necessary to take into account not only the speed of the transport, but the waiting time. To make the problem concrete, we suppose that a bus will go from the point 0 to the point B in 3 minutes, a streetcar in 10 minutes, and that one can walk it in 20 minutes. The intervals between buses are identically distributed independent random variables with an exponential distribution, so that, independently of the time at which we arrive at the bus stop, it is necessary to wait for the next bus for a time $\ge \tau$ with probability $e^{-\tau/c}$*. The constant c is equal to the mean interval between buses, as planned by the city bus service. We suppose the same about the streetcars, except that the mean interval between them is d. We suppose moreover that the buses and streetcars operate independently.

---

* As to the properties of the exponential distribution see, for example, Feller [1].

In order to have a maximization problem, we shall evaluate a path by the time which it costs, taken with a minus sign.

In the initial state 0, when we arrive at the waiting point, there are two actions: "Walk" or "Wait" (see figure 1.10). The reward for walking is equal to -20, and the reward for waiting is equal to

$$-\int_0^\infty e^{-\tau/c} e^{-\tau/d}\, d\tau = \frac{-cd}{c+d}.$$

After the wait, a bus arrives (state C) with probability $d/(c+d)$, and a streetcar (state D) with probability $c/(c+d)$. In each of the states C and D, one may go or wait. In state C the second solution is clearly unreasonable, and we will suppose that there is then only one decision—go*. The reward for this is equal to -3. The decision "Go" in state D yields the reward -10. The action "Wait" once again costs $-cd/(c+d)$, and we again arrive at the states C or D according to the previous probabilities $d/(c+d)$ or $c/(c+d)$. The decision to "Walk" in state 0 and the decision to "Go" in states C and D lead us to state B.

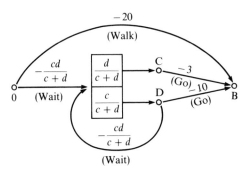

Figure 1.10

In state B, by the nature of the problem, the process stops. In order not to go out of our framework of definitions, we have to give in B a single action, with a zero running reward, carrying the system back into the state B. (We will call states with such actions *absorbing*.)

In order to pass to a nonhomogeneous model we need to fix the number of steps n and to define the terminal reward in the states 0, C, B, and D. The value of the terminal reward in the states 0, C and D must express the loss caused by the fact that we have not arrived at B at all. Put $r(B) = 0$ and $r(0) = r(C) = r(D) = -K$,

---

* We reject here also the absurd solution "Walk" in states C and D.

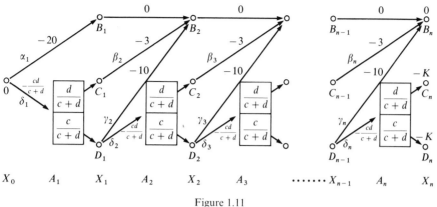

Figure 1.11

where K is a sufficiently large positive number*. Then we obtain the model depicted in figure 1.11. Here we have introduced some simplifications, dictated by the nature of the problem; we have omitted the states B, C, D at the time $t = 0$ and the state 0 at times $t > 0$ are omitted.

In order to write out the optimality equations, we introduce the following notations for the action at the tth step: α_t means "Walk", β_t means "Go by bus", γ_t means "Go by streetcar", and δ_t means "Wait". (Here $t = 1, \ldots ,n$.) Then the system (7.6) may be written in the form

$$v(0) = \max[u(\alpha_1),u(\delta_1)],$$

$$u(\alpha_1) = -20 + v(B_1),$$

$$u(\delta_t) = -\frac{cd}{c+d} + \frac{d}{c+d}v(C_t) + \frac{c}{c+d}v(D_t), \qquad t = 1, \ldots, n, \qquad (1)$$

$$v(C_t) = u(\beta_{t+1}), \qquad t = 1, \ldots, n-1,$$

$$v(D_t) = \max[u(\gamma_{t+1}),u(\delta_{t+1})], \qquad t = 1, \ldots, n-1,$$

$$u(\beta_t) = -3 + v(B_t), \qquad t = 2, \ldots, n,$$

$$u(\gamma_t) = -10 + v(B_t), \qquad t = 2, \ldots, n,$$

and the boundary conditions (7.7) in the form

$$v(B_t) = 0, \qquad t = 1, \ldots, n,$$

$$v(C_n) = -K,$$

$$v(D_n) = -K. \qquad (2)$$

* From what follows it will be clear that the optimal behavior for all $K \geq 20$ is the same. (This is true even for $K \geq 10$ if $n > 1$.)

First of all we are interested in $v(0)$. We find immediately from (1) and (2) that

$$u(\gamma_t) = -10, \qquad t = 2, \ldots, n,$$

$$u(\beta_t) = -3, \qquad t = 2, \ldots, n,$$

$$v(C_t) = -3, \qquad t = 1, \ldots, n-1,$$

$$u(\delta_n) = -\frac{cd}{c+d} - K, \qquad (3)$$

$$u(\alpha_1) = -20,$$

$$v(D_{n-1}) = \max\left[-10, -\frac{cd}{c+d} - K\right] = -10.$$

On substituting the values just found into (1), we find for the remaining unknowns the system

$$v(0) = \max[-20, u(\delta_1)],$$

$$u(\delta_t) = \frac{-3d + cv(D_t) - cd}{c+d}, \qquad t = 1, \ldots, n-1, \qquad (4)$$

$$v(D_t) = \max[-10, u(\delta_{t+1})], \qquad t = 1, \ldots, n-2,$$

where $v(D_{n-1}) = -10$.
From (4) we find that $u(\delta_{n-1}) = \kappa$, where

$$\kappa = \frac{-3d - 10c - cd}{c+d}.$$

The rest of the solution depends on whether κ is larger or smaller than -10. If $\kappa \leq -10$ we have

$$v(D_{n-2}) = \max[-10, \kappa] = -10,$$

$$u(\delta_{n-2}) = \frac{-3d - 10c - cd}{c+d} = \kappa,$$

$$\vdots$$

$$v(D_1) = \max[-10, \kappa] = -10, \qquad (5)$$

$$u(\delta_1) = \frac{-3d - 10c - cd}{c+d} = \kappa,$$

$$v(0) = \begin{cases} -20 & \text{if } \kappa \leq -20, \\ \kappa & \text{if } -20 \leq \kappa \leq -10. \end{cases}$$

If $\kappa \geq -10$, then we successively find that

$$u(\delta_{n-1}) = \kappa_1, \quad \text{where } \kappa_1 = \kappa,$$

$$v(D_{n-2}) = \kappa_1,$$

$$u(\delta_{n-2}) = \kappa_2, \quad \text{where } \kappa_2 = \frac{-3d + c\kappa_1 - cd}{c+d} \geq \kappa \geq -10,$$

$$v(D_{n-3}) = \kappa_2,$$

$$u(\delta_{n-3}) = \kappa_3, \quad \text{where } \kappa_3 = \frac{-3d + c\kappa_2 - cd}{c+d} \geq \kappa \geq -10, \qquad (6)$$

$$\vdots$$

$$v(D_1) = \kappa_{n-2},$$

$$u(\delta_1) = \kappa_{n-1}, \quad \text{where } \kappa_{n-1} = \frac{-3d + c\kappa_{n-2} - cd}{c+d} \geq \kappa \geq -10,$$

$$v(0) = \kappa_{n-1}.$$

Thus,

$$v(0) = \begin{cases} -20 & \text{if } \kappa_1 \leq -20, \\ \kappa_1 & \text{if } -20 \leq \kappa_1 \leq -10, \\ \kappa_{n+1} & \text{if } -10 \leq \kappa_1 \end{cases} \qquad (7)$$

where

$$\kappa_0 = -10, \qquad \kappa_{s+1} = \frac{-3d + c\kappa_s - cd}{c+d}, \qquad s = 0, 1, 2, \ldots.$$

We now turn to the simple optimal strategy. It suffices to give this strategy only in the states 0 and D_t, $t = 1, \ldots, n-1$, in which there is a choice. For $\kappa_1 \leq -20$, from formulas (3), (5)–(7) we have

$$v(0) = -20,$$

$$u(\alpha_1) = -20,$$

$$u(\delta_1) = \kappa \leq -20,$$

$$\left. \begin{array}{l} v(D_t) = -10, \\ u(\gamma_{t+1}) = -10, \\ u(\delta_{t+1}) = \kappa < -10, \end{array} \right\} \quad t = 1, \ldots, n-2,$$

$$v(D_{n-1}) = -10,$$

$$u(\gamma_n) = -10,$$

$$u(\delta_n) = \frac{-cd}{c+d} - K < -10,$$

which means that the optimal strategy is

$$\varphi(0) = \alpha_1,$$
$$\varphi(D_t) = \gamma_{t+1}, \qquad t = 1, \ldots, n-1.$$

If $-20 \leq \kappa_1 \leq -10$ we have

$$v(0) = \kappa,$$
$$u(\alpha_1) = -20 \leq \kappa$$
$$u(\delta_1) = \kappa,$$
$$\left. \begin{array}{l} v(D_t) = -10, \\ u(\gamma_{t+1}) = -10, \\ u(\delta_{t+1}) = \kappa \leq -10, \end{array} \right\} \quad t = 1, \ldots, n-2,$$
$$v(D_{n-1}) = -10,$$
$$u(\gamma_n) = -10,$$
$$u(\delta_n) = \frac{-cd}{c+d} - K < -10,$$

and the optimal strategy is

$$\varphi(0) = \delta_1,$$
$$\varphi(D_t) = \gamma_{t+1}, \qquad t = 1, \ldots, n-1.$$

Finally, if $\kappa \geq -10$

$$v(0) = \kappa_{n-1},$$
$$u(\alpha_1) = -20 < \kappa_{n-1},$$
$$u(\delta_1) = \kappa_{n-1},$$
$$\left. \begin{array}{l} v(D_t) = \kappa_{n-t-1}, \\ u(\gamma_{t+1}) = -10 \leq \kappa_{n-t-1}, \\ u(\delta_{t+1}) = \kappa_{n-t-1}, \end{array} \right\} \quad t = 1, \ldots, n-2,$$
$$v(D_{n-1}) = -10,$$
$$u(\gamma_n) = -10,$$
$$u(\delta_n) = \frac{-cd}{c+d} - K < -10,$$

and the optimal strategy is

$$\varphi(0) = \delta_1,$$

$$\varphi(D_t) = \delta_{t+1}, \qquad t = 1, \ldots, n-2,$$

$$\varphi(D_{n-1}) = \gamma_n.$$

Thus, for $\kappa \leq -20$ one should walk, if $-20 \leq \kappa \leq -10$ one should take the first transportation that comes along. If $-10 \leq \kappa$, one should wait for the bus, but take the streetcar at the last stage n. The natural conclusion suggests itself, that in the original homogeneous schema, if $-10 \leq \kappa$, one should wait for the bus, letting an unbounded number of streetcars go by. In order to justify this conclusion, we have to turn to models over an infinite time interval $[0,\infty)$.

§11. The Replacement Problem

Now we turn to the replacement problem which we formulated in §2. (The other problems which we considered in the Introduction and in §2, are best left aside until we have studied models with general state and action spaces.) The transition function and the payoffs in this model are given by formulas (2.4) and (2.5). In these formulas

$$q_0 \leq q_1 \leq \cdots \leq q_{K-1} \leq q_K = 1, \qquad p_x = 1 - q_x,$$

$$h_0 \geq h_1 \geq h_{K-1} > \alpha > \gamma, \tag{1}$$

$$r_0 \geq r_1 \geq \cdots \geq r_K.$$

We will suppose that the control is exercised over the time interval $[0,n]$. Note that if one adds the number C to all the parameters h_x, α, and γ, then for any strategy the total gain will increase by the same amount, nC. The value v of the model also increases by nC, and the optimal strategies remain as they were before. Therefore, without loss of generality, we may suppose that the parameter γ is equal to zero. The inequality $\alpha > \gamma$ now becomes the condition $\alpha > 0$. In order to return to the initial case we need in the following formulas to replace h_x and α by $h_x - \gamma$ and $\alpha - \gamma$.

In distinction from §8, we do not turn the model around in time, but rather use the optimality equations in the form (7.1)–(7.2). We have

$$v_{t-1}(x) = \max[u_t(xc), u_t(xd)],$$

$$u_t(xc) = p_x h_x + q_x v_t(0) + p_x v_t(x+1), \tag{2}$$

$$u_t(xd) = \alpha + v_t(0),$$

where $0 \leq x \leq K$, $t = 1, \ldots n$, while

$$v_n(x) = r_x, \tag{3}$$

$0 \le x \le K$. The simple optimal strategy is equal to $\varphi = \psi_1 \psi_2 \cdots \psi_n$, where

$$\psi_t(x) = \begin{cases} c & \text{if } u_t(xc) \ge u_t(xd), \\ d & \text{if } u_t(xc) < u_t(xd). \end{cases} \tag{4}$$

(If $u_t(xc) = u_t(xd)$, then either c or d will do; for definiteness we choose c.)

It is difficult to give an explicit calculation for v_t and ψ_t in the general case, but it is possible to give a qualitative description of the answer. It is natural to expect that new equipment is working better than old, so that for any t

$$v_t(0) \ge v_t(1) \ge \cdots \ge v_t(K). \tag{5}$$

Further, for each t we decompose the space X into the set C_t of those states in which the optimal strategy (4) prescribes that one should preserve the equipment in use, and the set D_t in which one must make a replacement. Common sense suggests that if at some time t it is advantageous to replace equipment that has served through a time x, then it is even more advantageous to replace older equipment. This means that D_t should have the form

$$D_t = \{k_t, k_t + 1, \ldots, K\}. \tag{6}$$

Here we note that the state K belongs to D_t, since $u_t(Kc) = v_t(0) < \alpha + v_t(0) = v_t(Kd)$. The set C_t is empty for $k_t = 0$, and for $k_t > 0$ it has the form

$$C_t = \{0, 1, \ldots, k_t - 1\}. \tag{7}$$

Inequalities (5) are verified by induction from t to $t - 1$, while at the same time one establishes stepwise the structures of the sets C_t and D_t. If $t = n$ inequalities (5) follow from the hypotheses (1) and from formula (3). Now suppose that these inequalities are valid for some t, $1 \le t \le n$. Inasmuch as, in view of (2) and (4),

$$C_t = \{x: u_t(xc) \ge \alpha + v_t(0)\}, \qquad D_t = \{x: u_t(xc) < \alpha + v_t(0)\},$$

$$v_{t-1}(x) = \begin{cases} u_t(xc) & \text{if } x \in C_t, \\ \alpha + v_t(0) & \text{if } x \in D_t, \end{cases} \tag{8}$$

then both the structure (6)–(7) of the sets C_t and D_t, and also inequalities (5) for the value v_t, will be proved if we verify that

$$x > 0, \ x \in C_t \Rightarrow u_t(x - 1, c) \ge u_t(xc).$$

Making use of the representation of both terms in the form (2), and taking account of the monotonicity of h_x and $v_t(x)$, we have

$$\begin{aligned} u_t(x-1,c) - u_t(xc) &= (q_{x-1} - q_x)v_t(0) + (1 - q_{x-1})[h_{x-1} + v_t(x)] \\ &\quad - (1 - q_x)[h_x + v_t(x+1)] \\ &\ge (q_{x-1} - q_x)v_t(0) + [(1 - q_{x-1}) - (1 - q_x)][h_x + v_t(x+1)] \\ &= (q_x - q_{x-1})[h_x + v_t(x+1) - v_t(0)]. \end{aligned}$$

Since $q_{x-1} \leq q_x$, then it remains to verify that

$$h_x + v_t(x + 1) \geq v_t(0). \tag{9}$$

Inasmuch as x belongs to the set C_t, then $u_t(xc) \geq \alpha + v_t(0)$ and $p_x > 0$, since if $p_x = 0$ one finds from (2) that $u_t(x) = v_t(0) < \alpha + v_t(0)$. Therefore (9) follows from the relation

$$v_t(0) < \alpha + v_t(0) \leq u_t(xc) = q_x v_t(0) + p_x[h_x + v_t(x + 1)]$$

and from the fact that $p_x + q_x = 1$.

In view of (6) and (8), k_t is the smallest value of x for which $u_t(xc) < \alpha + v_t(0)$, or, taking account of (2),

$$h_x + v_t(x + 1) < v_t(0) + \left(\frac{q_x}{p_x} + 1\right)\alpha. \tag{10}$$

§12. Countable Models: Optimality Equations and ε-Optimal Strategies

Let us say that a model is *finite* if the spaces X_t and A_t are all finite, and that it is *countable* if all of these spaces are finite or countable and at least one of them is countable. To this point we have been considering only finite models. Do their properties extend to countable models as well?

Formula (3.2), defining the distribution P in the space L of paths corresponding to a given initial distribution μ and a given strategy π, remains valid in a countable model as well, but now the space L is not finite but rather countable. Not every random variable has a mathematical expectation on a countable space L with a probability distribution P. In order to have an *apriori* assurance that the formulas (3.3)–(3.4) define a value w for any strategy π, it suffices to require that the running and terminal rewards should be bounded. One may weaken this requirement introducing either of the following two conditions:

   α) The running reward function q and the terminal reward function r are bounded above;
   α') The running reward function q and the terminal reward function r are bounded below.

Then formulas (3.3)–(3.4) will yield a definite quantity as the value w of any strategy π. This value may be equal to $-\infty$ under Condition α), and under Condition α') it may be equal to $+\infty$. In maximization problems, the value $+\infty$ is more inconvenient than $-\infty$. (See Example 13.2 below) Therefore in this section we assume that Condition α) is satisfied*.

Further, a function on a countable set may fail to have a largest value. Therefore in countable models there is no reason to expect that optimal strategies exist.

---

* In Chapter IV, which is devoted to the control over an infinite time interval, the results of this section will be extended to a wider class of models on the finite interval $[m,n]$, containing in particular the models which satisfy Condition α').

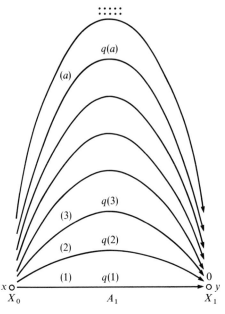

Figure 1.12

For example, if in the schema indicated in figure 1.12, $r = 0$ and $q(a) = a - 1/a$, $a = 1, 2, \ldots$, then for any strategy π

$$w(\pi) = \sum_{a=1}^{\infty} \frac{a-1}{a} \pi(a|x) < \sum_{a=1}^{\infty} \pi(a|x) = 1,$$

while at the same time

$$v(x) = \sup_{\pi} w(\pi) = \sup_{a} \frac{a-1}{a} = 1.$$

The difficulty just mentioned lies in the essence of the problem, and leads to the following modification of the concept of optimality. Suppose that $\varepsilon \geq 0$. The strategy π is said to be *ε-optimal for the process* Z_μ (or for the initial distribution μ) if

$$w(\mu,\pi) \geq v(\mu) - \varepsilon.$$

It is said to be *ε-optimal for the model* Z, or *uniformly ε-optimal*, if that relation is satisfied for all initial distributions μ. (The case $\varepsilon = 0$ brings us back to the preceding definition of optimality.)

We shall consider the changes undergone by the constructions of §§4–9 in passing from finite models to countable models.

If §4, for the finite model Z, we constructed a uniformly optimal (not necessarily simple) strategy. In place of this, in the countable model we will construct for each $\varepsilon > 0$ a uniformly *ε-optimal* strategy.

Suppose that π_x is an ε-optimal strategy for the process Z_x (the existence of such a strategy follows directly from the definition of the least upper bound, since $v(x) < \infty$). As in §4, we construct a combination $\bar{\pi}$ of the strategies π_x using formula (4.1). Since $w(x,\bar{\pi}) = w(x,\pi_x)$ for all $x \in X_m$, then

$$w(x,\bar{\pi}) \geq v(x) - \varepsilon \tag{1}$$

for all $x \in X_m$. We still have to prove that each strategy $\bar{\pi}$ for which the inequality (1) is satisfied is uniformly ε-optimal. As in the finite case, it follows from formulas (3.2)–(3.4) that for any initial distribution μ and any strategy π

$$w(\mu,\pi) = \sum_{X_m} \mu(x)w(x,\pi). \tag{2}$$

We find from (1) and (2) that

$$w(\mu,\pi) = \sum_{X_m} \mu(x)w(x,\pi) \leq \sum_{X_m} \mu(x)v(x)$$
$$\leq \sum_{X_m} \mu(x)[w(x,\bar{\pi}) + \varepsilon] = w(\mu,\bar{\pi}) + \varepsilon.$$

The left inequality above here shows that

$$\sup_{\pi} w(\mu,\pi) \leq \sum_{X_m} \mu(x)v(x), \tag{3}$$

and the right inequality that

$$w(\mu,\bar{\pi}) \geq \sum_{X_m} \mu(x)v(x) - \varepsilon. \tag{4}$$

It follows from (3) and (4) that

$$\sup_{\pi} w(\mu,\pi) = \sum_{X_m} \mu(x)v(x) \leq w(\mu,\bar{\pi}) + \varepsilon. \tag{5}$$

This means that the strategy $\bar{\pi}$ is uniformly ε-optimal.

It is clear from formula (5) that

$$v(\mu) = \sum_{X_m} \mu(x)v(x) = \mu v, \tag{6}$$

as at formula (4.3).

The contents of §5, including the deduction of the fundamental equation, carry over to countable models without change (it makes no difference whether Condition α) or Condition α') is satisfied; we will use this remark in §3 of Chapter 4).

In §6 the arguments are unchanged up to the derivation of the inequality

$$w(x,\pi) \leq Vu(x) \tag{7}$$

for an arbitrary strategy π and $x \in X_m$. As in §6, it follows from (7) that $v \leq Vu$. In order to prove that $v = Vu$, we will now, for any $\varepsilon > 0$, construct a strategy for which

$$w(x,\pi) \geq Vu(x) - \varepsilon. \tag{8}$$

As before, the fundamental equation for the product $\gamma\pi'$ has the form

$$w(x,\gamma\pi') = \sum_{A(x)} \gamma(a|x)[q(a) + w'(p_a,\pi')] \tag{9}$$

(see (6.5)). Suppose that the strategy π' is ε'-optimal for the derived model Z'; such a strategy exists for any $\varepsilon' > 0$. Then $w'(p,\pi') \geq v'(p_a) - \varepsilon'$, and it follows from (9) that

$$w(x,\gamma\pi') \geq \sum_{A(x)} \gamma(a|x)[q(a) + v'(p_a)] - \varepsilon'$$

$$= \sum_{A(x)} \gamma(a|x)u(a) - \varepsilon' \tag{10}$$

(here $u(a) = q(a) + v'(p_a)$ as in §6). The subset $\bar{A}(x)$ of the fibre $A(x)$ on which $u(a) = Vu(x) = \sup_{A(x)} u(a)$ may, in the countable case, turn out to be empty. Instead of it we consider the set

$$A_\kappa(x) = \{a : a \in A(x), u(a) \geq Vu(x) - \kappa\},$$

which is nonempty for any $x \in X_m$ and $\kappa > 0$, and take as $\gamma(\cdot|x)$ any probability distribution on $A(x)$ concentrated on $A_\kappa(x)$. For such a γ

$$\sum_{A(x)} \gamma(a|x)u(a) \geq Vu(x) - \kappa. \tag{11}$$

The inequality (8) follows from (10) and (11) for $\varepsilon' + \kappa \leq \varepsilon$.

Thus, the result 6.a) (i.e. the equations $v = Vu$, $u = Uv'$) remains valid. Instead of 6.b) we now have the following:

b′) *For any $\kappa > 0$ there exists a selector ψ of the correspondence $A(x)$ of X_m into A_{m+1} such that*

$$u(\psi) \geq v - \kappa. \tag{12}$$

The result 6.c), along with its derivation, also remains valid, but we cannot always make use of it, since there do not always exist an optimal strategy π and a selector ψ such that $u(\psi) = v$. The considerations presented above imply the following generalization of 6.c):

c′) *Suppose that ε' and κ are any nonnegative numbers. If the strategy π' is ε'-optimal for the model Z' and if the selector ψ satisfies inequality (12), then the strategy $\psi\pi'$ is $(\varepsilon' + \kappa)$-optimal for the model Z.*

Using the operator T_ψ defined by formula (6.10), we may rewrite condition (12) in the form

$$T_\psi v' \geq v - \kappa. \tag{13}$$

From the result 6.a) one obtains without any modifications all the versions of the optimality equation presented in §7.

It follows from c') that if $\kappa_{m+1}, \kappa_{m+2}, \ldots, \kappa_n$ are arbitrary nonnegative numbers and if

$$T_{\psi_t} v_t \geq v_{t-1} - \kappa_t \tag{14}$$

for $t = m+1, m+2, \ldots, n$, then the simple strategy $\varphi = \psi_{m+1}\psi_{m+2}\cdots\psi_n$ is uniformly ε-optimal for $\varepsilon = \kappa_{m+1} + \kappa_{m+2} + \cdots + \kappa_n$. By b'), for any positive κ_t there are such ψ_t.

Thus, for countable models:

    1) *The value v satisfies the optimality equations;*
    2) *For any $\varepsilon > 0$ there exists a simple uniformly ε-optimal strategy φ;*
    3) *Such a strategy $\varphi = \psi_1\psi_2\cdots\psi_n$ may be found at each step from inequality (14), independently of the values of φ at the other steps.*

Of course, if it is possible to choose all the ψ_t in such a way that (14) is satisfied with $\kappa_t = 0$, then there exists a uniformly optimal strategy $\varphi = \psi_{m+1}\cdots\psi_n$. This would be the case for example if all the fibres $A(x)$ were finite, since the supremum on a finite set is always attained.

The results of §§8,9 carry over completely to countable models.

§13. Countable Models: Sufficiency of Simple Strategies

Would we not lose something if we were to use only simple strategies? The results obtained so far do not yet give an answer to that question. It follows from them only that our losses can be made arbitrarily small. Now we shall prove that *for a fixed initial distribution μ, for each strategy there exists a simple strategy which is not worse*. That simple strategy will in general depend on μ.

This follows from the following two results:

    1. *For each μ and any strategy π there exists a Markov strategy σ such that*

$$w(\mu,\sigma) = w(\mu,\pi). \tag{1}$$

We shall say that such a strategy is *equivalent to π for the process Z_μ.*

    2. *For any Markov strategy σ there exists a simple strategy φ such that*

$$w(\mu,\varphi) \geq w(\mu,\sigma) \quad \text{for all } \mu. \tag{2}$$

We shall say that such a φ is *uniformly not worse than σ, or uniformly majorizes* σ.

$$* \quad * \quad *$$

In order to prove Result 1 we consider the Markov strategy

$$\sigma(a|x) = P\{a_t = a | x_{t-1} = x\} = \frac{P\{x_{t-1}a_t = xa\}}{P\{x_{t-1} = x\}}, \tag{3}$$

where $a \in A_t$, $x \in X_{t-1}$, $t = m+1, \ldots, n$, and P denotes the measure in the space L of paths corresponding to the initial distribution μ and the strategy π. (If it should happen for some x that $P\{x_{t-1} = x\} = 0$, the right hand side in (3) loses its meaning. For such x one may choose as $\sigma(a|x)$ any distribution on $A(x)$.)

The probability distribution Q in the space L corresponding to the initial distribution μ and the strategy σ in general does not coincide with P. But since

$$w(\mu,\pi) = P\left[\sum_{m+1}^{n} q(a_t) + r(x_n)\right] = \sum_{m+1}^{n} Pq(a_t) + Pr(x_n)$$

and analogously

$$w(\mu,\sigma) = \sum_{m+1}^{n} Qq(a_t) + Qr(x_n),$$

then it is sufficient for (1) that each of the elements $x_m, a_{m+1}, \ldots, a_n, x_n$ should have the same probability distribution relative to P and Q.

This assertion is proved by induction. It is true for x_m, because the distribution of x_m relative to both P and Q is equal to μ. Suppose that it is valid for x_{t-1}, where $m + 1 \leq t \leq n$. Since the strategy σ is Markovian, then

$$Q\{x_{t-1}a_t = xa\} = Q\{x_{t-1} = x\}\sigma(a|x), \tag{4}$$

where $a \in A_t$, $x \in X_{t-1}$ (as one finds formally on summing in (3.2)). Using (3) and (4), we get

$$P\{a_t = a\} = \sum_{x \in X_{t-1}} P\{x_{t-1}a_t = xa\} = \sum_{x \in X_{t-1}} P\{x_{t-1} = x\}\sigma(a|x)$$

$$= \sum_{x \in X_{t-1}} Q\{x_{t-1} = x\}\sigma(a|x) = \sum_{x \in X_{t-1}} Q\{x_{t-1}a = xa\} = Q\{a_t = a\}.$$

Thus our assertion is valid for a_t as well. It remains to show that if it is valid for a_t, then it is valid for x_t as well. By the meaning of the transition function

$$P\{a_t x_t = ax\} = P\{a_t = a\}p(x|a), \tag{5}$$

$$Q\{a_t x_t = ax\} = Q\{a_t = a\}p(x|a), \tag{6}$$

where $a \in A_t$, $x \in X_t$ (formally, this also can be obtained by summing on (3.2)). From (5) and (6) we finally obtain

$$P\{x_t = x\} = \sum_{a \in A_t} P\{a_t x_t = ax\} = \sum_{a \in A_t} P\{a_t = a\}p(x|a)$$

$$= \sum_{a \in A_t} Q\{a_t = a\}p(x|a) = \sum_{a \in A_t} Q\{a_t x = ax\} = Q\{x_t = x\}.$$

for all $x \in X_t$.

* * *

The proof of Result 2 is based on the following general lemma.

Lemma 1. *Let f be a function and v a probability distribution on a countable space E. If vf < +∞, then the set*

$$\Gamma = \{x : f(x) \geq vf\}$$

has positive measure v(Γ).*

Proof. Put $c = vf$. If $c = -\infty$, then $\Gamma = E$, $v(\Gamma) = 1$, and the conclusion holds. If c is finite, put $g(x) = f(x) - c$. Then

$$vg = vf - c = 0. \tag{7}$$

But also

$$vg = \sum_{x \in \Gamma} g(x)v(x) + \sum_{x \in E\backslash\Gamma} g(x)v(x). \tag{8}$$

On combining (7) and (8) we get

$$\sum_{x \in E\backslash\Gamma} g(x)v(x) = -\sum_{x \in \Gamma} g(x)v(x). \tag{9}$$

Now suppose $v(\Gamma) = 0$. Then the right hand side of (9) is zero. On the other hand, then $v(E\backslash\Gamma) = 1$, and, since $g(x) < 0$ at every point of $E\backslash\Gamma$, the left hand side of (9) is strictly negative. This contradiction proves the lemma.

In view of (12.2) condition (2) is equivalent to the requirement that

$$w(x,\varphi) \geq w(x,\sigma)$$

for all $x \in X_m$. Following what we did in §6, we decompose the Markov strategy σ into a product $\gamma\sigma'$, γ being the restriction of σ to X_m and σ' its restriction to $X_{m+1} \cup \cdots \cup X_n$. According to formula (5.1)

$$w(x,\sigma) = \gamma_x f,$$

where $\gamma_x(\cdot) = \gamma(\cdot|x)$ is a probability distribution on $A(x)$ and

$$f(a) = q(a) + w'(p_a,\sigma'),$$

$a \in A_{m+1}$. By Lemma 1, the subset of the fibre $A(x)$ on which $f(a) \geq \gamma_x f = w(x,\sigma)$ has positive γ_x-measure, and so is not empty. If $\psi(x)$ is any point of this subset, then $f[\psi(x)] \geq w(x,\sigma)$. But in view of the fundamental equation (5.1), $f[\psi(x)] = w(x,\psi\sigma')$. Hence

$$w(x,\psi\sigma') \geq w(x,\sigma). \tag{10}$$

---

* The lemma is valid, and its proof almost unchanged, for probability distributions and functions on arbitrary measure spaces. We shall make use of this remark in §8 of Chapter 3.

Now suppose that Result 2 is valid for the derived model Z'. Then there exists in this model a simple strategy φ', uniformly majorizing the Markov strategy σ'. In view of (5.1) and our assumption we have

$$w(x,\psi\varphi') = q[\psi(x)] + w'(p_{\psi(x)},\varphi')$$
$$\geq q[\psi(x)] + w'(p_{\psi(x)},\sigma')$$
$$= w(x,\psi\sigma') \geq w(x,\sigma).$$

Accordingly, in the model Z the simple strategy $\varphi = \psi\varphi'$ uniformly majorizes σ, so that Result 2 is valid for the model Z as well.

With obvious modifications, our argument remains valid for the one-step model as well, and thus yields a starting point for the induction. (There are no φ' and σ' and by $w'(p_a,\cdot)$ one understands $p_a r$, r being the terminal reward.)

* * *

The following example shows that the strategy σ in Result 1 can in general not be chosen independently of μ.

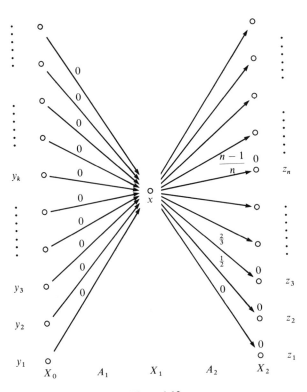

Figure 1.13

EXAMPLE 1. Consider the two-step model depicted in figure 1.13. In this model the value of any Markov strategy σ is equal to

$$w(y_k,\sigma) = \sum_{s=1}^{\infty} \sigma(z_s|x) \frac{s-1}{s},$$

$k = 1, 2, \ldots$. It is constant on X_0 and is smaller than 1. On the other hand, the value for the non-Markovian strategy π, prescribing that in the initial state y_k one should go to z_k, is equal to

$$w(y_k,\pi) = \frac{k-1}{k},$$

$k = 1, 2, \ldots$, and takes on values on X_0 arbitrarily close to 1. This means that here any Markov strategy is worse than π for certain initial states.

The exceptions are the uniformly optimal strategies π for which there always exist Markovian, and even simple strategies φ, equivalent to π for any μ. Since $w(\mu,\pi) = v(\mu)$, then the equivalence of φ and π for any μ implies, simply, the uniform optimality of φ. Thus we assert that *from the existence of any uniformly optimal strategy whatever, it follows that there exists a simple uniformly optimal strategy.*

This assertion follows from Results 1 and 2 and from the fact that in countable models uniform optimality of a strategy π is the same thing as the optimality of π for some fixed initial distribution μ and indeed for any μ satisfying the condition

$$\mu(x) > 0 \quad \text{for every } x \in X_m. \tag{11}$$

To prove this fact we note that according to formulas (12.2) and (12.6), for a strategy π which is optimal for the initial distribution μ,

$$\sum_{X_m} \mu(x)[v(x) - w(x,\pi)] = v(\mu) - w(\mu,\pi) = 0. \tag{12}$$

Since $w(x,\pi) \leq v(x)$, then it follows from (10) and (11) that $w(x,\pi) = v(x)$ for all $x \in X_m$.

* * *

In conclusion we note that if in the definition of a model we had supposed the reward bounded below rather than bounded above (i.e. if we had adopted Condition 12.α') instead of Condition 12.α)) then we would not have Result 2.

EXAMPLE 2. Suppose in the model depicted in figure 1.12, $r = 0$ and $q(a) = 2^a$, $a = 1, 2, \ldots$. Then any simple strategy has a finite value, and the Markov strategy σ defined by the formula $\sigma(a|x) = 2^{-a}$ has the value $w(x,\pi) = +\infty = v(x)$.

Chapter 2

Semicontinuous Models

§1. On the Concept of Measurability

In Chapter 1 we were always dealing with probability distributions in various spaces, such as the state space, action space, and path space. For the finite and countable spaces, which we have been considering to this point, the probability distribution is a very simple concept. In the case of uncountable spaces the situation is more complex. In distinction from the discrete case, here it does not suffice to give the probabilities of individual points. On the other hand, as a rule, it is not possible to define probabilities for all sets in a consistent way. Therefore probabilities are given only for certain classes of sets, called measurable sets.

We shall recall the basic definitions connected with the concept of measurability.

A system of subsets of a space E is called a σ-algebra, if it includes E, contains along with any set its complement, and contains along with any finite or countable collection of sets their union and intersection. We will say that E is a *measurable space* if a σ-algebra in E is fixed (we shall denote it by $\mathcal{B}(E)$). A set is considered to be measurable if and only if it belongs to $\mathcal{B}(E)$.

In a finite or countable space E, one usually adopts the set of all subsets of that space as $\mathcal{B}(E)$. If E is a real line, then we will understand by $\mathcal{B}(E)$ the minimal σ-algebra containing all intervals. The elements of this σ-algebra will be called *Borel sets.*

A mapping i of a measurable space E into a measurable space E' will be said to be *measurable*, if the preimage of any set of $\mathcal{B}(E')$ is in $\mathcal{B}(E)$.

It is clear that from the measurability of the mappings $E \xrightarrow{i} E'$, $E' \xrightarrow{k} E''$ it follows that their product $E \xrightarrow{ki} E''$ is measurable.

A real-valued function f is said to be *measurable* if it defines a measurable mapping into the real line. For this it is necessary and sufficient that all the sets $\{x: f(x) \geq c\}$, or all the sets $\{x: f(x) > c\}$, should be measurable, where c is any constant. All the usual operations over finite or countable sets of functions, such as addition, multiplication, passage to the limit, taking the supremum and infimum, lead once again to measurable functions.

Every measurable subset $\tilde{E}$ of a measurable space E also becomes a measurable space if one puts into $\mathcal{B}(\tilde{E})$. all the subsets of $\tilde{E}$ which belong to $\mathcal{B}(E)$ (these form a σ-algebra in $\tilde{E}$).

We will denote by $E_1 \times E_2 \times \cdots \times E_k$ the set of all ordered k-tuples $x_1 x_2 \cdots x_k$, where $x_t \in E_t$, $t = 1, \ldots, k$. If $E_1, E_2, \ldots, E_k$ are measurable spaces, then $E_1 \times$

$E_2 \times \cdots \times E_k$ may also be considered as a measurable space, where one takes as $\mathscr{B}(E_1 \times E_2 \times \cdots \times E_k)$ the minimal σ-algebra containing all the "rectangle" sets $\Gamma_1 \times \Gamma_2 \times \cdots \times \Gamma_k$, where $\Gamma_t \in \mathscr{B}(E_t)$, $t = 1, 2, \ldots, k$.

In the case when $E_1, E_2, \ldots, E_k$ are real lines, and $\mathscr{B}(E_1), \mathscr{B}(E_2), \ldots, \mathscr{B}(E_k)$ the σ-algebras of their Borel subsets, this construction leads to the k-dimensional coordinate measurable space, the elements of the system $\mathscr{B}(E_1 \times E_2 \times \cdots E_k)$ also being called the *Borel sets*. In what follows, *when we speak of measurable sets in a k-dimensional space, we will always mean Borel sets**.

A *measure* v in a measurable space E is a nonnegative function on $\mathscr{B}(E)$ satisfying the following condition: if Γ is representable in the form of a union of a finite or countable number of disjoint measurable sets Γ_n, then $v(\Gamma)$ is equal to the sum of the $v(\Gamma_n)$. If moreover $v(E) = 1$, then v is said to be a *probability measure* or a *probability distribution*.

If v is a measure in a measurable space E, then to each nonnegative measurable function f on E there corresponds a nonnegative number

$$vf = \int f \, dv = \int_E f(x)v(dx) = \lim_{n \to \infty} \sum_{k=0}^{\infty} \frac{k}{2^n} v\left\{x : \frac{k}{2^n} \le f(x) < \frac{k+1}{2^n}\right\}$$

called the *integral* of f relative to the measure v. Here vf can take on not only finite values but also the value $+\infty$. For any measurable function f one puts

$$vf = vf^+ - vf^-, \tag{1}$$

where $f^+ = \max(f, 0)$, $f^- = \max(-f, 0)$. For the integral vf to have a meaning it is necessary and sufficient that at least one of the numbers vf^+, vf^- should be finite.

We note that if f is the characteristic function[†] χ_Γ of a set $\Gamma \in \mathscr{B}(E)$, then $vf = v(\Gamma)$. Therefore the measure v is uniquely defined if the values of the integral vf are known for all bounded measurable functions.

§2. The General Definition of a Model

The space L of paths consists of all possible collections

$$l = x_m a_{m+1} x_{m+1} \cdots a_n x_n,$$

where

$$x_m \in X_m, \qquad a_{m+1} \in A_{m+1}, \qquad x_{m+1} \in X_{m+1}, \qquad \ldots, \qquad a_n \in A_n, \qquad x_n \in X_n$$

$$j(a_{t+1}) = x_t \qquad (m \le t \le n-1). \tag{1}$$

---

* The class of Borel sets in k-dimensional coordinate space is narrower than the class of sets measurable in the Lebesgue sense. (They are also frequently called measurable sets.) In this connection see the footnote on page 82.

† The characteristic function χ_Γ of a set Γ is equal to 1 on Γ and 0 off Γ.

It is a subset of the product

$$R_n = X_m \times A_{m+1} \times X_{m+1} \times \cdots \times A_n \times X_n.$$

If the factors X_t and A_t are measurable spaces, then R_n is also a measurable space. If L is a measurable subset in R_n, then L may also be considered as a measurable space.

We will suppose that:

α) *The set X of states and the set A of actions are measurable spaces. In addition $X_m, X_{m+1}, \ldots, X_n$ are disjoint measurable subsets of X and $A_{m+1}, \ldots, A_n$ are disjoint measurable subsets of A.*

β) *The mapping j is measurable.*

γ) *The set of all pairs xx, where $x \in X_t$, belongs to $\mathscr{B}(X_t \times X_t)$ for each $t = m, \ldots, n$.*

The conditions $\alpha)$–$\gamma)$ guarantee the measurability of L, and also the measurability of the set H_t of all histories $h = x_m a_{m+1} x_{m+1} \cdots a_t x_t$ at the time t, in the space $R_t = X_m \times A_{m+1} \times X_{m+1} \times \cdots \times A_t \times X_t$.

Indeed, condition (1) may be written in the form

$$\prod_{t=m}^{n-1} \delta_t(x_t, j(a_{t+1})) = 1, \tag{2}$$

where $\delta_t(x,y)$, defined for $(x,y) \in (X_t \times X_t)$ (the "Kronecker delta") is equal to 1 when $x = y$ and 0 when $x \neq y$. It follows from $\gamma)$ that δ_t is measurable on $X_t \times X_t$, so that $\delta_t(x_t, j(a_{t+1}))$ is measurable on R_n. In view of (2), L is a measurable subset of R_n. The measurability of H_t in R_t is proved analogously.

In the finite and countable cases the probability distribution in the space of paths was given by formula (1.3.2). In the general case the corresponding formula is given by

$$P(dx_m \, da_{m+1} \, dx_{m+1} \, da_{m+2} \cdots dx_{n-1} \, da_n \, dx_n)$$
$$= \mu(dx_m)\pi(da_{m+1}|x_m)p(dx_{m+1}|a_{m+1})\pi(da_{m+2}|x_m a_{m+1} x_{m+1})$$
$$\cdots p(dx_{n-1}|a_{n-1})\pi(da_n|x_m a_{m+1} x_{m+1} a_{m+2} \cdots x_{n-1})p(dx_n|a_n) \tag{3}$$

This notation means that

$$Pf = \int_{X_m} \mu(dx_m) \int_{A(x_m)} \pi(da_{m+1}|x_m) \int_{X_{m+1}} p(dx_{m+1}|a_{m+1})$$
$$\times \int_{A(x_{m+1})} \pi(da_{m+2}|x_m a_{m+1} x_{m+1}) \cdots \int_{X_{n-1}} p(dx_{n-1}|a_{n-1})$$
$$\times \int_{A(x_{n-1})} \pi(da_n|x_m a_{m+1} x_{m+1} a_{m+2} \cdots x_{n-1})$$
$$\times \int_{X_n} p(dx_n|a_n)f(x_m a_{m+1} \cdots x_n) \tag{4}$$

for all functions f for which the right side has a meaning.

Consider first the inside integral (over X_n). For it to have a meaning it is necessary that f be measurable. We need to apply formula (4) to the case when f is the value of the path l. Therefore we have to require that the reward functions q and r are also measurable. For the next integral (over A_{n-1}) to exist, we need the first integral to be a measurable function of a_n. This leads us to the necessity of including in the definition of the transition function $p(\cdot|a)$ the requirement that it should be measurable in a. For all subsequent integrals to be meaningful, we need to be concerned with the measurability of the second integral (over $A(x_{n-1})$) relative to the variables $x_m a_{m+1} x_{m+1} \cdots x_{n-1}$. Therefore it is necessary to include in the definition of the strategy $\pi(\cdot|h)$ the requirement that it should be measurable in h.

Of course in the general case it is necessary to retain the hypothesis that q and r are bounded above, introduced in §12 of Chapter 1 for the countable models.

Taking account of all this, we will say that *the elements enumerated in the points* a)–f) *of* §2 *of Chapter 1 define a model, provided conditions* α)–γ) *are satisfied, along with the following requirements*:

δ) *The transition function p is measurable in a, i.e. $p(\Gamma|a)$ is a measurable function on A_t for any Γ of $\mathcal{B}(X_t)$, $t = m+1, \ldots, n$.*

ε) *The running reward function q and the terminal reward function r are measurable and bounded above**.*

In the definition of a strategy we include the following measurability requirement: $\pi(\Gamma|h)$ *is a real-valued measurable function on H_t for any $\Gamma \in \mathcal{B}(A_t)$, $t = m, \ldots, n-1$.* For a simple strategy $\varphi = \psi_1\psi_2 \cdots \psi_n$ this requirement reduces to the condition that ψ_t be a measurable selector of the correspondence $A(x)$ of X_{t-1} into A_t, $t = 1, \ldots, n$, and for a Markov strategy σ to the condition that the functions $\sigma(\Gamma|x)$ should be measurable on X_{t-1} for any $\Gamma \in \mathcal{B}(A_t)$.

* * *

For any pair of measurable spaces E and E', one may speak of a *transition function from E into E'*. This is a function $v(x|\Gamma)$ of the point $x \in E$ and the measurable set $\Gamma \subset E'$, for each x being a probability measure on E', and for each Γ a measurable function on E. In these terms one may say that p is a transition function of A into X, and π a transition function of the space H of all histories into A. We require in addition that the measure $p(\cdot|a)$ should be concentrated on X_t for $a \in A_t$, and $\pi(\cdot|h)$ on the fibre $A(x)$ where x is the endpoint of the history h.

* * *

We have already mentioned in Chapter 1 that it is sometimes convenient to consider the transition function and the running reward function as depending

---

* Instead of the boundedness above of the reward function, we could require that it be bounded below (cf. the analogous remark in §12 of Chapter 1). A more general class of models including both cases is considered in Chapter 5.

not only on a_t but on x_{t-1} as well. This case reduces to the basic one if one adopts the pair $x_{t-1}a_t$ as the action. Such a reduction is applicable to general models; it is necessary only to introduce a measurable structure into the spaces of pairs $x_{t-1}a_t$. We will suppose that a_t and x_t take on values from measurable spaces X_t and A_t, and that the pair $x_{t-1}a_t$ belongs to a measurable subset $\tilde{A}_t$ of the product $X_{t-1} \times A_t$. The projection j is defined here by the formula $j(x_{t-1}a_t) = x_{t-1}$. In order that one can act in any state, we require that j *maps* $\tilde{A}_t$ *onto the entire space* X_{t-1}. Condition β) is obviously satisfied automatically. Conditions α) and γ) do not change, and conditions δ) and ε) are modified in the obvious way.

$$* \quad * \quad *$$

Instead of giving a transition function, one can give a recurrence equation

$$x_t = F_t(x_{t-1}, a_t, s_t), \quad x_{t-1}a_t \in \tilde{A}_t, \quad s_t \in S_t,$$

and a probability distribution Π_t on S_t for the random parameters s_t (cf. the end of §2 of Chapter 1). We will suppose, as we did there, that the parameters s_t are mutually independent. The transition function may be constructed according to the formula

$$p(\Gamma|xa) = \Pi_t\{s_t : F_t(x,a,s_t) \in \Gamma\}, \quad xa \in \tilde{A}_t. \tag{5}$$

For (5) to have a meaning and for it to yield a transition function it is sufficient to require that the function F_t should be measurable in the totality of its arguments.

> This last requirement means by definition that the preimage $F_t^{-1}(\Gamma)$ of any measurable set $\Gamma \subset X_t$ is measurable. The right hand side of (5) describes the measure of an xa-section of $F_t^{-1}(\Gamma)$. It remains to refer to the following well-known facts:
> (a) If C is a measurable set in the product $Y \times Z$, then all the z-sections C_z are measurable sets in Y;
> b) If P is any measure on Y, then $P(C_z)$ is a measurable function on Z.
> See for instance Halmos [1], §§34,35. Also, it is not hard to deduce this from the lemma of §3 in Appendix 4.

§3. Is It Possible to Extend to General Models the Methods Applied in the Study of Finite and Countable Models?

The investigation of finite and countable models in Chapter 1 was based on the fundamental equation (1.5.1). For general models that equation takes on the form

$$w(x,\pi) = \int_{A(x)} \pi(da|x)[q(a) + w'(p_a,\pi_a)], \tag{1}$$

where $x \in X_m$ and π is any strategy. It is proved in the same way as in the finite case; it is necessary only to replace the sums by integrals. Instead of the equation

$P(xal') = \pi(a|x)P_a(l')$ we have to use the formula

$$Pf(x_{m+1}a_{m+1} \cdots x_n) = \int_{A(x)} P_a f(x_{m+1}a_{m+1} \cdots x_n)\pi(da|x),$$

which follows from (2.4). Here the measure P corresponds to the initial state x and the strategy π in the model Z, and the measure P_a to the initial distribution p_a and the strategy π_a in the model Z'.

For the value w' appearing in the fundamental equation, we now have the expression

$$w'(p_a,\pi_a) = \int_{X_{m+1}} w'(y,\pi_a)p(dy|a) \tag{2}$$

(cf. (1.5.3)). This follows from the general formula

$$w(\mu,\pi) = \int_{X_m} w(x,\pi)\mu(dx), \tag{3}$$

which one deduces from (2.3) in the same way as (1.4.2) was deduced from (1.3.2) in the finite case. As in §4 of Chapter 1, it follows from (3) that if $w(x,\pi) = v(x)$ for all $x \in X_m$, then the strategy π is uniformly optimal*. Analogously one carries the Markov property of Chapter 1, §8 over to the general case.

In Chapter 1 we indicated a recurrence procedure for constructing the value v and simple optimal strategies, using the operators U and V. In the general case the operator U is given by the formula

$$Uf(a) = q(a) + \int_X f(y)p(dy|a) \tag{4}$$

(cf. (1.6.3)). For the operator V formula (1.1.5) remains valid. The essential difficulty is connected with the fact that V may carry measurable functions into nonmeasurable functions.

Indeed, suppose that j is an orthogonal projection of a square A onto its side X. As is known, there exists a Borel subset C of the square A such that $j(C)$ is not a Borel subset of X (see §5 of Appendix 2). If $g(a) = \chi_C(a)$, then $Vg(x) = \chi_{j(C)}(x)$, and the latter function is not measurable.

In view of what has been said, the function v_{n-1}, calculated according to the formulas $v_{n-1} = Vu_n$, $u_n = Ur$, may be nonmeasurable. Then the expression Uv_{n-1}, containing an integral, has no meaning, and therefore the recurrence formulas $v_{t-1} = Vu_t$, $u_t = Uv_t$ of Chapter 1, §6, cannot be applied.

One of the ways of dealing with these difficulties is that of dealing only with measurable functions from some class $\mathscr{L}$ which is invariant relative to the operators V and U. Such a method will be investigated in this chapter for the case of semicontinuous models. It is possible to carry over to this case all the results proved for finite models.

---

* We note that all the preceding constructions remain valid as well under the assumption that the reward function is bounded below, a fact of which we shall make use in §2 of Chapter 5.

The construction of a uniformly optimal strategy for finite models made use also of the principle of combination of strategies. For formula (1.4.1) to define a strategy in the general case, it is necessary to guarantee the measurability of $\pi(\cdot|h)$ in h. For this purpose it is necessary to prove special theorems on the possibility of a measurable selection.

A very general class of measurable models, including as special cases countable and semicontinuous models, will be studied in the following chapter. However, to do that it is necessary to apply new, more delicate methods of constructing measurable strategies. (At the same time the results obtained are weaker than for countable and semicontinuous models.)

§4. Definition of a Semicontinuous Model

Our starting point is the analogue between properties of functions defined on finite sets and continuous functions defined on compacta. In particular both attain largest and smallest values. However, for us it is essential only to have the largest value. The proof that a continuous function f on a compact set takes on its supremum uses only the fact that for each c the set $\{x: f(x) \geq c\}$ is closed.

> Indeed, making use of compactness, one constructs a convergent sequence $\{x_n\}$ such that $f(x_n)$ converges to $b = \sup f$. For each $\varepsilon > 0$ the set $\{x: f(x) \geq b - \varepsilon\}$ contains all the x_n from some index onward, and therefore contains the limit x' of the sequence $\{x_n\}$. Inasmuch as $f(x') \geq b - \varepsilon$ for each $\varepsilon > 0$, then $f(x') \geq b$. Since from the definition of b we have $f(x') \leq b$, then $f(x') = b$.

Suppose that E is any metric space*. A function given in E is said to be *semicontinuous* if all the sets $\{x: f(x) \geq c\}$ are closed[†].

Any nonincreasing sequence of semicontinuous functions f_n converges to a semicontinuous function f (which may take on the value $-\infty$). This follows from the obvious relation $\{x: f(x) \geq c\} = \bigcap_n \{x: f_n(x) \geq c\}$. In particular, all limits of nonincreasing sequences of continuous functions are semicontinuous. The converse is also true: *any semicontinuous function f is the limit of a nonincreasing sequence of continuous functions f_n converging to it.*[‡]

---

* The set E is said to be a *metric space* if to any $x, y \in E$ one can assign a nonnegative number $\rho(x, y)$, called the *distance* between x and y, satisfying the conditions
  1) $\rho(x, y) = \rho(y, x)$;
  2) $\rho(x, y) = 0$ if and only if $x = y$;
  3) $\rho(x, z) \leq \rho(x, y) + \rho(y, z)$ for any triple $x, y, z \in E$ (triangle inequality).
The convergence of a sequence $\{x_n\}$ to a point $x_0 \in E$ is defined by the requirement that $\rho(x_n, x_0) \to 0$ as $n \to \infty$.

† It is usual to call such functions upper semicontinuous. Lower semicontinuous functions are those for which the sets $\{x: f(x) \leq c\}$ are all closed. A function is continuous if and only if it is both upper and lower semicontinuous. Since we will not encounter lower semicontinuous functions, we will make use of the abbreviation "semicontinuous function" instead of the complete expression "upper semicontinuous function".

‡ If f is bounded then f_n may be taken bounded as well.

Here is a simple proof of this assertion for a semicontinuous function f which is bounded above (we will not encounter any other cases). Put

$$f_n(x) = \int_0^1 \max\left[-n, F_x\left(\frac{r}{n}\right)\right] dr, \qquad F_x(r) = \sup_{\rho(x,y) \le r} f(y).$$

For each x the function F_x is nonincreasing and tends to $f(x)$ as $r \downarrow 0$. Therefore $f_n \downarrow f^*$. From the triangle inequality it follows that

$$\{y : \rho(x_0, y) \le r - \delta\} \subseteq \{y : \rho(x, y) \le r\} \subseteq \{y : \rho(x_0, y) \le r + \delta\}$$

for $\rho(x_0, x) < \delta < r$. This means that

$$F_{x_0}(r - \delta) \le F_x(r) \le F_{x_0}(r + \delta)$$

for $\rho(x_0, x) < \delta < r$. Accordingly, $F_x(r) \to F_{x_0}(r)$ as $x \to x_0$ for all values of r where $F_{x_0}(r)$ is continuous, i.e. on the entire segment $[0,1]$ except perhaps for a countable number of points. Since

$$-n \le \max\left[-n, F_x\left(\frac{r}{n}\right)\right] \le \sup_x f(x),$$

one may pass to the limit in the expression for $f_n(x)$ under the integral sign as $x \to x_0$, and the function f_n is continuous.

Each metric space E may be considered as a measurable space, by taking as $\mathscr{B}(E)$ the minimal σ-algebra containing all open and closed sets. The elements of this σ-algebra are called *Borel sets* in the space E[†]. We shall denote by $\mathscr{L}(E)$ the collection of all semicontinuous and bounded above functions on E. The model Z is said to be *semicontinuous* if the following four conditions are satisfied.

A. *The set X of states and the set A of actions are separable metric spaces*[‡]. *In addition $X_m, X_{m+1}, \ldots, X_n$ are closed subsets of X and $A_{m+1}, \ldots, A_n$ are closed subsets of A.*

B. *If $x_k \to x \in X$ and $a_k \in A(x_k)$, then the sequence $\{a_k\}$ has a limit point point lying in $A(x)$.* (We will call this property *quasi-continuity* of the correspondence $A(x)$ in x.)[§]

C. *If $f \in \mathscr{L}(X_t)$ and*

$$g(a) = \int_{X_t} p(dx \,|\, a) f(x) \qquad (a \in A_t), \tag{1}$$

then $g \in \mathscr{L}(A_t)$, $t = m + 1, \ldots, n$.

D. *The running reward function q restricted to the set A_t belongs to the set $\mathscr{L}(A_t)$, and the terminal reward function r belongs to $\mathscr{L}(X_n)$.*

---

[*] It follows from the theorem on the monotone passage to the limit under the integral sign that if the functions f_n are measurable, f_1 is bounded above and $f_n \downarrow f$, then $\mu f_n \downarrow \mu f$ for any probability measure μ (see Halmos [1], §27, Theorem 2).

[†] For k-dimensional coordinate space this is consistent with the definition of Borel sets given in §1.

[‡] A metric space E is called *separable* if it is possible to construct a countable subset C with the property: for any x of E and any $\varepsilon > 0$ there exists a y of C such that $\rho(x, y) < \varepsilon$. Such a set C is said to be *everywhere dense*.

[§] A correspondence $A(x)$ is called *closed* if the relations $a_k \in A(x_k)$, $x_k \to x$, $a_k \to a$ imply that $a \in A(x)$. Every quasi-continuous correspondence is closed. The converse is not true.

Conditions $A-D$ are satisfied automatically if the spaces X and A are finite. Thus all finite models are semicontinuous.

Condition C is equivalent to the following simpler requirement:

C_1. *If the function f is continuous and bounded, then the function g defined by* (1) *is also continuous.*

In order to deduce C_1 from C, it suffices to note that the function f is continuous if and only if f and $-f$ are upper semicontinuous. On the other hand, C follows from C_1, since every function of the class $\mathscr{L}$ is the limit of a nonincreasing sequence of bounded continuous functions*.

We note that conditions α and $\gamma)-\varepsilon)$ of §2 follow from conditions $A-D$. Condition β of §2 also follows from $A-D$ if the space X can be represented as the union of a countable number of its compact subsets.

Indeed, $\alpha)$ follows from A and the definition of the σ-algebra $\mathscr{B}(E)$ in the metric space E. It follows from B that the preimage $j^{-1}(C)$ of a compactum $C \subset X$ is a compactum in A; indeed, for any sequence $\{a_n\} \subset j^{-1}(C)$, the sequence $\{j(a_n)\} \subset C$ has a limit point $x_0 \in C$, and then, in view of B, the sequence $\{a_n\}$ has a limit point $a_0 \in A(x_0) = j^{-1}(x_0) \subset j^{-1}(C)$. This means that if C is a compactum, then $j^{-1}(C) \in \mathscr{B}(A)$. If $X = \bigcup_1^\infty C_n$, where the C_n are compact, then any closed set $D \subset X$ is the sum of compacta $D_n = D \cap C_n$, and $j^{-1}(D) = \bigcup_1^\infty j^{-1}(D_n) \in \mathscr{B}(A)$. If the preimages of all closed sets are measurable, then the preimages of all Borel sets are measurable, and we have condition $\beta)$.

The diagonal $\{(x, y) : x = y\}$ of the product $X \times X$ is defined by the equation $\rho(x, y) = 0$. Since ρ is continuous we have $\gamma)$.

In a metric space a semicontinuous function is measurable, since its level sets $\{x : f(x) \geq c\}$, c being a real number, are closed. Therefore $\varepsilon)$ follows from D.

In order to deduce condition $\delta)$, we denote by $\mathscr{R}$ the class of all functions f corresponding to measurable functions g under the formula (1). In view of $C)$, $\mathscr{R}$ contains all continuous bounded functions. Obviously it is closed relative to addition, multiplication by scalars, and bounded passage to the limit. By Lemma 1 of Appendix 5, the class $\mathscr{R}$ contains all bounded measurable functions, and in particular the characteristic functions of all measurable sets (the lemma is applied to the set $\mathscr{C}$ of all continuous bounded functions).

* * *

Now we shall discuss the generalization which we spoke of at the end of §2. It is not hard to verify that under the reduction described there one obtains a semicontinuous model, if the generalized model satisfies conditions A and B in the preceding form, and also the following modifications of conditions C_1 and D:

C_1'. *If f is a continuous bounded function on X_t, then*

$$g(xa) = \int_{X_t} f(y) p(dy \,|\, xa) \tag{2}$$

is a continuous function on $\tilde{A}_t$.

* See footnotes ‡ on p. 50 and * on p. 51.

D′. *The reward function q on the set $\tilde{A}_t$ belongs to $\mathscr{L}(\tilde{A}_t)$, and the reward function r belongs to $\mathscr{L}(X_n)$.*

* * *

Now we turn to the recurrence equation

$$x_t = F_t(x_{t-1}, a_t, s_t) \qquad (x_{t-1}a_t \in \tilde{A}_t,\ s_t \in S_t), \tag{3}$$

considered in §2 of Chapter 1 and §2 of Chapter 2. When does this equation define a semicontinuous model?

We need to impose the previous requirements A, B, *and* D′ *on the spaces* X_t, A_t *and* $\tilde{A}_t$, *and also on the reward functions q and r. The functions F_t should be measurable in the totality of their arguments, and continuous in* $x_{t-1}a_t$*.* The first of these conditions makes it possible to define the transition function $p(\cdot\,|x_{t-1}a_t)$ (see §2). Property C′₁ follows from the second of these conditions. Indeed, suppose that f is a bounded continuous function on X_t. The transition function (2.5) carries f into the function

$$g(xa) = \int_{X_t} f(y)p(dy\,|xa) = \int_{S_t} f[F_t(x,a,s,)]\Pi_t(ds)$$
$$= Ef[F_t(x,a,s_t)] \qquad (xa \in \tilde{A}_t). \tag{4}$$

Here, when f is the indicator of Γ, the equality of the integrals follows from the definition of the transition function p; it then extends to all bounded measurable functions with the aid of Lemma 1 of §1 of Appendix 5. Under the hypotheses we have made, the integrand is bounded and continuous in xa for each $s \in S_t$. The continuity of g follows from Lebesgue's theorem on the passage to the limit under the integral sign[†]. Thus, under the above italicized conditions, equation (3) yields a semicontinuous model.

§5. Optimality Equations and Simple Optimal Strategies

We agree to say that a *correspondence admits a measurable selection* (uniformization), if a measurable selector (see §3 of Chapter 1) exists for it. An example of a mapping not admitting a uniformization will be considered in §1 of Chapter 3 (Example 1).

---

[*] In fact, measurability in the triple x_{t-1}, a_t, and s_t follows from measurability in s_t and the continuity in $x_{t-1}a_t$.

[†] See Halmos [1], §26, Theorem 4.

We shall make use of the following general theorem.

Theorem A. *Suppose that E and E' are separable metric spaces and that $Q(x)$ is a quasi-continuous correspondence of E into E' (see §4,B). If $f \in \mathcal{L}(E')$, then the function*

$$g(x) = \sup_{y \in Q(x)} f(y) \qquad (x \in E)$$

belongs to $\mathcal{L}(E)$, the sets

$$\bar{Q}(x) = \{y : y \in Q(x), f(y) = g(x)\} \qquad (x \in E)$$

are nonempty, and the correspondence $\bar{Q}(x)$ admits a measurable selection.

This theorem will be proved in the next section.

Making use of Theorem A, we may extend the results of §§6,7 of Chapter 1 on the value v and on simple optimal strategies to semicontinuous models. In order to derive these results we need the following properties:

1) *The value v belongs to $\mathcal{L}(X_m)$, and hence it is a measurable function;*
2) *$v(\mu) = \mu v$ for any initial distribution μ;*
3) *There exists a uniformly optimal strategy.*

For finite models Property 1 is trivial, and Properties 2) and 3) were deduced before we started to investigate the connection between the model Z and its derived model Z'. In the semicontinuous case Properties 1)–3) will be proved by induction from Z' to Z, simultaneously with the optimality equation.

Under the assumption that Properties 1)–3) are satisfied for the derived model Z', we show that:

a) *The value v of the model Z may be expressed in terms of the value v' of the model Z' by the equations*

$$v = Vu, \qquad u = Uv', \tag{1}$$

where the operators U and V are defined by the formulas

$$Uf(a) = q(a) + \int_X f(x)p(dx \,|\, a) \qquad (a \in A), \tag{2}$$

$$Vg(a) = \sup_{A(x)} g(a) \qquad (x \in X \backslash X_n); \tag{3}$$

b) *There exists a measurable selector ψ of the correspondence $A(x)$ of X_m into A_{m+1} such that*

$$u[\psi(x)] = v(x); \tag{4}$$

    c) *If π' is an optimal strategy for the model Z', and ψ the selector of* b) *above, then the strategy $\psi\pi'$ is optimal for the model Z;*
    d) *The model Z also has properties* 1)–3).

For a degenerate model consisting of one column X_n, Properties 1)–3) are satisfied in a trivial way ((1) follows from 4.D). By induction, Properties 1)–3) will be valid for any semicontinuous model, and along with them results a)–c) as well.

Just as in §6 of Chapter 1 we deduce from the fundamental equation (3.1) that

$$w(x,\pi) \le Vu(x), \qquad x \in X_m, \tag{5}$$

$$u(a) = q(a) + v'(p_a), \qquad a \in A_{m+1}, \tag{6}$$

and π is any strategy. By the hypotheses 1)–2), $v' \in \mathscr{L}(X_{m+1})$ and

$$v'(p_a) = \int_{X_{m+1}} v'(y)p(dy\,|\,a). \tag{7}$$

Therefore it follows from conditions 4.C–4.D that $u \in \mathscr{L}(A_{m+1})$. Comparing (2) and (6)–(7), we have

$$u = Uv'.$$

Now we will construct a strategy π for which formula (5) is satisfied with equality. Suppose that π' is an optimal strategy for Z' (hypothesis 3)). Then, in view of the fundamental equation and formula (6), for any product $\gamma\pi'$ (see §6 of Chapter 1)

$$w(x,\gamma\pi') = \int_{A(x)} \gamma(da\,|\,x)[q(a) + w'(p_a,\pi')]$$

$$= \int_{A(x)} \gamma(da\,|\,x)[q(a) + v'(p_a)] = \int_{A(x)} u(a)\gamma(da\,|\,x).$$

For the right side to be equal to $Vu(x) = \sup_{A(x)} u(a)$, it suffices that the distribution $\gamma(\cdot\,|\,x)$ should be concentrated at any point $\psi(x)$ of the set

$$\bar{A}(x) = \{a : a \in A(x), u(a) = Vu(x)\};$$

In addition, for $\psi\pi'$ to be a strategy, we need to choose the selector ψ of the correspondence $\bar{A}(x)$ to be measurable. Since $u \in \mathscr{L}(A_{m+1})$, this can be done by Theorem A. From (5) and the equation $w(x,\psi\pi') = Vu(x)$ it follows that $v = Vu$. We have proved a).

It is clear that the selector of the correspondence $A(x)$ satisfies condition b) if and only if ψ is a measurable selector of the correspondence $\bar{A}(x)$. Therefore the considerations of the preceding paragraph prove b) and c).

It remains to be shown that Properties 1)–3) remain valid for the model Z. Property 3) is proved by the construction of the strategy $\psi\pi'$. By Theorem A, Property 1) follows from the inclusion $u \in \mathscr{L}(A_{m+1})$ and the equation $v = Vu$. For the proof of 2) we note that if π is a uniformly optimal strategy for Z, then

$$v(\mu) = w(\mu,\pi) = \int_{X_m} w(x,\pi)\mu(dx) = \int_{X_m} v(x)\mu(dx) = \mu v.$$

As in §6 of Chapter 1, the results a)–b) may be formulated in terms of the operators T_ψ and T, transforming functions on X_{m+1} into functions on X_m according to the formulas

$$T_\psi f(x) = q[\psi(x)] + \int_{X_{m+1}} f(y)p(dy\,|\,\psi(x)) \tag{8}$$

and

$$Tf(x) = \sup_{A(x)} \left[q(a) + \int_{X_{m+1}} f(y)p(dy\,|\,a) \right]. \tag{9}$$

It is clear that both of these operators have meaning for functions f of $\mathscr{L}(X_{m+1})$, and it is clear from conditions 4.C, 4.D and Theorem A that $T\mathscr{L}(X_{m+1}) \subset \mathscr{L}(X_m)$.

In Result a) equations (1) may be replaced by the equation

$$v = Tv',$$

and in Result b) formula (4) may be replaced by the equation

$$T_\psi v' = v.$$

As in §7 of Chapter 1, one deduces from Results a)–c) the optimality equations

$$\begin{aligned} v &= Tv \quad \text{on } X\backslash X_n, \\ v &= r \quad \text{on } X_n, \end{aligned} \tag{10}$$

the existence of a simple strategy $\varphi = \psi_{m+1} \cdots \psi_n$ for which

$$T_\varphi v = v \quad \text{on } X\backslash X_n, \tag{11}$$

and the uniform optimality of that strategy.

* * *

In the case of a generalized model, when the state x_{t-1} is not defined uniquely by the action a_t (see §§2,4), the operators T and T_ψ have to be defined by the formulas

$$Tf(x) = \sup_{a \in A(x)} \left[q(xa) + \int_{X_t} f(y)p(dy\,|\,xa) \right] \quad (x \in X_{t-1}) \tag{12}$$

and

$$T_\psi f(x) = q(x\psi(x)) + \int_{X_t} f(y)p(dy\,|\,x\psi(x)). \tag{13}$$

Here the optimality equations and the condition for a simple optimal strategy φ retain the form (10) and (11).

* * *

If a semicontinuous model is given by equation (4.3), then, in view of (4.4),

$$T_t f(x) = \sup_{a \in A(x)} \left\{ q(xa) + \int_{S_t} f[F_t(x,a,s)] \Pi_t(ds) \right\}$$

$$= \sup_{a \in A(x)} \{ q(xa) + Ef[F_t(x,a,s_t)] \}. \tag{14}$$

The operator T_ψ on the step t may be written out in the form

$$T_{\psi_t} f(x) = q(x\psi(x)) + \int_{S_t} f[F_t(x,\psi(x),s)] \Pi_t(ds)$$

$$= q(x\psi(x)) + Ef[F_t(x,\psi(x),s_t)]. \tag{15}$$

* * *

The calculations of the value of a model and of the optimal strategies relative to formulas (10)–(11) do not present an easy problem. There exist methods of numerical solution using computers, on which we shall not touch. It is possible to obtain simple and explicit expressions in examples, where T transforms into itself a certain family of functions depending on a small number of parameters. We shall use this in the analysis of concrete examples in §§7–11.

* * *

In applications one encounters cases in which the model is not semicontinuous, but explicit calculations show that:

A′, *There exist measurable functions $\tilde{v}_t$ on the spaces X_t, and measurable selectors ψ_t of the correspondences $A(x)$ of X_{t-1} into A_t such that $\tilde{v}_n = r$ and*

$$T_{\psi_t} \tilde{v}_t = T_t \tilde{v}_t = \tilde{v}_{t-1} \qquad (t = m+1, \ldots, n).$$

Then one may assert that $\tilde{v}_t = v_t$ for all t and that the simple strategy $\varphi = \psi_{m+1} \cdots \psi_n$ is uniformly optimal. Indeed, the condition A′ completely replaces Theorem A in carrying out the induction of the first part of this section; in Property 1) semicontinuity of the value v is replaced by measurability.

§6. Theorems on Measurable Selection

Theorem A will be deduced from the following more general theorem.

Theorem B. *Suppose that to each x of a measurable space E there corresponds a nonempty compact subset $Q_x = Q(x)$ of a separable metric space E', and suppose* *that for each y of E' the function $F(x) = \rho(Q_x, y)$ is measurable*. Then the correspondence Q(x) admits a measurable selection.*

Every correspondence $Q(x)$, subject to the conditions of Theorem B, will be called *measurable*.

We note that when $Q(x)$ consists of one point $\psi(x)$, this definition coincides with the usual definition of the measurability of $\psi^\dagger$.

The proof of Theorem B and the deduction of Theorem A from Theorem B are based on the following proposition.

Measurability Criterion. *For the correspondence $Q(x)$ to be measurable it is necessary and sufficient that there exist a sequence of open sets $Q^1(x) \supseteq Q^2(x) \supseteq \cdots \supseteq Q^n(x) \supseteq \cdots \supseteq Q(x)$ with the following properties:*

a) *For any n and y the set*

$$\{x : y \in Q^n(x)\}$$

is measurable;

b) *Every sequence of points $y_n \in Q^n(x)$ has a limit point in the set $Q(x)$.*

For the proof of necessity it suffices to put

$$Q^n(x) = \left\{ y : \rho(y, Q_x) < \frac{1}{n} \right\}.$$

Here the set

$$\{x : y \in Q^n(x)\} = \left\{ x : \rho(Q_x, y) < \frac{1}{n} \right\}$$

is measurable for any n and y, and as a limit point $\bar{y} \in Q(x)$ for a sequence of points $y_n \in Q^n(x)$ one may take a limit point of the sequence $\{y'_n\}$, y'_n being a point of the compactum $Q(x)$ closest to y_n.

In order to prove sufficiency, consider a sequence $\{y_m\}$ which is everywhere dense in E'. Fix y and put

$$f_{mn}(x) = \begin{cases} \rho(y, y_m) & \text{if } y_m \in Q^n(x), \\ +\infty & \text{if } y_m \notin Q^n(x). \end{cases}$$

Then

$$\rho(y, Q^n(x)) = \inf_{y' \in Q^n(x)} \rho(y, y') = \inf_{y_m \in Q^n(x)} \rho(y, y_m) = \inf_m f_{mn}(x).$$

Choose a point y'_n in $Q^n(x)$ so that

$$\rho(y, y'_n) < \rho(y, Q^n(x)) + \frac{1}{n}.$$

---

* The distance $\rho(Q, y)$ from the point y to the set Q is defined as the greatest lower bound of $\rho(x, y)$ over all $x \in Q$.

† Indeed, if U is an ε-neighborhood of the point y, then $\{x : \psi(x) \in U\} = \{x : \rho(Q_x, y) < \varepsilon\}$.

By b), the sequence $\{y_n'\}$ has a limit point $\bar{y} \in Q(x)$. Obviously

$$\rho(y,Q_x) \leq \rho(y,\bar{y}) \leq \lim_{n \to \infty} \rho(y,Q^n(x)) \leq \rho(y,Q_x),$$

so that

$$\rho(y,Q_x) = \lim_{n \to \infty} \rho(y,Q^n(x)).$$

The functions f_{mn} are measurable in view of a). Since measurability is preserved on taking the lower bound and the limit of a sequence of functions, then the function $\rho(y,Q_x)$ is measurable in x. This is true for each y, which means that the correspondence $Q(x)$ is measurable.

Corollary. *If the correspondence $Q(x)$ is measurable, then for any $y' \in E'$ the correspondence*

$$\bar{Q}(x) = \{y : y \in Q(x), \rho(y,y') = \rho(Q(x),y')\}$$

is also measurable.

From the compactness of $Q(x)$ it follows that $\bar{Q}(x)$ is nonempty and compact. Put

$$Q^n(x) = \left\{ y : \rho(y,Q_x) < \frac{1}{n}, \; \rho(y,y') < \rho(y',Q_x) + \frac{1}{n} \right\}.$$

Obviously these sets are open, and they satisfy the needed inclusions as well as condition a) of the measurability criterion. We shall verify condition b). If $y_n \in Q^n(x)$, then there exists a point $y_n' \in Q(x)$ with $\rho(y_n,y_n') < 1/n$. Since $Q(x)$ is compact, the sequence y_n' has a limit point $\bar{y} \in Q(x)$. It is clear that $\bar{y}$ is a limit point for the sequence y_n as well, and that $\rho(\bar{y},y') = \rho(Q_x,y')$.

Now we shall prove Theorem B on measurable selection. We take an everywhere dense sequence $\{y_m\}$ in E' and put

$$Q_0(x) = Q(x),$$

$$Q_m(x) = \{y : y \in Q_{m-1}(x), \rho(y,y_m) = \rho(Q_{m-1}(x),y_m)\},$$

$$(m = 1,2, \ldots; x \in E).$$

In view of the Corollary the sets $Q_m(x)$ are compact and measurable in x. Obviously $Q_1(x) \supseteq \cdots \supseteq Q_m(x) \supseteq \cdots$, and the intersection $Q_\infty(x)$ of these compact sets is nonempty. If $y \in Q_\infty(x)$, then

$$\rho(y_m,y) = \rho(y_m,Q_{m-1}(x)) \qquad (m = 1,2, \ldots). \tag{1}$$

Therefore for any two points $y', y'' \in Q_\infty(x)$

$$\rho(y_m,y') = \rho(y_m,y'')$$

for all y_m. Hence $y' = y''$. Thus $Q_\infty(x)$ consists of a single point, which we shall denote by $\psi(x)$. Since the sets $Q_{m-1}(x)$ are measurable, and, in view of (1),

$$\rho(y_m, \psi(x)) = \rho(y_m, Q_{m-1}(x)),$$

then the function $\rho(y_m, \psi(x))$ is measurable in x for all y_m. Since the convergence $y_m \to y$ implies the convergence

$$\rho(y_{m_k}, \psi(x)) \to \rho(y, \psi(x)),$$

and since $\{y_m\}$ is everywhere dense in E', then the function $\rho(y, \psi(x))$ is measurable in x for any $y \in E'$. By the remark made just before the measurability criterion, this last condition is equivalent to the measurability of the function $\psi(x)$.

Theorem B is proved. Now we shall deduce Theorem A of the preceding section from it.

First of all we note that from the quasi-continuity of $Q(x)$ it follows that any sequence $y_k \in Q(x)$ has a limit point in $Q(x)$. Accordingly, $Q(x)$ is compact.

The set $\bar{Q}(x)$ of Theorem 5.A is nonempty, since the semicontinuous function f attains its maximum $g(x)$ on the compactum $Q(x)$. The set $\bar{Q}(x)$ is compact, as the intersection of a closed set $\{y: f(y) \geq g(x)\}$ with the compactum $Q(x)$.

From the boundedness above of f it follows that g is bounded above. We shall show that g is semicontinuous. Suppose that $x_n \to x$ and $g(x_n) \geq c$. Choosing a point $y_n \in \bar{Q}(x_n)$ for each n, we will have $f(y_n) \geq c$. In view of the quasi-continuity of Q the sequence y_n has a limiting point $y \in Q(x)$. In view of the semicontinuity of f we have $f(y) \geq c$, which means that $g(x) \geq f(y) \geq c$.

By Theorem B, it remains to verify that $\bar{Q}(x)$ is measurable. It follows from the quasi-continuity of Q that for any $y \in E'$ the function $F(x) = -\rho(y, Q_x)$ is semi-continuous and therefore measurable. That means that the correspondence $Q(x)$ is measurable. In order to deduce the measurability of $\bar{Q}(x)$ in x from the measurability of $Q(x)$, we make use of the measurability criterion. Suppose that f_n are continuous functions on E', monotonically converging from above to f. The open sets

$$Q^n(x) = \left\{ y: \rho(y, Q_x) < \frac{1}{n}, \; f_n(y) > g(x) - \frac{1}{n} \right\}$$

are nested and contain $\bar{Q}(x)$. They satisfy condition a) of the measurability criterion, since the mapping $Q(x)$ and the function $g(x)$ are measurable. We shall verify condition b). Suppose that $y_n \in Q^n(x)$ and that y'_n is a point of the compactum $Q(x)$ closest to y_n. Then $\rho(y_n, y'_n) \to 0$. Therefore the point $\bar{y} \in Q(x)$ which is limiting for $\{y'_n\}$ is also limiting for $\{y_n\}$. For $n > m$

$$f_m(y_n) \geq f_n(y_n) > g(x) - \frac{1}{n}.$$

Letting $n \to \infty$, we conclude that

$$f_m(\bar{y}) \geq g(x).$$

As $m \to \infty$ we find that $f(\bar{y}) \geq g(x)$, which means that $\bar{y} \in \bar{Q}(x)$. Thus condition b) of the measurability criterion is also satisfied, and the proof of Theorem A is complete.

§7. The Model for Allocation of a Resource Between Production and Consumption

Armed now with our general results on semicontinuous models, we continue our study of the concrete problems posed in the Introduction (see also §2 of Chapter 1).

We begin with the problem of the distribution of one good between production and consumption. In this problem the output x_t is connected with the input a_t by the equation

$$x_t = F(a_t, s_t), \tag{1}$$

s_t being a random parameter. For X_t and A_t we may take the halfline $[0, +\infty)$. According to the meaning of the problem the action a_t may be chosen from the segment $[0, x_{t-1}]$, so that the fibre $A(x)$ is the segment $[0,x]$, and the set $\tilde{A}_t$ is the region between the x axis and the line $a = x$ indicated in figure 2.1. Conditions A and B of §4 are satisfied; the quasi-continuity of the correspondence $A(x)$ follows from the compactness of the union of the $A(x)$ over all $x \leq c < \infty$. The income over n steps is equal to

$$q_1(x_0 - a_1) + q_2(x_1 - a_2) + \cdots + q_n(x_{n-1} - a_n).$$

By what was said at the end of §4, for semicontinuity of a model it suffices that the functions q_t should be semicontinuous and bounded above, and that the function F should be measurable in the pair (a,s), and continuous in a.

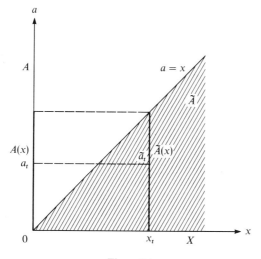

Figure 2.1

It is natural to suppose that if the input does not exceed a constant c, then for any random situation the output cannot exceed some constant $\Phi(c)$. In other words: the function $F(a,s)$ is bounded on each set $[0,c] \times S_t$. Under these conditions one may dispense with the requirement that the functions q_t be bounded above. Indeed, consider the sequence $c_0 = x_0$, $c_t = \Phi(c_{t-1})$. Obviously, given the initial state x_0 and any action, the inequality $x_t \leq c_t$ is satisfied, and we may replace the state space at the time t by the segment $X_t = [0,c_t]$. Here the set $\tilde{A}$ is replaced by the triangle $\tilde{A}_t = \{(a,x) : 0 \leq a \leq x \leq c_t\}$, and the boundedness of the function q_t on $\tilde{A}_t$ follows from its continuity.

Now we apply to our problem the general results of §5. Suppose that v_t is the value of the model on the interval $[t,n]$. According to formulas (5.10) and (5.14),

$$v_n = 0,$$

$$v_{t-1}(x) = T_t v_t(x) = \max_{0 \leq a \leq x} \{q_t(x - a) + Ev_t[F(a,s_t)]\}. \tag{2}$$

A simple optimal strategy $\varphi = \psi_1 \psi_2 \cdots \psi_n$ is obtained if one defines $\psi_t(x)$ as that value of a for which the maximum in formula (2) is attained.

$$* \quad * \quad *$$

We shall carry out some calculations for one very special case, which is however of interest for economics. Suppose that

$$q_t(c) = c^\alpha, \qquad 0 \leq c < \infty \tag{3}$$

for some $\alpha \in (0,1)$, the s_t being independent identically distributed positive random variables, and

$$F(a,s) = as. \tag{4}$$

(Formula (3) describes all homogeneous concave functions. Formula (4) expresses the hypothesis that for any random situation the output is proportional to the input.) Our model is homogeneous. In view of (2)

$$v_t = T^{n-t}0,$$

where

$$Tf(x) = \max_{0 \leq a \leq x} [(x - a)^\alpha + Ef(as_t)] \qquad (x \geq 0). \tag{5}$$

We have

$$T0(x) = \max_{0 \leq a \leq x} (x - a)^\alpha = x^\alpha$$

and

$$\psi_n(x) = 0.$$

The next step should be the application of the operator T to the function $f = x^\alpha$. We solve a slightly more general problem, and calculate $T(bx^\alpha)$, where $b \geq 0$. We have

$$T(bx^\alpha)(x) = \max_{0 \leq a \leq x} [(x - a)^\alpha + \lambda b a^\alpha], \tag{6}$$

where

$$\lambda = E s_t^\alpha;$$

we are supposing that this mathematical expectation is finite. By ordinary calculus methods, we find that the maximum in (6) is achieved at the point $\xi(b)x$ and is equal to $\chi(b)x^\alpha$, where

$$\xi(b) = \frac{(\lambda b)^{\frac{1}{1-\alpha}}}{1 + (\lambda b)^{\frac{1}{1-\alpha}}},$$

$$\chi(b) = [1 + (\lambda b)^{\frac{1}{1-\alpha}}]^{1-\alpha}. \tag{7}$$

Accordingly,

$$v_{n-k}(x) = b_k x^\alpha,$$
$$\psi_{n-k}(x) = d_k x, \qquad (x \geq 0), \tag{8}$$

the numbers b_k and d_k being found from the formulas

$$b_0 = 0,$$

$$b_{k+1} = [1 + (\lambda b_k)^{\frac{1}{1-\alpha}}]^{1-\alpha}, \tag{9}$$

$$d_{k+1} = \frac{(\lambda b_k)^{\frac{1}{1-\alpha}}}{1 + (\lambda b_k)^{\frac{1}{1-\alpha}}} \tag{10}$$

(these numbers are independent of n). It follows from (9) that the numbers

$$c_k = b_k^{\frac{1}{1-\alpha}} \tag{11}$$

are connected by the relation

$$c_{k+1} = 1 + \mu c_k,$$

where

$$\mu = [E s_t^\alpha]^{\frac{1}{1-\alpha}}. \tag{12l}$$

Therefore

$$c_0 = 0,$$
$$c_k = 1 + \mu + \mu^2 + \cdots + \mu^{k-1}. \tag{13}$$

In view of (10) and (11), the coefficients b_k and d_k are expressed in terms of c_k by the formulas

$$b_k = c_k^{1-\alpha}, \qquad d_k = \frac{\mu c_k}{1 + \mu c_k}. \tag{14}$$

Formulas (8) and (12)–(14) yield the complete solution to the problem.

Now we consider the behavior of the optimal action ψ_t at fixed t as the number n of steps tends to infinity. According to (8) and (14)

$$\psi_t(x) = d_{n-t}x = \frac{\mu c_{n-t}}{1 + \mu c_{n-t}} x.$$

If $\mu < 1$, then $c_k \to 1/(1 - \mu)$ as $k \to \infty$, and in the limit

$$\psi_t(x) = \mu x,$$

so that one has to invest a constant share of the output. If on the other hand $\mu \geq 1$, then $c_k \to \infty$ and $\psi_t(x) = x$. This means that at the beginning of the control period one has to invest almost all the resource in expansion of production. (Near the end of that period consumption rapidly increases.) One should however not overestimate the practical value of this last result. It happens due to terms which correspond to very large values of c, and evaluating the utility of such values of c by an unbounded function $q(c) = c^\alpha$ is rather doubtful.

* * *

The explicit solution of the last problem was obtained because we managed to guess at the very beginning a simple set $\mathscr{L}$ of functions $f(x)$, invariant relative to the operators T_t and containing the terminal reward function r. Such a set obviously contains all the functions $v_t = T_{t+1}T_{t+2}\cdots T_n r$, so that, on solving the optimal control problem we need only to consider functions from $\mathscr{L}$. (In our case $\mathscr{L}$ consisted of the functions

$$f(x) = bx^\alpha,$$

where $b \geq 0$, $r = 0$ and the operators T_t did not depend on t.) This trick will help us to solve some other concrete problems.

§8. The Water Regulation Problem

In this problem, described in the Introduction and in §2 of Chapter 1,

$$x_t = \min[x_{t-1} - a_t + s_t, M], \tag{1}$$

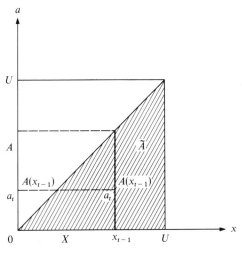

Figure 2.2

where x_t is the quantity of water in a reservoir at the end of the tth period, a_t is the demand for water over that period, s_t is the random influx of water, and M is the volume of the reservoir. The objective function has the form

$$q(a_1) + q(a_2) + \cdots + q(a_n).$$

If we suppose that the s_t are independent identically distributed random variables with distribution Π, then we obtain a homogeneous model. The state space here is the segment $[0,M]$. The same segment serves as the action space A. The fibre $A(x)$ consists of the points $a \in [0,x]$. The set $\tilde{A}$ is now a triangle (see fig. (2.2)). Since the function (1) is continuous in $x_{t-1}a_t$ for each value of s_t, then for semi-continuity of the model it is sufficient that the function q should be semicontinuous and bounded above.

We shall write out the operator T for this model. Since

$$\min(x - a + s, M) = \begin{cases} x - a + s & \text{for } s \leq M - x + a \\ M & \text{for } s > M - x + a \end{cases},$$

then formula (5.14) takes the form

$$Tf(x) = \sup_{0 \leq a \leq x} \left[q(a) + \int_0^{M-x+a} f(x - a + s)\Pi(ds) \right.$$
$$\left. + f(M)\Pi(M - x + a, +\infty) \right] \qquad (0 \leq x \leq M),$$

and the optimality equations become

$$v_n = 0,$$

$$v_{t-1}(x) = \max_{0 \le a \le x}\left[q(a) + \int_0^{M+x-a} v_t(x - a + s)\Pi(ds) \right.$$

$$\left. + v_t(M)\Pi(M - x + a, +\infty) \right] \qquad (0 \le x \le M, 1 \le t \le n).$$

§9. The Problem of the Allocation of Stakes

According to §1.2, this problem is described by the recurrence equation

$$x_t = [a_t\sigma_t + (1 - a_t)\tau_t]x_{t-1}, \tag{1}$$

where $x_{t-1} = a_t x_{t-1} + (1 - a_t)x_{t-1}$ is the allocation of stakes at the tth step and σ_t and τ_t are random coefficients. We will suppose that the pairs $(\sigma_1,\tau_1), (\sigma_2,\tau_2), \ldots,$ $(\sigma_t,\tau_t), \ldots,$ are independent and identically distributed. The objective function coincides with the terminal gain $r(x_n)$.

The model is homogeneous. For the state and action spaces X and A we take the halfline $[0,\infty)$ and the segment $[0,1]$ respectively; see figure 2.3. The fibre $A(x)$ is the whole space A. The spaces X and A satisfy the conditions 4.A and 4.B, and the function F given by formula (1) satisfies the requirement of continuity in $x_{t-1}a_t$. Therefore the model will be semicontinuous if the terminal reward function r is semicontinuous and bounded above. As in the case of the one-good model, the condition that the reward function r should be bounded above may be replaced by the assumption that the random variables σ_t and τ_t should be bounded.

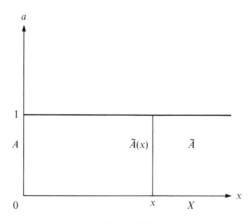

Figure 2.3

In this case the operator T has the form

$$Tf(x) = \sup_{0 \le a \le 1} Ef[a\sigma x + (1 - a)\tau x] \tag{2}$$

Here we have omitted the subscript t on the random variables σ_t, τ_t, since by hypothesis the above expression does not depend on t. The value v_t on the interval $[t,n]$ is found from the relations

$$v_n = r, \qquad v_{t-1} = Tv_t, \qquad t = 1, 2, \dots, n.$$

$$* \ * \ *$$

As in §7, the operator T leaves invariant the set $\mathscr{L}$ of functions of the form

$$f(x) = bx^\alpha \qquad (b \ge 0) \tag{3}$$

(α here being a fixed positive number). Indeed, for the function (3)

$$Tf(x) = \sup_{0 \le a \le 1} Eb[a\sigma x + (1 - a)\tau x]^\alpha = \lambda bx^\alpha, \tag{4}$$

where λ is the supremum on the segment $[0,1]$ of the function

$$\Phi(a) = E[a\sigma + (1 - a)\tau]^\alpha. \tag{5}$$

Therefore it is easy to obtain the solution of the problem for a terminal reward function r lying in $\mathscr{L}$.

It follows from (4)–(5) that for the terminal reward (3) the value of our model on the time interval $[t,n]$ is equal to

$$v_t(x) = b\lambda^{n-t}x^\alpha.$$

Let us suppose that the integral (5), and the integrals obtained from it by differentiating with respect to a, converge uniformly relative to a. (For this it would suffice for example to require that σ and τ take on values in some segment $[\varepsilon, 1/\varepsilon]$, where $\varepsilon > 0$.) Then the function $\Phi(a)$ is continuous, attains its maximal value λ at some point a^*, and

$$\Phi'(a) = \alpha E\{[a\sigma + (1 - a)\tau]^{\alpha - 1}(\sigma - \tau)\},$$
$$\Phi''(a) = \alpha(\alpha - 1)E\{[a\sigma + (1 - a)\tau]^{\alpha - 2}(\sigma - \tau)^2\}. \tag{6}$$

Clearly the optimal action at each step consists in investing the fixed proportion a^* of the means at hand into the first sector, and $1 - a^*$ into the second.

If follows from (6) that $\Phi''(a) < 0$ for $0 < \alpha < 1$ and $\Phi''(a) \geq 0$ for $\alpha \geq 1$. In the second case the function $\Phi(a)$ is convex (linear if $\alpha = 1$), and takes on its largest values at an endpoint of the segment $[0,1]$. That endpoint is $a^* = 0$ or 1 depending on which of the quantities

$$\Phi(0) = E\tau^\alpha$$

or

$$\Phi(1) = E\sigma^\alpha$$

is the larger.

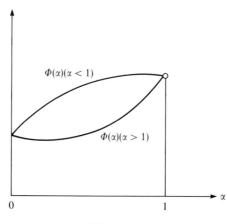

Figure 2.4

In the first case, $\alpha \in (0,1)$, the function $\Phi(a)$ is concave, and the position of the point a^* depends on the signs of the first derivatives

$$\Phi'(0) = \alpha E(\sigma\tau^{\alpha-1} - \tau^\alpha)$$

and

$$\Phi'(1) = \alpha E(\sigma^\alpha - \sigma^{\alpha-1}\tau).$$

Since $\Phi''(0) < 0$, then $\Phi'(1) < \Phi'(0)$. If $0 \leq \Phi'(1)$, then $a^* = 1$. If $\Phi'(0) \leq 0$, then $a^* = 0$. Finally, if $\Phi'(1) < 0 < \Phi'(0)$, then $0 < a^* < 1$ (see figure 2.4). In this last case the point a^* is found from the equation $\Phi'(a) = 0$ or, in expanded form,

$$E\{[a\sigma + (1-a)\tau]^{\alpha-1}(\sigma - \tau)\} = 0. \tag{7}$$

We note the special case when the coefficient τ is not random. Then the double inequality $\Phi'(1) < 0 < \Phi(1)$, under which one should direct the resources into both sectors, takes the form

$$E\sigma^\alpha - \tau E\sigma^{\alpha-1} < 0 < \tau^{\alpha-1}E\sigma - \tau^\alpha$$

or

$$\frac{E\sigma^\alpha}{E\sigma^{\alpha-1}} < \tau < E\sigma \tag{8}$$

* * *

Now suppose that we desire to achieve with the largest possible probability a certain level of accumulation c: we would be equally satisfied with any sum larger than or equal to c, and anything less is worth nothing to us. Without loss of generality, we may suppose that $c = 1$ and that the function r has the form

$$r(x) = \begin{cases} 0 & \text{if } 0 \le x < 1, \\ 1 & \text{if } 1 \le x. \end{cases} \tag{9}$$

We consider only the simplest case, when the coefficient τ is equal to 1 and the coefficient σ takes on the values 2 or 0 with probabilities p and $q = 1 - p$. One may imagine a game where the gambler wins back his stake and as much more again with probability p, and with probability q he loses his stake. He can wager any amount ax, where $0 \le a \le 1$ if his fortune is equal to x. His aim is to obtain at the end of the game a sum not less than 1 with maximum probability.

The optimal behavior of the gambler essentially depends on the relation between p and q. If $p > q$, then the conditions of the game are favorable for the player; according to the law of large numbers, he will achieve the point $x = 1$ with a probability close to 1 if the numbers of games is large enough and the bets are sufficiently small. In the limit, as $n \to \infty$, the value is equal to 1 for all $x > 0$. For $x = 0$, of course, $v = 0$. Determining the value v and the optimal strategy for a fixed number n of steps is a problem. But we will not take up this problem, rather turning to the case $p \le q$, which is nontrivial even when one admits an arbitrarily long interval of play.

In the case $p < q$, by the same law of large numbers, small bets will, with high probability, lead the player to ruin. Therefore the idea occurs that one ought to make the largest possible stakes consistent with the fortune at hand but avoiding unnecessary risk. This means that if $x \le \frac{1}{2}$ one should stake the entire capital x at hand, and if $\frac{1}{2} \le x \le 1$ one should stake the deficit $1 - x$. If $x \ge 1$ one does not play at all. The corresponding strategy is given at each stage by the same selector:

$$\psi(x) = \begin{cases} 1 & \text{for } 0 \le x \le \frac{1}{2}, \\ \dfrac{1-x}{x} & \text{for } \frac{1}{2} \le x \le 1, \\ 0 & \text{for } 1 \le x. \end{cases} \tag{10}$$

We call this the *bold strategy*. We will show that *the bold strategy is optimal for any number n of steps.*

The value v of the model is equal to $T^n r$, and the value $w(\cdot,\psi)$ of the bold strategy ψ is equal to $T^n_\psi r^*$. Therefore the problem reduces to the proof of the equality

$$T^n_\psi r = T^n r. \tag{11}$$

In the present case T operates according to the formula

$$Tf(x) = \sup_{0 \le a \le 1} \left[pf(x + ax) + qf(x - ax) \right]$$
$$= \sup_{0 \le y \le x} \left[pf(x + y) + qf(x - y) \right], \tag{12}$$

gotten from (2) for the coefficients σ and τ at hand. In accordance with (10), the operator T_ψ is given by the formula

$$T_\psi f(x) = \begin{cases} pf(2x) + qf(0) & \text{for } 0 \le x \le \frac{1}{2}, \\ pf(1) + qf(2x - 1) & \text{for } \frac{1}{2} \le x \le 1, \\ f(x) & \text{for } 1 \le x. \end{cases} \tag{13}$$

---

* Formula (1.7.5), being a direct consequence of the fundamental equation, is true also for general models.

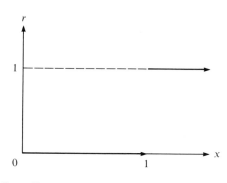

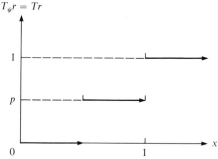

Figure 2.5

We shall prove (11) by induction on n. For $n = 0$ it is trivial: $r = r$. For $n = 1$, starting from (9), we find by a direct calculation that

$$T_\psi r(x) = Tr(x) = \begin{cases} 0 & \text{for } 0 \le x < \tfrac{1}{2}, \\ p & \text{for } \tfrac{1}{2} \le x < 1, \\ 1 & \text{for } 1 \le x; \end{cases}$$

(see figure 2.5)). Now suppose that (11) is true for some $n \ge 1$. We shall prove then that it is true for $n + 1$ as well.

To abbreviate the notation we will write

$$f_t = T_\psi^t r. \tag{14}$$

In view of the induction hypothesis, $f_n = T^n r$, so that the relation $T_\psi^{n+1} r = T^{n+1} r$, which we need, reduces to the equation $f_{n+1} = T f_n$. Since $f_{n+1} = T_\psi f_n \le T f_n$, then it is sufficient to prove the inequality $f_{n+1} \ge T f_n$, which, by virtue of (12), can be rewritten out in the form

$$f_{n+1}(x) \ge p f_n(x + y) + q f_n(x - y) \qquad (0 \le y \le x). \tag{15}$$

It follows immediately from (9), (13), and (14) that

$$0 \le f_t \le 1, \qquad f_t(0) = 0, \qquad f_t(x) = 1 \quad \text{for } x \ge 1$$
$$(t = 0,1,2,\ldots). \tag{16}$$

Hence inequality (15) is valid for $x \ge 1$ and in we need consider now only values $x \le 1$. Further, it is easy to see that the value $v(x) = T^n r(x) = f_n(x)$ is a non-decreasing function of x: the outcome with a large fortune is not worse than that with a smaller one (it suffices to make the same stakes.) Thus if $x + y > 1$ and $x + y' = 1$, then

$$p f_n(x + y') + q f_n(x - y') = p + q f_n(x - y') \ge p + q f_n(x - y)$$
$$= p f_n(x + y) + q f_n(x - y).$$

Therefore inequality (15) will be true for $x + y > 1$ given that it is true for $x + y = 1$. So we may suppose that $x + y \le 1$.

It follows from the induction hypothesis that (15) is valid when n is replaced by $n - 1$:

$$f_n(x) \ge p f_{n-1}(x + y) + q f_{n-1}(x - y) \qquad (0 \le y \le x) \tag{17}$$

(here we are using the fact that $n \ge 1$). In order to pass from (17) to (15) we note that, from (14), $f_{t+1} = T_\psi f_t$. Taking account of (13) and (16), on writing this equation out in detail we have

$$f_{t+1}(x) = \begin{cases} p f_t(2x) & 0 \le x \le \tfrac{1}{2}, \\ p + q f_t(2x - 1) & \text{for } \tfrac{1}{2} \le x \le 1. \end{cases} \tag{18}$$

In the remainder of the derivation of (15) we have to consider four possible cases: 1) $x + y \leq \frac{1}{2}$, 2) $x \leq \frac{1}{2} \leq x + y$, 3) $x - y \leq \frac{1}{2} \leq x$, 4) $\frac{1}{2} \leq x - y \leq x$. In all of these cases $0 \leq y \leq x, x + y \leq 1$.

In the first case it follows from (18) that

$$f_{n+1}(x) = pf_n(2x), \qquad f_n(x \pm y) = pf_{n-1}(2x \pm 2y)$$

Equation (15) is then obtained from (17) by replacing x and y by $2x$ and $2y$ and multiplying by p.

The second case is more complicated. Applying (18) repeatedly, we have

$$f_{n+1}(x) = pf_n(2x) = p^2 + pqf_{n-1}(4x - 1) = p^2 + qf_n(2x - \tfrac{1}{2}),$$
$$f_n(x + y) = p + qf_{n-1}(2x + 2y - 1),$$
$$f_n(x - y) = pf_{n-1}(2x - 2y)$$

Here we have taken into account the inequalities $2x \geq \frac{1}{2}$, $2x - \frac{1}{2} \leq \frac{1}{2}$, which follow from the conditions of the second case. Therefore the desired inequality (15) reduces to

$$f_n(2x - \tfrac{1}{2}) \geq pf_{n-1}(2x + 2y - 1) + pf_{n-1}(2x - 2y)$$

or, using the notations $z = 2x - \frac{1}{2}$, $u = |2y - \frac{1}{2}|$, to

$$f_n(z) \geq p[f_{n-1}(z + u) + f_{n-1}(z - u)]. \tag{19}$$

Since $q \geq p$, this last inequality follows from (17) (we check that $u \leq z$ if $y \leq x$ and $\frac{1}{2} \leq x + y$).

In the third case we find from (18) that

$$f_{n+1}(x) = p + qf_n(2x - 1) = p + qpf_{n-1}(4x - 2)$$
$$= pq + pf_n(2x - \tfrac{1}{2}),$$
$$f_n(x + y) = p + qf_{n-1}(2x + 2y - 1),$$
$$f_n(x - y) = pf_{n-1}(2x - 2y)$$

here we have used the inequalities $2x - 1 \leq \frac{1}{2} \leq 2x - \frac{1}{2}$, holding in this case. That means that (15) now reduces to the inequality

$$q + f_n(2x - \tfrac{1}{2}) \geq p + qf_{n-1}(2x + 2y - 1) + qf_{n-1}(2x - 2y)$$

or, with the same notations as in (19), to

$$f_n(z) \geq p - q + qf_{n-1}(z + u) + qf_{n-1}(z - u) \tag{20}$$

as before, $0 \le u \le z$. Since $p < q$ and $f_{n-1} \le 1$, then

$$p - q \le (p - q)f_{n-1}(z + u).$$

Hence the right side of (20) does not exceed the right side of (17). Therefore (20) follows from (17).

Finally, in the fourth case

$$f_{n+1}(x) = p + qf_n(2x - 1),$$
$$f_n(x \pm y) = p + qf_{n-1}(2x \pm 2y - 1)$$

and (15) easily reduces to (17) on replacing x by $2x - 1$ and y by $2y$.

The optimality of the bold strategy is proved.

·   * * *

Now we make a remark which will be useful in the case of an infinite time interval.

We note that the bold strategy maximizes the probability of the event $C_n = \{$there exists $t \le n$ such that $x_t \ge 1\}$. Indeed, the value of any strategy is equal to the probability of the event $\{x_n \ge 1\}$. For the bold strategy ψ it is equal to f_n and, by the definition of ψ, coincides with the probability of the event C_n. Therefore it is sufficient to prove that for any strategy π

$$P_x^\pi(C_n) \le f_n(x).$$

Suppose that $\bar\pi$ is the strategy obtained from π by the following natural modification: we stop betting as soon as we achieve some state with $y \ge 1$. Obviously

$$P_x^\pi(C_n) = P_x^{\bar\pi}\{x_n \ge 1\},$$

and in view of the optimality of the bold strategy the right side does not exceed $f_n(x)$.

We have $C_1 \subseteq C_2 \subseteq \cdots \subseteq C_n \subseteq \cdots$. Therefore it follows from the above remark that the sequence of functions f_n is nondecreasing and accordingly has a limit f_∞.

* * *

Formula (18) shows that the graph of the function f_{n+1} can be obtained from the graph of f_n as follows. First we contract the graph of f_n along the x axis with the coefficient $\frac{1}{2}$. Then we contract twice the graph of f_n along the y axis, first, with coefficient p and, then, with coefficient q. Then we place the two resulting graphs in the southwest and northeast corners of the unit square; the two parts are joined at the point $x = \frac{1}{2}, y = p$ (see figure 2.6, where for clarity f_n is represented by a continuous convex curve). One

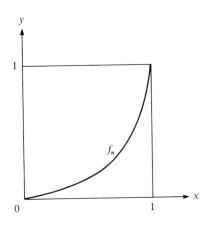 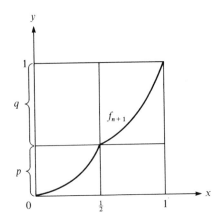

Figure 2.6

proves by induction on n that the function f_n is constant on each halfinterval $[(k-1)/2^n, k/2^n)$, $k = 1, \ldots, 2^n$, and that the magnitude of each jump of f_n lies between p^n and q^n. Further, if the gambler starts with a dyadic-rational fortune of $k/2^n$, and is playing boldly, he wins either 0 or 1 after n games. This fortune will not change. Hence $f_\infty(k/2^n) = f_n(k/2^n)$. Relying on the properties of $f_\infty(x)$ described above, one easily proves that $f_\infty(x)$ is strictly increasing and continuous on the interval $[0,1]$.

§10. The Problem of Allocation of a Resource Among Consumption and Several Productive Sectors

In the problem of allocation of resources among two productive sectors and consumption

$$x_t = i_t[\gamma_t \sigma_t + (1 - \gamma_t)\tau_t], \tag{1}$$

where $x_{t-1} = i_t + c_t$ is the allocation of resources over the period t between production and consumption, γ_t and $1 - \gamma_t$ being the proportions of the resource i_t invested into the first and second sector and σ_t and τ_t being random efficiency coefficients, which we suppose here to be mutually independent. The utility is measured by the quantity

$$q_1(x_0 - i_1) + q_2(x_1 - i_2) + \cdots + q_n(x_{n-1} - i_n). \tag{2}$$

It is natural here to take as X_t the ray $[0, +\infty)$, and as A_t the product $I_t \times \Gamma = [0, +\infty) \times [0,1]$. Since $i_t \le x_{t-1}$, the fibre $A(x)$ is $[0,x] \times [0,1]$. It is easy to see that the spaces X_t and A_t satisfy the conditions 4.A, 4.B, and since the function (1) is continuous in $a_t = (i_t, \gamma_t)$, then for the model to be semicontinuous it suffices that the function q_t should be semicontinuous and bounded above. We note that, as in §7, one can, instead of requiring that the functions q_t be bounded above, require that the random variables σ_t and τ_t be bounded.

In this model the operators T_t are defined by the formulas

$$T_t f(x) = \sup_{\substack{0 \le i \le x \\ 0 \le \gamma \le 1}} \{q_t(x - i) + Ef[i\gamma\sigma_t - i(1 - \gamma)\tau_t]\}$$

$$= \sup_{0 \le i \le x} \{q_t(x - i) + \sup_{0 \le \gamma \le 1} Ef[i\gamma\tau_t + i(1 - \gamma)\tau_t]\}. \tag{3}$$

* * *

We shall consider in detail the homogeneous case, when q_t and the distributions of the pairs (σ_t, τ_t), and accordingly, the operator T_t, do not depend on t. If

$$q(c) = c^\alpha, \tag{4}$$

where $\alpha \in (0,1)$, then these operators, just as they did in §§7,9, leave invariant the set $\mathscr{L}$ of functions of the form

$$f(x) = bx^\alpha \tag{5}$$

with $b \ge 0$. Indeed, for the function (5) we have

$$Tf(x) = \sup_{0 \le i \le x} \{(x - i)^\alpha + \sup_{0 \le \gamma \le 1} Ebi^\alpha[\gamma\sigma + (1 - \gamma)\tau]^\alpha\}$$

$$= \sup_{0 \le i \le x} [(x - i)^\alpha + \lambda bi^\alpha], \tag{6}$$

where

$$\lambda = \sup_{0 \le \gamma \le 1} E[\gamma\sigma + (1 - \gamma)\tau]^\alpha \tag{7}$$

Here we have omitted the subscript t on the random variables σ and τ, because the expectations in (6) and (7) do not depend on t. We have already calculated the expression (6) in §7. There we showed that

$$Tf(x) = \chi(b)x^\alpha \tag{8}$$

and that the supremum in (6) is achieved for

$$i = i(b,x) = \xi(b)x, \tag{9}$$

where $\chi(b)$ and $\xi(b)$ are given by formulas (7.7). The expression (7) was investigated in §9.

Thus we see that our problem breaks into two problems which we have already solved. The optimal proportions γ^* and $1 - \gamma^*$ of each of the productive sectors and the number λ are calculated as in §9 (independent of t and the allocation of resources between consumption and production). Then, as in §7, one finds the optimal allocation of resources between production and consumption. Here the two sectors with efficiency coefficients σ and τ are replaced by a single sector with efficiency coefficient s such that

$$Es^\alpha = E[\gamma^*\sigma + (1 - \gamma^*)\tau]^\alpha = \lambda$$

or $Eq(s) = Eq[\gamma^*\sigma + (1 - \gamma^*)\tau]$.

In view of formulas (7.8) and (7.12)–(7.14) we have

$$v_t(x) = [1 + \lambda^{\frac{1}{1-\alpha}} + \lambda^{\frac{2}{1-\alpha}} + \cdots + \lambda^{\frac{n-t-1}{1-\alpha}}]x^\alpha;$$

The optimal amount of investment in production at the tth step, given a resource x_{t-1}, is equal to

$$i_t^*(x_{t-1}) = \frac{\lambda^{\frac{1}{1-\alpha}} + \lambda^{\frac{2}{1-\alpha}} + \cdots + \lambda^{\frac{n-t}{1-\alpha}}}{1 + \lambda^{\frac{1}{1-\alpha}} + \lambda^{\frac{2}{1-\alpha}} + \cdots + \lambda^{\frac{n-t}{1-\alpha}}} x_{t-1};$$

the optimal shares γ_t^* and $1 - \gamma_t^*$ of the first and second sectors are equal to γ^* and $1 - \gamma^*$, and are calculated as in §9.

* * *

The splitting of the problem described above holds also in the nonhomogeneous case, under the assumption that $q_t(c) = B_t c^\alpha$. (Of course if the distributions of the efficiency coefficients σ_t and τ_t change with t, so does also the optimal proportion γ_t of the first sector.) This splitting holds also if one considers not just two, but an arbitrary number of sectors.

§11. The Stabilization Problem

In the stabilization problem

$$x_t = x_{t-1} - a_t + s_t, \qquad q(x_{t-1}a_t) = -b(x_{t-1} - a_t)^2 - ca_t^2,$$

where b and c are positive constants and the s_t are independent identically distributed random variables. (The terminal reward function is taken to be zero.) We shall consider only the case when $Es_t = 0$, i.e. when there are no systematic perturbations.

The model is homogeneous, and as the spaces X and A we take the real lines $-\infty < x < +\infty$ and $-\infty < a < +\infty$. The model is not semicontinuous, since the fibres $A(x)$ are not compact, and the quasi-continuity condition 4B is violated. Therefore we cannot assert in advance that the value of the model satisfies the optimality equations, or, moreover, that there exists an optimal strategy. However, we shall show that the assertions of the last part of §5 are applicable, i.e. that condition A' is satisfied.

In the case at hand the operator T is given by the formula

$$Tf(x) = \sup_{-\infty < a < +\infty} [-b(x - a)^2 - ca^2 + Ef(x - a + s_t)]. \tag{1}$$

We shall show that the set $\mathscr{L}$ of functions

$$f(x) = -lx^2 - m, \tag{2}$$

where l and m are nonnegative, is invariant relative to the operator T. We have

$$Tf(x) = \sup_a \{-b(x-a)^2 - ca^2 - E[l(x - a + s_t)^2 + m]\}$$

$$= \sup_a [-(b+l)(x-a)^2 - ca^2 - l\sigma^2 - m],$$

where

$$\sigma = Es_t^2$$

is the variance of the random variables s_t. Differentiating with respect to a, we find that the maximum is achieved for

$$a = \frac{b+l}{b+c+l} x \tag{3}$$

and is equal to

$$Tf(x) = -l'x^2 - m',$$

where

$$l' = \frac{c(b+l)}{b+c+l}, \qquad m' = m + l\sigma^2.$$

It therefore follows that the value v_t of the model over the control interval $[t,n]$ is equal to

$$v_t = -(l_{n-t}x^2 + m_{n-t}), \qquad t = 0, \ldots, n, \tag{4}$$

where the coefficients l_k and m_k are calculated recurrently according to the formulas

$$l_0 = 0, \qquad l_{k+1} = \frac{cl_k + bc}{l_k + b + c}, \tag{5}$$

$$m_0 = 0, \qquad m_{k+1} = \sigma^2 l_k + m_k \tag{6}$$

and that the simple strategy

$$a_t = \frac{l_{n-t} + b}{l_{n-t} + b + c} x_{t-1} \tag{7}$$

is optimal on the control interval $[0,n]$.

It remains to express l_k and m_k from equations (5) and (6). We note that $l_{k+1} = g(l_k)$, where $w = g(z)$ is the linear-fractional transformation

$$w = \frac{cz + bc}{z + (b + c)}. \tag{8}$$

The transformation (8) has two fixed points

$$z_{1,2} = \frac{-b \pm \sqrt{b^2 + 4bc}}{2}$$

and it may be rewritten in the form

$$\frac{w - z_1}{w - z_2} = \lambda \frac{z - z_1}{z - z_2},$$

where

$$\lambda = \frac{c - z_1}{c - z_2} = \frac{b + 2c - \sqrt{b^2 + 4bc}}{b + 2c + \sqrt{b^2 + 4bc}}$$

Therefore formula (5) may be put into the form

$$\frac{l_{k+1} - z_1}{l_{k+1} - z_2} = \lambda \frac{l_k - z_1}{l_k - z_2}$$

and therefore

$$\frac{l_k - z_1}{l_k - z_2} = \lambda^k \frac{l_0 - z_1}{l_0 - z_2} = \lambda^k \frac{z_1}{z_2}.$$

Hence

$$l_k = \frac{(1 - \lambda^k)z_1}{1 - \lambda^k \dfrac{z_1}{z_2}}. \tag{9}$$

It follows from (6) that

$$m_k = \sigma^2(l_0 + l_1 + \cdots + l_{k-1}). \tag{10}$$

Since $|\lambda| < 1$, then in the limit as $n - t = k \to \infty$

$$l_k \to l_\infty = z_1 = \frac{\sqrt{b^2 + 4bc} - b}{2}, \tag{11}$$

and the optimal action at each fixed step t becomes

$$a_t = \frac{l_\infty + b}{l_\infty + b + c} x_{t-1}. \tag{12}$$

Chapter 3

General (Borel) Models

§1. Introduction. The Main Results

Lebesgue's measure and integration theory looks equally simple in any measurable space E. However more elaborate constructions, such as conditional distributions, the construction of measures in infinite products and so forth, cannot be carried out in every measurable space. Therefore it is necessary to have a concept of what is to be meant by a "good" measurable space. The class of such spaces must be sufficiently narrow so as to exclude the possibility of pathological examples, and sufficiently broad so as to be invariant to passing to measurable subsets and products spaces.

Two measurable spaces E_1 and E_2 are said to be *isomorphic* if there exists a 1–1 measurable mapping of E_1 onto E_2 such that the inverse mapping is also measurable. A measurable space is said to be *Borelian*, or to be a *Borel space*, if it is isomorphic to a measurable subset of a complete* separable metric space (called for short a *Polish* space).

It is obvious that a measurable subset of a Borel space is also a Borel space.

In Appendix 1 we prove that every Borel space is isomorphic either to a finite set, or to a countable set, or to the unit interval. (In the first two cases the σ-algebra of measurable sets coincides with the system of all subsets, and in the third case with the σ-algebra of Borel sets on the interval.) It follows easily from this that a product of Borel spaces is a Borel space.

In this chapter we will study general models (see §2 of Chapter 2) *under the sole additional assumption that the state space X and the action space A are Borelian.*

In this case the condition γ) of measurability of the diagonal, given in §2 of Chapter 2, is satisfied automatically. In view of the isomorphism of Borel spaces it suffices to consider the cases of a finite set, countable set, and the interval. For the finite and countable cases the measurability of the diagonal is trivial, since all sets are measurable. For the interval it was proved in §4 of Chapter 2.

---

* A metric space is said to be *complete* if every Cauchy sequence in it converges. A sequence $\{x_n\}$ is said to be a *Cauchy sequence* provided $\rho(x_m, x_n) \to 0$ as m and n independently tend to infinity.

For discrete and semicontinuous models we have established three main results:

I. *The value v of the model satisfies the optimality equations*

$$v = Vu \quad \text{on } X \backslash X_n,$$
$$u = Uv \quad \text{on } A, \tag{1}$$

where the operators U and V are defined by the formulas

$$Uf(a) = q(a) + \int_X p(dx\,|\,a)f(x) \qquad (a \in A), \tag{2}$$

$$Vg(x) = \sup_{A(x)} g(a), \tag{3}$$

and the boundary condition

$$v = r \quad \text{on } X_n. \tag{4}$$

II. *For each $\varepsilon > 0$ there exists a simple uniformly ε-optimal strategy. In the finite and semicontinuous cases this is valid with $\varepsilon = 0$ as well.*

III. *For a fixed initial distribution μ and for each strategy π there exists a simple strategy φ which is no worse than π^*.*

We have already spoken in §3 of Chapter 2 of the difficulties encountered in carrying over the methods of Chapter 1 to general models. But is it perhaps possible to achieve the same results by different methods?

EXAMPLE 1. Consider the one-step model Z depicted in figure 3.1. Here X_0 is the segment $0 \leq x \leq 1$, A is a Borel subset of the square $X_0 \times Y$, where Y is the

---

* In the finite and semicontinuous case III is a trivial consequence of II, since one may take as φ a simple uniformly optimal strategy.

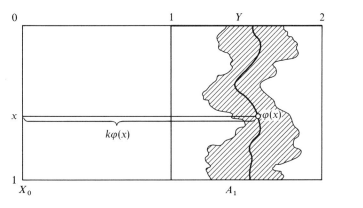

Figure 3.1

segment $1 \leq y \leq 2$, j is the orthogonal projection of A onto X_0; X_1 consists of one point. The transition function is uniquely determined by the condition $p(X_1 | a) = 1$, $a \in A$. The reward functions are arbitrary.

Here the simple strategies are given by measurable selectors ψ of the correspondence j^{-1}. If k is the orthogonal projection of $X_0 \times Y$ onto Y, then for any such selector the composite function $y = k(\varphi(x))$ will be measurable and will have a graph belonging to A. However, it is known, that there exists a Borel subset Q of the square $X_0 \times Y$ which projects onto X_0 and contains no graph of any measurable function $y = f(x)(x \in X_0)$ with values in Y^*. If $A = Q$, then there are no simple strategies in our model, and assertion II is false for it.

The following proposition will be proved in §2.

Proposition 1. *If the correspondence j^{-1} in the model Z does not admit a measurable selection then there is no strategy in this model at all.*

In other words, the existence of a simple strategy is equivalent to the existence of *any* strategy.

Models in which there are no strategies at all will be called *trivial*. For such models the very statement of a control problem loses its meaning and we exclude them from consideration.

Further, the expression (2) for the operator U has a meaning only for measurable functions f. On the other hand, the optimality equations contain the function Uv, and v may be non-measurable, as is shown by the following example.

EXAMPLE 2. Consider a one-step model Z with the same elements X_0, Y, X_1, j, and p as in Example 1, and with $A = X_0 \times Y$ (see figure 3.2). We choose a subset D of A and we put the running reward function q equal to 1 if $a \in D$ and equal to 0 otherwise. To fix ideas, we put the terminal reward function r equal to 0. Obviously,

---

* See §3 of Appendix 3.

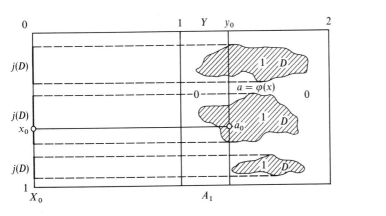

Figure 3.2

if x_0 does not belong to the projection $j(D)$ of the set D into X_0, then $v(x_0) = 0$. If however the point $a_0 = (x_0, y_0)$ of D projects into x_0, then $w(x_0, \varphi) = 1$ for the simple strategy $\varphi(x) = (x, y_0)$, $x \in [0,1]^*$, and therefore $v(x_0) = 1$. Thus,

$$v(x) = \begin{cases} 1 & \text{for } x \in j(D), \\ 0 & \text{otherwise.} \end{cases}$$

It is known[†] that there exists a Borel subset D of the square $X_0 \times Y$ for which the orthogonal projection into the side X_0 is not a Borel set. For such a D the value v will be non-measurable.

Fortunately, for a wide class of functions we can give a meaning to expression (2) for the operator U. In fact, if μ is an arbitrary measure on a measurable space E, then an integral with respect to this measure can be defined not only for measurable functions, but also for all functions f having the following property, which we call μ-measurability: there exists a measurable function $\tilde{f}$ such that $f(x) = \tilde{f}(x)$ (a.s. μ)[‡]. The integral $\mu\tilde{f}$ does not depend on the choice of $\tilde{f}$, and it may be taken as the value of $uf^{\tilde{}}$[§]. The set Γ is said to be μ-measurable if its indicator function χ_Γ is μ-measurable. It is not hard to verify that a function f is μ-measurable if and only if for any number c the set $\{x : f(x) > c\}$ is μ-measurable[¶].

A function f is called universally measurable if it is μ-measurable relative to every probability measure μ. This is equivalent to the requirement that for any c the set $\{x : f(x) > c\}$ is universally measurable, i.e. μ-measurable for any μ.

If the function f is universally measurable, then the integral with respect to any measure has a meaning, and hence the expression (2).

It will be proved that in a nontrivial model the value v is universally measurable, and that in such a model Results I and III are valid.

Result II is false in the general case. In fact, in Example 2 the value of any strategy π is obviously equal to

$$w(x, \pi) = 1 \cdot \pi(D \,|\, x) + 0 \cdot \pi(A \backslash D \,|\, x) \qquad (x \in X_0),$$

and is a measurable function not exceeding the function $v(x)$ (see figure 3.3). For any $\varepsilon \in (0,1)$, the measurable set $\Gamma = \{x : x \in X_0, w(x, \pi) \geq 1 - \varepsilon\}$ is contained in the nonmeasurable set $\{x : x \in X_0, v(x) \geq 1 - \varepsilon\} = j(D)$. Therefore there exists a point $x_0 \in j(D)$ which is not in Γ, and at that point $w(x_0, \pi) < 1 - \varepsilon = v(x_0) - \varepsilon$. Thus no strategy π is ε-optimal for any $\varepsilon < 1$.

---

* We leave it to the reader as an exercise to verify that such a mapping φ of the segment X_0 into the square A is measurable.

† See §5 in Appendix 2.

‡ If $I(x)$ is some property of the point x, the statement $I(x)$ (a.s. μ) means that there exists a measurable set Γ such that $\mu(\Gamma) = 0$ and $I(x)$ is true for all $x \notin \Gamma$. The definition of μ-measurability is often given in another form: f is called μ-measurable if $f_1 \leq f \leq f_2$ everywhere, f_1 and f_2 are measurable and $f_1 = f_2$ (a.s. μ). Both definitions are equivalent. (If $\mu(\Gamma) = 0$ and $f = \tilde{f}$ outside Γ we can put $f_1 = f_2 = f$ outside Γ and $f_1 = -\infty$, $f_2 = +\infty$, on Γ.)

§ It is assumed that at least one of the numbers $\mu f_+ = \mu\tilde{f}_+$ or $\mu f_- = \mu\tilde{f}_-$ is finite (see §1 of Chapter 2).

¶ If E is n-dimensional coordinate space and μ is a measure on the Borel subsets of E equal, for any n-dimensional parallelepiped, to its volume, then μ-measurability of a set or of a function coincides with measurability in the sense of Lebesgue.

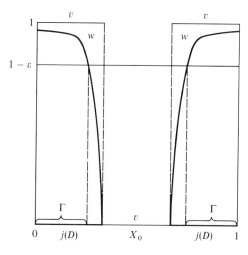

Figure 3.3

We shall prove the following weakened version of Result II.

IIa. *In a nontrivial model, for any $\varepsilon > 0$, and for any initial distribution μ, there exists a simple ε-optimal (a.s. μ) strategy φ.*

(A strategy π is said to be *ε-optimal* (a.s. μ) if for μ-almost all x it is ε-optimal for Z_x, i.e. $w(x,\pi) \geq v(x) - \varepsilon$ (a.s. μ).)

In Chapters 1 and 2 Result II was separated into two assertions, as follows.

II_1. For any $\varepsilon > 0$ there exists a simple strategy φ such that

$$T_{\varphi} v \geq v - \varepsilon \quad \text{on} \quad X \backslash X_n, \tag{5}$$

where the operator T_{φ} is defined by the formula

$$T_{\varphi} f(x) = q(\varphi(x)) + \int_X f(y) p(dy \mid \varphi(x)).$$

(for the finite and semicontinuous models this is true for $\varepsilon = 0$ as well).

II_2. If the simple strategy φ satisfies (5), then it is ε'-optimal for $\varepsilon' = (n - m)\varepsilon$.

Assertion II_1 is invalid in general models (Example 2), and we shall prove Result IIa without making use of II_2. Therefore, in general models, Result II_2 loses its utility, although it remains valid (see the end of §7).

§2. Proof of Main Results: Outline

In order to prove Proposition 1, it is sufficient to construct a simple strategy φ starting from an arbitrary strategy π. We do this in two stages. First we construct a Markov strategy σ, putting

$$\sigma(\cdot \mid x) = \pi(\cdot \mid h),$$

where $h = x_m^0 a_{m+1}^0 \cdots x_{t-1}^0 a_t^0 x$ and $x_m^0 a_{m+1}^0 \cdots x_n^0$ is any fixed path $(x \in X_t,$ $t = m, \ldots, n-1)$.

It is clear that $\sigma(\cdot \,|\, x)$ is a probability distribution on A_t concentrated on $A(x)$. The measurability of $\sigma(\Gamma \,|\, x)$ in x for any $\Gamma \in \mathcal{B}(A_t)$ follows from the fact that the set

$$\{x : \sigma(\Gamma \,|\, x) > c\}$$

is a section of the measurable set

$$\{x_m a_{m+1} \cdots x_{t-1} a_t x : \pi(\Gamma \,|\, x_m a_{m+1} \cdots x_{t-1} a_t x) > c\}$$

for $x_m = x_m^0, a_{m+1} = a_{m+1}^0, \ldots, x_{t-1} = x_{t-1}^0, a_t = a_t^0$; it is well-known that a section of a measurable set in a product space is measurable in the corresponding factor-space.

The second stage is the construction of a simple strategy φ starting from the Markov strategy σ. This stage is based on the following general theorem on measurable selection (see Appendix 3, §2).

Theorem A. *Suppose that i is a measurable mapping of the Borel space E onto the Borel space E', and suppose that $v(\cdot \,|\, x')$, where $x' \in E'$, is a finite measure on E such that:*

1) *For each measurable set Γ of E, the quantity $v(\Gamma \,|\, \cdot)$ is a measurable function on E'.*
2) *The measure $v(\cdot \,|\, x')$ is concentrated on the fibre $E(x') = i^{-1}(x')$, and $v(E \,|\, x') > 0$ for all $x' \in E'$.*

Then the correspondence i^{-1} admits a measurable selection, i.e. there exists a measurable mapping φ of the space E' into the space E such that $i(\varphi(x')) = x'$ for all $x' \in E'$.

We apply this theorem to $E = A$, $E' = X \backslash X_n$, $i = j$, $v = \sigma$.

$$* \quad * \quad *$$

The proof of the universal measurability of the value function v is based on the following representation for it:

$$v(x) = \sup_\pi w(x, \pi) = \sup_{P \in S(x)} PI, \qquad x \in X_m,$$

where I is the value of the path and $S(x)$ is the family of measures in the space of paths corresponding to all possible strategies π and to the initial distribution μ, concentrated at the point x.

Put $kP = x$ if P belongs to $S(x)$. This defines a mapping k of the set

$$S_0 = \bigcup_{X_m} S(x)$$

onto X_m*. The universal measurability of the function v follows from the following general theorem.

Theorem B. *If i is a measurable mapping of the Borel space E onto the Borel space E', and if f is a measurable function on E, then the function*

$$f'(x') = \sup_{i(x) = x'} f(x) \qquad (x' \in E')$$

is universally measurable.

We wish to apply this theorem to $E = S_0$, $E' = X_m$, $i = k$, $f(P) = PI$. To this end, we have to construct a σ-algebra $\mathscr{B}(S_0)$ in S_0 such that S_0 becomes a Borel space, and mapping k and the function $f(P) = PI$ become measurable. This will be done in §§3–6.

Theorem B is a consequence of a still more general result:

Theorem C. *A measurable mapping of a Borel space E into a Borel space E' transforms measurable sets of the space E into universally measurable sets of the space E'* [†].

Theorem C is proved in Appendix 2. In order to deduce Theorem B from it, it suffices to note that for any number c

$$\{x' : f'(x') > c\} = i\{x : f(x) > c\}.$$

* * *

In the construction of ε-optimal strategies of general form we rely on the following theorem on measurable selection, proved in §1 of Appendix 3.

Theorem D. *Suppose that i is a measurable mapping of a Borel space E onto a Borel space E', and μ a probability measure on E'. Then there exists a measurable mapping ψ of E' into E such that*

$$i(\psi(x')) = x' \quad \text{(a.s. } \mu\text{).} [‡] \tag{1}$$

Such a ψ will be called an *measurable* (a.s. μ) *selector of the correspondence* i^{-1}.

Suppose that μ is any initial distribution and ε any positive number. In view of the universal measurability of the function v there exists a measurable function

---

* The mapping k is related to the system $S(x)$ in exactly the same way as the mapping j to the system $A(x)$.

[†] Example 2 of §1 shows that the image of a measurable set under a measurable mapping may fail to be measurable.

[‡] Theorem D loses its validity if one requires (1) to be valid for all x' of E'; this is clear from Example 1 of §1.

$\tilde{v}$ and a measurable subset E' of the set X_m such that $\mu(E') = 1$ and $v(x) = \tilde{v}(x)$ for all x of E'. Denote by E the collection of all measures P of S_0 satisfying the conditions

$$kP \in E', \qquad PI \geq \tilde{v}(kP) - \varepsilon. \qquad (2)$$

It is clear that E' belongs to $\mathscr{B}(X_m)$ and E to $\mathscr{B}(S_0)$, so that E and E' may be considered as Borel spaces. The mapping k induces a measurable mapping i of the space E into the space E'. By the definition of v and $\tilde{v}$, for each x of E' there exists a measure P of $S(x)$ for which

$$PI \geq v(x) - \varepsilon = \tilde{v}(x) - \varepsilon = \tilde{v}(kP) - \varepsilon,$$

i.e. a measure P of $S(x)$ lying in E. Hence, i maps E onto E'.

Put $P^x = \psi(x)$, where ψ is measurable (a.s. μ) selector as in Theorem D. In view of (1) and (2)

$$P^x I \geq \tilde{v}(kP^x) - \varepsilon = \tilde{v}(x) - \varepsilon = v(x) - \varepsilon \quad \text{(a.s. } \mu\text{)}. \qquad (3)$$

The formula

$$\bar{P}(\Gamma) = \int_{X_m} P^x(\Gamma)\mu(dx)$$

defines a probability measure $\bar{P}$ in the space L of paths. We shall prove at the end of §6 that: 1) The measure $\bar{P}$ may be obtained by formula (2.2.3) from the initial distribution μ and some strategy $\bar{\pi}$; 2) The strategy $\bar{\pi}$ is an (a.s. μ)-*combination* of the strategies π^x corresponding to the measures P^x, in the sense that

$$P_x^{\bar{\pi}} = P^x \quad \text{(a.s. } \mu\text{)}, \qquad (4)$$

where $P_x^{\bar{\pi}}$ is the measure in the space L corresponding to the initial state x and the strategy $\bar{\pi}$. It is clear from (3) and (4) that $\bar{\pi}$ is an (a.s. μ) ε-optimal strategy. Knowing that there exists such a strategy, one can deduce the optimality equations (Result I) in roughly the same way as in the countable case. However now this deduction does not lead to the construction of a simple ε-optimal strategy (see §7).

* * *

In the general case we will deduce Result IIa from Result III. As in the countable models, Result III is a consequence of the following two propositions.

III.1. *For any initial distribution μ and any strategy π there exists a Markov strategy σ equivalent to π in the process Z_μ.*

III.2. *For any Markov strategy σ there exists a simple strategy φ uniformly majorizing σ.*

Both of these propositions are proved in about the same way as for the countable models. However in the proof of III.1 one uses a more general concept of conditional probabilities, and in the proof of III.2 special attention must be paid to the question of the measurability of φ (see §8).

In order to deduce IIa from III, we must, starting from a simple strategy which is ε-optimal for the process Z_μ, construct a simple strategy which is ε-optimal (a.s. μ). This is done in §9 with the aid of a lemma which makes possible, starting from any sequence of simple strategies φ_k and any $\varepsilon > 0$, to construct a simple strategy φ such that $w(x,\varphi) \geq w(x,\varphi_k)$-$\varepsilon$ for all $x \in X_m$ and all $k = 1, 2, \ldots$. There we also present an example which shows that in general models—in distinction from the countable models—the existence of a uniformly optimal strategy does not imply the existence of a simple (and therefore Markov) uniformly optimal strategy.

§3. The Space of Probability Measures

In order to complete the proof of the universal measurability of the value function v, outlined at the beginning of §2, we have to investigate the class S_0 of all measures in the space L of paths corresponding to the initial distributions concentrated at one point, and to all possible strategies. In preparation, we study some general properties of the class of all probability measures on any Borel space.

The class $\mathscr{M} = \mathscr{M}(E)$ of all probability measures μ on any measurable space E will be considered as a measurable space. To this end, we introduce the family of functions

$$F(\mu) = \mu f$$

on $\mathscr{M}(E)$ corresponding to all bounded (or nonnegative) measurable functions f on E and denote by $\mathscr{B}(\mathscr{M})$ the minimal σ-algebra relative to which all the functions $F(\mu)$ are measurable. The following result is proved in §2 of Appendix 5.

Theorem 1. *If E is a Borel space, then the class $\mathscr{M}(E)$ is also a Borel space.*

§4. Measures on Product Spaces, and Transition Functions

Before moving onward, we need to discuss how to construct measures in product spaces with the help of transition functions.

To give a transition function for a sequence $E_0, E_1, \ldots, E_{s-1}, E_s$ of measurable spaces means to give, for each $t = 0, \ldots, s-1$, a transition function from $E_0 \times E_1 \times \cdots \times E_t$ to E_{t+1} (see Chapter 2, §2). In other words, it means to assign to each $h = x_0 x_1 \cdots x_t$, where $x_0 \in E_0$, $x_1 \in E_1$, $\ldots$, $x_t \in E_t$, a probability measure $v(\cdot|h)$ on the space E_{t+1} such that $v(\Gamma|h)$ is a measurable function of h for each measurable subset Γ of the space E_{t+1}.

Any probability measure P on the product $E_0 \times E_1 \times \cdots \times E_s$ induces a probability measure on each of the products $E_0 \times E_1 \times \cdots \times E_t$ with $t < s$, defined for any measurable subset C of $E_0 \times E_1 \times \cdots \times E_t$ by the formula

$$P(C) = P(C \times E_{t+1} \times \cdots \times E_s).$$

The following two theorems establish a connection between transition functions and probability measures in product spaces.

Theorem E. *Suppose that v is a transition function for the sequence $E_0, E_1, \ldots,$ E_{s-1}, E_s of measurable spaces, and let μ be a probability measure on E_0. Then there exist a unique probability measure P on the product $E_0 \times E_1 \times \cdots \times E_{s-1} \times E_s$ such that*

$$P(dx_0) = \mu(dx_0), \tag{1}$$

$$P(dx_0\, dx_1 \cdots dx_t\, dx_{t+1}) = v(dx_{t+1} | x_0 x_1 \cdots x_t) P(dx_0\, dx_1 \cdots dx_t) \tag{2}$$
$$(t = 0, 1, \ldots, s-1).$$

The converse assertion requires more rigid restrictions on the spaces E_t.

Theorem F. *For any probability measure P on the product $E_0 \times E_1 \times \cdots \times E_{s-1} \times E_s$ of Borel spaces, there exist a probability measure μ on E_0 and a transition function v such that equations (1)–(2) are satisfied.*

For the proof of Theorem E we note that the measure P defined by the formula

$$P(dx_0\, dx_1 \cdots dx_s) = \mu(dx_0) v(dx_1 | x_0) \cdots v(dx_s | x_0 x_1 \cdots x_{s-1}), \tag{3}$$

satisfies conditions (1)–(2). On the other hand, (1)–(2) implies (3), and therefore the measure P is determined uniquely by the conditions of Theorem E.

Theorem F is proved in Appendix 4.

* * *

Suppose that B is a measurable subset of the product $E_0 \times E_1 \times \cdots \times E_{s-1} \times E_s$. Under what conditions on the initial distribution and transition function will the corresponding measure P be concentrated on B?

Suppose that $t < s$. Let B_t consist of all collections $x_0 x_1 \cdots x_t$ such that $x_0 x_1 \cdots x_t x_{t+1} \cdots x_s \in B$ for some $x_{t+1}, \ldots, x_s$ (B_t is the projection of B into the product $E_0 \times E_1 \times \cdots \times E_t$). For each $h \in B_t$, we denote by $E[h]$ the set of $x \in E_{t+1}$ such that $hx \in B_{t+1}$.

Theorem 1. *Suppose that all the sets B_t are measurable. Consider the measure P in the space $E_0 \times E_1 \times \cdots \times E_{s-1} \times E_s$ corresponding to the initial distribution μ and to the transition function v. For P to be concentrated on B it is sufficient that*

$\mu(B_0) = 1$ and

$$v(E[h]|h) = 1 \qquad (4)$$

for any $h \in B_t$, $t = 0, 1, \ldots, s - 1$.

Put $B_s = B$. We shall prove by induction that

$$P(B_t) = 1 \qquad (5)$$

for all $t = 0, 1, \ldots, s$. For $t = 0$ this assertion follows from the condition $\mu(B_0) = 1$ and formula (1). Suppose that (5) is valid for some $t < s$. By (2) and (4)

$$\begin{aligned}
P(B_{t+1}) &= \int_{B_{t+1}} P(dx_0\, dx_1 \cdots dx_t\, dx_{t+1}) \\
&= \int_{B_t} P(dx_0\, dx_1 \cdots dx_t)v(E[x_0 x_1 \cdots x_t]|x_0 x_1 \cdots x_t) \\
&= \int_{B_t} P(dx_0\, dx_1 \cdots dx_t) = P(B_t),
\end{aligned}$$

so that (5) is valid for $t + 1$ as well.

As with Theorem F, the converse result can only be established under more restrictive conditions.

Theorem 2. *Suppose that the spaces $E_0, E_1, \ldots, E_{s-1}, E_s$ are Borelian, the sets $B_0, B_1, \ldots, B_{s-1}, B_s = B$ are measurable, and that the correspondence $E[h]$ of B_t into E_{t+1} admits a measurable selection, $t = 0, \ldots, s - 1$. Let P be a probability measure in the product $E_0 \times E_1 \times \cdots \times E_{s-1} \times E_s$, concentrated on B. Then it is possible to choose a measure μ and a transition function v, corresponding to P as in Theorem F, such that $\mu(B_0) = 1$ and condition (4) is satisfied, $t = 0, \ldots, s - 1$.*

Indeed, from the equation $P(B_s) = 1$ we obtain formula (5) for all $t = 0, \ldots, s$. For $t = 0$ we find from (1) that $\mu(B_0) = 1$.

Further, it follows from (5) and (2) that

$$\begin{aligned}
1 &= P(B_{t+1}) \\
&= \int_{B_t} P(dx_0\, dx_1 \cdots dx_t)v(E[x_0 x_1 \cdots x_t]|x_0 x_1 \cdots x_t). \qquad (6)
\end{aligned}$$

At the same time

$$1 = P(B_t) = \int_{B_t} P(dx_0\, dx_1 \cdots dx_t). \qquad (7)$$

Subtracting (6) from (7), we get

$$\int_{B_t} [1 - v(E[h]|h)]P(dh) = 0. \qquad (8)$$

Since $v(\cdot|h)$ is a probability measure, then $v(E[h]|h) \leq 1$. It therefore follows from (8) that

$$v(E[h]|h) = 1 \quad \text{(a.s. } P\text{) on } B_t.\tag{9}$$

Suppose that φ is a measurable selection of the correspondence $E[h]$, and let δ_x be a probability measure concentrated at the point x. In view of (9) there exists a measurable subset Γ_t of the set B_t such that $P(\Gamma_t) = 1$ and $v(E[h]|h) = 1$ for all $h \in \Gamma_t$. The formula

$$\bar{v}(\cdot|h) = \begin{cases} v(\cdot|h) & \text{for } h \in \Gamma_t \text{ or } h \in (E_0 \times E_1 \times \cdots \times E_t)\backslash B_t, \\ \delta_{\varphi(h)} & \text{for } h \in B_t\backslash\Gamma_t, \end{cases}$$

$t = 0, 1, \ldots, s-1$, yields a transition function identically satisfying (4). Since the measure P in the product $E_0 \times E_1 \times \cdots \times E_t$ is concentrated on Γ_t, then

$$\bar{v}(\cdot|x_0x_1 \cdots x_t) = v(\cdot|x_0x_1 \cdots x_t) \quad \text{(a.s. } P\text{) on } E_0 \times E_1 \times \cdots \times E_t.$$

Therefore equation (2) remains valid on the replacement of v by $\bar{v}$. Thus, the transition function $\bar{v}$ satisfies all the requirements of Theorem 2.

$$* \quad * \quad *$$

The general construction of a measure with the help of transition functions, as described in this section, may be used for constructing a measure in the space L of paths. In this case, we have a sequence of spaces $X_m, A_{m+1}, X_{m+1}, \ldots, A_n, X_n$ and the transition laws are given in turn by the strategy π and the transition function p of the model. At the even steps, for any collection $x_m a_{m+1} x_{m+1} \cdots a_t \in X_m \times A_{m+1} \times \cdots \times A_t$

$$v(dx_t|x_m a_{m+1} x_{m+1} \cdots a_t) = p(dx_t|a_t),$$

$t = m+1, \ldots, n$. At the odd steps

$$v(da_{t+1}|h) = \pi(da_{t+1}|h):$$

this last formula defines v only for histories $h = x_m a_{m+1} \cdots x_t$, i.e. for collections satisfying the condition $j(a_{m+1}) = x_m, \ldots, j(a_t) = x_{t-1}$; we extend v arbitrarily to the remaining elements h (for example, we may concentrate the measure $\pi(\cdot|x_m a_{m+1} \cdots x_t)$ at a fixed point of the space A_{t+1}). The transition function v satisfies the conditions of Theorem 1 for $B = L$, and, given the initial distribution μ, there corresponds to it a measure P in $X_m \times A_{m+1} \times \cdots \times X_n$ concentrated on L^*. It is easy to see that this is the same measure on L as the measure constructed in

---

* For $t = 2k$ the set B_t coincides with the space H_{m+k} of histories at the time k; for $t = 2k+1$ B_t consists of all collections ha such that $h \in H_{m+k}$ and $j(a) = x$, where x is the end of the history h, $k = 0, 1, \ldots, n-m-1$.

Chapter 2, §2 (compare formula (3) with formula (2.2.3)). For this measure, formulas (1)–(2) take on the form

$$P(dx_m) = \mu(dx_m), \tag{10}$$

$$P(dx_m \, da_{m+1} \cdots dx_t \, da_{t+1}) = \pi(da_{t+1} | x_m a_{m+1} \cdots x_t) P(dx_m \, da_{m+1} \cdots dx_t), \tag{11}$$

$$P(dx_m \cdots da_{t+1} \, dx_{t+1}) = p(dx_{t+1} | a_{t+1}) P(dx_m \cdots da_{t+1}), \tag{12}$$
$$t = m, m+1, \ldots, n-1.$$

§5. Strategic Measures

Suppose that Z is a nontrivial model. Any measure P on the space L of paths, corresponding to an initial distribution μ and a strategy π, will be called a *strategic measure*. In this section we shall prove the measurability and convexity of the class S of all strategic measures. From the measurability of S we will deduce the measurability of the class S_0 and of the mapping $S_0 \overset{k}{\to} X_m$. We will need the convexity of the class S in order to extend the operation of combination of strategies to general models.

We deduce the properties of the class S of strategic measures from the following description of them.

Theorem 1. *In a nontrivial model, a probability measure P on the space L of paths is a strategic measure if and only if, for each $t = m, \ldots, n-1$ and for any bounded measurable function f on H_{t+1},*

$$Pf(ha_{t+1}x_{t+1}) = P \int_{X_{t+1}} f(ha_{t+1}x)p(dx | a_{t+1}) \tag{1}$$

($h \in H_t$ is considered as a function on L, i.e. as a random point). It is possible to select a countable system W of bounded measurable functions on H_{t+1} such that if (1) is satisfied for $f \in W$, then it is satisfied for all bounded measurable functions $f \in H_{t+1}$.

The necessity of condition (1) follows from formula (4.12).

In order to prove the sufficiency, we apply Theorem 2 of the preceding section to a probability measure P satisfying the condition (1) and to the subset $B = L$ of $X_m \times A_{m+1} \times \cdots \times X_n$. According to that theorem

$$P(dx_m) = \mu(dx_m), \tag{2}$$

$$P(dx_m \, da_{m+1} \cdots dx_t \, da_{t+1}) = v(da_{t+1} | x_m a_{m+1} \cdots x_t) P(dx_m \, da_{m+1} \cdots dx_t), \tag{3}$$

$$P(dx_m \cdots da_{t+1} \, dx_{t+1}) = v(dx_{t+1} | x_m \cdots a_{t+1}) P(dx_m \cdots da_{t+1})$$
$$(t = m, m+1, \ldots, n-1), \tag{4}$$

while the transition function v satisfies the condition (4.4). We note that for any history $h = x_m a_{m+1} \cdots x_t$ the set $E[h]$ on which the measure $v(\cdot|h)$ is concentrated coincides with the fibre $A(x_t)$. Therefore the formula

$$\pi(\cdot|h) = v(\cdot|h)$$

defines some strategy π.

We will not need formula (4). Instead of it we make use of the formula

$$P(dx_m \cdots da_{t+1} dx_{t+1}) = p(dx_{t+1}|a_{t+1})P(dx_m \cdots da_{t+1}). \tag{5}$$

which follows from (1). Formulas (2), (3) and (5), which coincide with (4.10)–(4.12), show that P is a strategic measure and that it corresponds to the initial distribution μ and strategy π.

We know that the space H_{t+1} is Borelian. If it is finite or countable, then as W one may choose a system of functions differing from 0 on a finite set and taking on only rational values. If H_{t+1} is uncountable, then we start with an isomorphism q of H_{t+1} onto the interval $[0,1]$. We put $W = \{q^n\}$ $n = 0, 1, 2, \ldots$. If (1) is satisfied for all functions $f(h) = q^n(h)$, $h \in H_{t+1}$, then (1) is valid also for any function $f = \varphi(q)$, where φ is a polynomial, and therefore for any function $f = \varphi(q)$, where φ is any continuous function (this follows from the Weierstrass theorem). Since equation (1) holds under a bounded point-by-point passage to the limit, then it is satisfied for $f = \varphi(q)$, where φ is any bounded measurable function on the interval $[0,1]$. (This follows from Lemma 1, Appendix 5, §1). Any bounded measurable function f on H_{t+1} can be represented in such a form; indeed $f(h) = \varphi(q(h))$ where $\varphi(y) = f(q^{-1}(y))$. The theorem is completely proved.

By the definition of measurable space $\mathcal{M}(L)$, the functions of P in both sides of equation (1) are measurable. A countable number of conditions of type (1) distinguish a measurable subset of the space $\mathcal{M}(L)$; therefore the class S of all strategic measures is measurable.

Further, it follows from Theorem 1 that the class S is convex, i.e. that if v is a probability measure in the space S, then the measure P^* defined by the formula

$$P^*(\cdot) = \int_S P(\cdot)v(dP)$$

is also strategic. Indeed, condition (1) is linear in P and therefore is preserved on integration with respect to P.

§6. Universal Measurability of the Value of the Model and Almost-Surely (a.s.) ε-Optimal Strategies

Now we have all the necessary tools to deduce the measurability of the class S_0 and of the mapping $S_0 \overset{k}{\to} X_m$ used in §2 for the proof of the universal measurability of the value $v(x)$, $x \in X$.

We note that for a measure P of $\mathcal{M}(L)$ to belong to the class S_0, it is necessary and sufficient that P should belong to the class of strategic measures S and that for any rational number c

$$P\{q(x_m) < c\} = 0 \text{ or } 1, \tag{1}$$

q being an isomorphic mapping of the Borel space X_m onto the interval $[0,1]$. Indeed, (1) holds if and only if the probability distribution for the random variable $q(x_m)$ is concentrated at some single point y of $[0,1]$. And this last is equivalent to having the initial distribution μ corresponding to P concentrated at the point $x = q^{-1}(y)$. The function in the left side of (1) is a measurable function of P, and therefore a countable number of conditions (1) distinguishes from the measurable class S a measurable subclass S_0.

Further, for any set Γ of the space X_m

$$k^{-1}(\Gamma) = \{P : P \in S_0, P\{x_m \in \Gamma\} = 1\},$$

from which it follows that $k^{-1}(\Gamma)$ is measurable for any measurable Γ, which means that the mapping k is measurable.

The gaps that remained in the proof of the universal measurability of the function v in §2 are now filled in.

* * *

Now we turn to the question of the existence of an (a.s. μ) ε-optimal strategy. Here it remains to justify constructing the combination of strategies described in §2. We have strategic measures P^x, $x \in X_m$, such that

$$P^x\{x_m = x\} = 1 \quad (\text{a.s. } \mu) \tag{2}$$

and such that $x \to P^x$ is a measurable mapping of the space X_m into the space S, and we define

$$\bar{P}(\cdot) = \int_{X_m} P^x(\cdot)\mu(dx). \tag{3}$$

A change of variables transforms (3) into the integral

$$\int_S P(\cdot)v(dP),$$

where v is the measure on the space S induced by the measure μ on X_m and the measurable mapping $x \to P^x$. In view of the convexity of the class S, the measure $\bar{P}$ is also strategic.

It follows from (3) and (2) that for $\Gamma \in \mathscr{B}(X_m)$

$$\bar{P}\{x_m \in \Gamma\} = \int_{X_m} P^x\{x_m \in \Gamma\}\mu(dx) = \int_{X_m} \chi_\Gamma(x)\mu(dx) = \mu(\Gamma),$$

so that μ is the initial distribution for $\bar{P}$.

Let π be the strategy corresponding to $\bar{P}$, and P_x^π the distribution in the space of paths corresponding to the initial state x and the strategy π. We show that for any bounded measurable function ξ on the space L

$$P_x^\pi \xi = P^x \xi \quad \text{(a.s. } \mu\text{)}. \tag{4}$$

If f is a bounded measurable function on X_m, then, by virtue of (3) and (2),

$$\bar{P}(f \cdot \xi) = \int_{X_m} P^x(f(x_m) \cdot \xi)\mu(dx) = \int_{X_m} (f(x) \cdot P^x \xi)\mu(dx). \tag{5}$$

On the other hand, we have

$$P_x^\pi \{x_m = x\} = 1$$

for each $x \in X_m$, and by formula (2.2.3)

$$\bar{P}(f \cdot \xi) = \int_{X_m} P_x^\pi(f(x_m) \cdot \xi)\mu(dx) = \int_{X_m} (f(x) \cdot P_x^\pi \xi)\mu(dx). \tag{6}$$

In view of the arbitrariness of f, (4) follows from (5) and (6).

Clearly, for any countable collection of functions $\{\xi_n\}$ one can choose a set Γ from X_m of μ-measure 1, such that equation (4) is satisfied for every $x \in \Gamma$ and any function ξ_n. The set Ξ of those functions ξ for which (4) is true for all $x \in \Gamma$ is closed relative to linear operations and bounded passage to the limit. Choosing an appropriate system $\{\xi_n\} = W$, we find that Ξ contains all bounded measurable functions (cf. the analogous argument in §5). This means that

$$P_x^\pi = P^x$$

for all $x \in \Gamma$, and the strategy π is an (a.s. μ) combination of the strategies π^x corresponding to the measures P^x.

§7. The Optimality Equations

As in the discrete models (cf. Chapter 1, §§4–7 and 12), the derivation of the optimality equations (Result I) is preceded by the proof of the formula

$$v(\mu) = \mu v \left[= \int_{X_m} v(x)\mu(dx) \right]. \tag{1}$$

Since the function $v(x)$ is universally measurable and, along with the reward functions, bounded above, the integral in (1) has a meaning for any initial measure μ. For any strategy π we have $w(x,\pi) \leq v(x)$, so that

$$w(\mu,\pi) = \int_{X_m} w(\mu,\pi)\mu(dx) \leq \int_{X_m} v(x)\mu(dx) = \mu v. \tag{2}$$

On the other hand, for any $\varepsilon > 0$ there exists an (a.s. μ) ε-optimal strategy π for which $w(x,\pi) \geq v(x) - \varepsilon$ (a.s. μ). Hence

$$w(\mu,\pi) = \int_{X_m} w(x,\pi)\mu(dx) \geq \int_{X_m} v(x)\mu(dx) - \varepsilon = \mu v - \varepsilon. \tag{3}$$

Now (2) and (3) imply (1).

It is clear from (3) that if the strategy π is (a.s. μ) ε-optimal, then it is ε-optimal for the process Z_μ. We will find this remark useful in §9.

Now we turn to the proof of the relations

$$v = Vu, \qquad u = Uv', \tag{4}$$

connecting the values v and v' of the model Z and its derived model Z' (the operators U and V were defined by formulas (1.2) and (1.3)).

The same as in Chapter 1, §6, one deduces from the fundamental equation (2.3.1) that

$$w(x,\pi) \leq Vu(x) \tag{5}$$

for all $x \in X_m$ and any strategy π, where

$$u(a) = q(a) + V'(p_a), \tag{6}$$

$a \in A_{m+1}$. It follows from (6), (1) and the definition (1.2) of the operator U that

$$u = Uv'.$$

In order to deduce equation (4), it remains for any $\varepsilon > 0$ and any $\bar{x} \in X_m$, to construct a strategy $\bar{\pi}$ in the model Z such that

$$w(\bar{x},\bar{\pi}) \geq Vu(\bar{x}) - \varepsilon. \tag{7}$$

(In distinction from the discrete case (cf. Chapter 1, §12) we cannot choose $\bar{\pi}$ here so that (7) is satisfied simultaneously for all $x \in X_m$.) By the definition of the supremum, it is possible to find an action $\bar{a}$ in the fibre $A(x)$ such that

$$u(\bar{a}) \geq Vu(\bar{x}) - \frac{\varepsilon}{2}. \tag{8}$$

Let γ be any measurable selector of the correspondence $A(x)$, $x \in X_m$. Clearly the function

$$\psi(x) = \begin{cases} \bar{a} & \text{for } x = \bar{x} \\ \gamma(x) & \text{for } x \neq \bar{x}, \, x \in X_m \end{cases}$$

is also a measurable selector of that correspondence. Further, suppose that π' is a strategy in the model Z' which is ($\varepsilon/2$)-optimal for the initial distribution

$p_{\bar{a}}$, so that

$$w'(p_{\bar{a}},\pi') \geq v'(p_{\bar{a}}) - \frac{\varepsilon}{2},$$

and, hence,

$$q(\bar{a}) + w'(p_{\bar{a}},\pi') \geq u(\bar{a}) - \frac{\varepsilon}{2}. \tag{9}$$

The strategy $\bar{\pi} = \psi\pi'$ in the model Z, consisting of the application of the simple strategy ψ at the first step and then the strategy π', satisfies condition (7). Indeed, applying formula (5) to $\bar{\pi}$ and taking into account (8) and (9), we have

$$w(\bar{x},\bar{\pi}) = q(\bar{a}) + w'(p_a,\pi') \geq u(\bar{a}) - \frac{\varepsilon}{2} \geq Vu(\bar{x}) - \varepsilon.$$

The expanded form of the optimality equations and their expression in terms of of the operator T, as introduced in Chapter 1, §7, are obtained from (4) in the same way as they were for the finite models.

$$* \quad * \quad *$$

The arguments presented above show that if π' is a uniformly ε_1-optimal strategy in the model Z', and if for the measurable selector ψ of the correspondence $A(x)$ of X_m into A_{m+1} we have $T_\psi v' \geq v - \varepsilon_2$, then $w(x,\psi\pi') \geq v(x) - (\varepsilon_1 + \varepsilon_2)$ for all $x \in X_m$. Hence Result II$_2$ of §1 can be obtained by an obvious induction.

§8. Sufficiency of the Simple Strategies

We shall show that the proposition III.1 stated at the end of §2 extends to general models.

As in the countable case, the Markov strategy σ which is equivalent to π for the initial distribution μ can be constructed with the aid of the conditional distributions of a_t given x_{t-1} ($t = m, \ldots, n-1$). In the general case the existence of such distributions follows from Theorems F and 2 of §4. These theorems are applied with $s = 1$ to the spaces $E_0 = X_t$, $E_1 = A_{t+1}$, to the sets $B_0 = X_t$, $B_1 = \{xa : x \in X_t, a \in A_{t+1}, j(a) = x\}$, and to the measure P on $X_t \times A_{t+1}$ defined by the formula

$$P(\Gamma) = P\{x_t a_{t+1} \in \Gamma\} \qquad (\Gamma \in \mathcal{B}(X_t \times A_{t+1})).$$

According to Theorem F

$$P(dx_t\, da_{t+1}) = P(dx_t)\sigma(da_{t+1}|x_t), \tag{1}$$

where $\sigma(\cdot|\cdot)$ is the transition function of X_t into A_{t+1}. Since $j(a_{t+1}) = x_t$ for any path then the measure P on the product $X_t \times A_{t+1}$ is concentrated on B_1. Therefore, by Theorem 2, one may choose σ in such a way that

$$\sigma(E[x] \mid x) = 1$$

for all $x \in X_t$. In our case

$$E[x] = \{a : a \in A_{t+1},\, xa \in B_1\} = \{a : j(a) = x\} = A(x),$$

so that the measure $\sigma(\cdot \mid x)$ is concentrated on the fibre $A(x)$.

Formula (1) replaces formula (1.13.3) which we used in the discrete case. Instead of formulas (1.13.4), (1.13.5), and (1.13.6), we have in the general case

$$Q(dx_t\, da_{t+1}) = Q(dx_t)\sigma(da_{t+1} \mid x_t), \qquad (2)$$

$$P(da_{t+1}\, dx_{t+1}) = P(da_{t+1})p(dx_{t+1} \mid a_{t+1}), \qquad (3)$$

$$Q(da_{t+1}\, dx_{t+1}) = Q(da_{t+1})p(dx_{t+1} \mid a_{t+1}) \qquad (4)$$

where Q is the strategic measure corresponding to μ and σ. Formulas (3) and (4) are obtained from (4.12) by integration with respect to $x_m a_{m+1} \cdots x_t$, and formula (2) by integration with respect to $x_m a_{m+1} \cdots a_t$ of formula (4.11), which in the case of a Markov strategy takes on the form

$$Q(dx_m\, da_{m+1} \cdots dx_t\, da_{t+1}) = \sigma(da_{t+1} \mid x_t)Q(dx_m\, da_{m+1} \cdots dx_t).$$

Formulas (1)–(4) make it possible to prove the coincidence of the distributions of a_t and x_t relative to the measures P and Q by means of the same induction as in the countable case.

$$* \quad * \quad *$$

Now we extend to general models the proof of the proposition III.2 on the existence of a simple strategy φ uniformly majorizing the Markov strategy σ (cf. Chapter 1, §13). The sole difference between the general case and the countable case consists in that we must ensure the measurability of the selector ψ of $A(x)(x \in X_m)$ with the property $f(\psi(x)) \geq \gamma_x f = w(x,\sigma)$. This can be done by applying Theorem A of §2 to

$$E = \{a : a \in A_{m+1},\, f(a) \geq w(j(a),\sigma)\}, \qquad E' = X_m, \qquad i = j,$$

$$v(\cdot \mid x) = \gamma_x(\cdot).$$

It is easy to see that E is a measurable subset of the space A_{m+1} and therefore is a Borel space. By Lemma 1 of §13, Chapter 1 (see also the footnote on page 41), $\gamma_x(E) > 0$ for all $x \in X_m$, which means that all the hypotheses of Theorem A are fulfilled.

§9. Simple a.s. ε-Optimal Strategies

We need the following lemma for the proof of Result IIa.

Lemma 1. *For any sequence $\{\varphi_k\}_{k=1,2,\ldots}$ of simple strategies and any $\varepsilon > 0$ there exists a simple strategy φ such that*

$$w(x,\varphi) \geq \sup_k w(x,\varphi_k) - \varepsilon, \qquad x \in X_m.$$

Proof. We shall show that our assertion is valid for a model Z if it is valid for the derived model Z'.

Let φ_k' denote the simple strategy in the derived model Z' which is gotten from φ_k by striking out the X_m and A_{m+1} columns. By hypothesis, there exists a simple strategy φ' in the model Z' such that

$$w'(x,\varphi') \geq w'(x,\varphi_k') - \frac{\varepsilon}{2} \qquad (x \in X_{m+1}, k = 1,2,\ldots), \tag{1}$$

where w' denotes the value in the model Z'. Consider in the model Z strategies ψ_k, consisting in the use at the first step of the strategy φ_k, and thereupon the strategy φ'. By formula (1.7.5.), expressing w in terms of w', it follows from (1) that

$$w(x,\psi_k) \geq w(x,\varphi_k) - \frac{\varepsilon}{2},$$

and therefore

$$\sup_k w(x,\psi_k) \geq \sup_k w(x,\varphi_k) - \frac{\varepsilon}{2}.$$

We obtain the desired strategy if we apply, at the initial point x, any strategy ψ_k for which

$$w(x,\psi_k) \geq \sup_k w(x,\psi_k) - \frac{\varepsilon}{2}. \tag{2}$$

We need only make sure that the dependence on x should be measurable. To this end it suffices to choose the strategy ψ_k with the smallest subscript k satisfying (2) (it is essential that the right side of (2) be a measurable function of x).

Our arguments apply also to the one-step model, except that in this case we need to start directly with (2), replacing ψ_k by φ_k. Lemma 1 is proved.

* * *

Suppose that μ is any initial distribution and k any integer. If the model Z is nontrivial, then there exists a strategy which is $(1/k)$-optimal for the process Z_μ. By Result III, there exists for Z_μ a simple $(1/k)$-optimal strategy φ_k. Clearly

$$\sup_k w(x,\varphi_k) \leq v(x) \qquad (x \in X_m). \tag{3}$$

Applying formulas (7.1) and (7.2) to $v(\mu)$ and $w(\mu,\varphi_k)$, we have

$$v(\mu) - \frac{1}{k} \le w(\mu,\varphi_k) = \int_{X_m} w(x,\varphi_k)\mu(dx)$$

$$\le \int_{X_m} \left[\sup_k w(x,\varphi_k) \right] \mu(dx) \le \int_{X_m} v(x)\mu(dx) = v(\mu).$$

It therefore follows that as $k \to \infty$

$$\int_{X_m} \left[\sup_k w(x,\varphi_k) \right] \mu(dx) = \int_{X_m} v(x)\mu(dx). \qquad (4)$$

It follows from (3) and (4) that

$$\sup_k w(x,\varphi_k) = v(x) \quad (\text{a.s. } \mu).$$

Applying Lemma 1 to the sequence $\{\varphi_k\}$, we find that *for any $\varepsilon > 0$ and any initial distribution μ there exists a simple strategy φ such that*

$$w(x,\varphi) \ge v(x) - \varepsilon \quad (\text{a.s. } \mu).$$

(Result IIa).

* * *

In the countable case, the existence of any optimal strategy for Z implies the existence of a simple strategy optimal for Z (see the end of §13 in Chapter 1). We present here an example showing that this result is not valid in the general case.

EXAMPLE 1. Consider the model Z depicted in figure 3.4. Here X_0 is a Borel set Q in the square $0 \le z \le 1, 0 \le y \le 1$, which projects orthogonally onto the segment

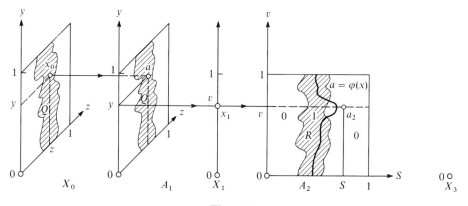

Figure 3.4

$0 \le y \le 1$ and does not contain the graph of even one measurable function $z = f(y)$ $(0 \le y \le 1)$ (see Example 1.1). The space A_1 is equal to X_0 and $j(y,z) = (y,z)$. The space X_1 is the segment $0 \le u \le 1$, the space A_2 the square $0 \le u \le 1$, $0 \le s \le 1$, and $j(u,s) = u$. The space X_3 consists of the single point 0.

The system passes deterministically to the point $u = y$ from each point (y,z) of the space A_1, and to each point 0 from each point $(u.s)$ of the space A_2.

The terminal reward is equal to zero. Denote by R the set into which Q goes if one superimposes the yz-plane onto the us-plane in such a way that the y-axis coincides with the u-axis and the z-axis with the s-axis. The running reward is equal to 1 on R and equal to 0 on $A \backslash R$.

It is clear that here $v(x) = 1$ for each $x \in X$.

As in Example 1.1, a simple strategy φ is given by a measurable function $s = f(u)$ $(0 \le u \le 1)$. Hence, for any simple strategy φ there exists a u such that $\varphi(u)$ does not belong to R, and we have $w(x,\varphi) = 0$ for any initial point $x = (y,z)$ with $y = u$. Thus no simple strategy φ can be optimal (even ε-optimal for $\varepsilon < 1$) in the model Z.

Since any Markov strategy is uniformly majorized by some simple strategy, then there exists no optimal (or ε-optimal for $\varepsilon < 1$) Markov strategy. However, a non-Markovian optimal strategy does exist.

Indeed, let π prescribe for the history $x_0 a_1 x_1 = (y,z)(y,z)u$ the action $a_2 = (u,s)$ with $s = z$ $(0 \le y, z, u \le 1)$ (the choice of a_1 is unique for any x_0). Such a function a_2 is measurable in $x_0 a_1 x_1$, and hence defines a deterministic strategy. By the construction, $w(x,\pi) = 1$ for all $x \in X_0$.

Part II

Control on an Infinite Time Interval

Chapter 4

Discrete Models

§1. Passage to an Infinite Interval of Control

When there is no natural moment for stopping a process, it is appropriate to consider control over an infinite time interval.

The problem of optimal control over an infinite time interval may be posed in various ways. One may seek a maximization of the average gain per unit time; to this we have devoted Chapter 7. In this chapter we maximize the total mean value of the reward I across an infinite time. Such an approach is interesting in the first place when the values of I are bounded above.

In this chapter we deal with discrete, i.e. finite and countable, models. The general case, requiring a more thorough acquaintance with measurability problems, and making use of the material of Chapters 2 and 3, will be taken up in Chapter 5.

§2. Summable Models

The definitions of a controlled Markov process and of a model do not change for an infinite control interval $[m,\infty)$, except that now the state spaces $X_m, X_{m+1}, \ldots$ and the action spaces $A_{m+1}, A_{m+2}, \ldots$ form infinite sequences, and there is no terminal payoff. It is also necessary to give strategies for histories of arbitrarily long length.

In Chapter 1 the value of a strategy π with the initial distribution μ was defined by the formula

$$w(\mu,\pi) = P\left[\sum_{m+1}^{n} q(a_t) + r(x_n)\right] = \sum_{m+1}^{n} Pq(a_t) + Pr(x_n)$$

where P is the measure in the space of paths defined by equation (1.3.2). In the case of an infinite interval it is natural to put

$$w(\mu,\pi) = \sum_{m+1}^{\infty} Pq(a_t). \tag{1}$$

Here $Pq(a_t)$ can be calculated according to formulas (1.3.2)–(1.3.3), breaking off the trajectory $x_m a_{m+1} x_{m+1} \cdots a_n x_n \cdots$ at x_n for an arbitrary $n \geq t$ (one easily sees that the value of $Pq(a_t)$ does not depend on n).

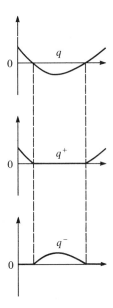

Figure 4.1

Let us put $q^+ = \max\{q,0\}$, $q^- = \max\{-q,0\}$ (see figure 4.1). The sum of the series (1) may fail to exist. But the series

$$\sum_{m+1}^{\infty} Pq^+(a_t) = w^+(\mu,\pi) \tag{2}$$

and

$$\sum_{m+1}^{\infty} Pq^-(a_t) = w^-(\mu,\pi), \tag{3}$$

always have definite sums, finite or $+\infty$. We will say that the model is *μ-summable above* if $w^+(\mu,\pi) < +\infty$ for all π, and that it is *μ-summable below* if $w^-(\mu,\pi) < +\infty$ for all π. The formulations for both cases often turn out to be quite symmetric. In such situations we shall frequently speak of "μ-summability", dropping the qualification "above" or "below". Formulations of this kind may be understood in two ways, either always with the word "above", or always with the word "below"

If the model is μ-summable, then

$$\sum_{m+1}^{\infty} Pq^+(a_t) - \sum_{m+1}^{\infty} Pq^-(a_t) = \sum_{m+1}^{\infty} [Pq^+(a_t) - Pq^-(a_t)]$$

$$= \sum_{m+1}^{\infty} P[q^+(a_t) - q^-(a_t)]. \tag{4}$$

The validity of these equalities is a consequence of the following general property of numerical series.

Property S. *If the sum of the positive terms, or the sum of the negative terms in a series, is finite, then the sum of the series has a meaning (finite or equal to − ∞ or + ∞) and does not change if the terms are rearranged, or grouped in parentheses in any manner (the number of parentheses and the number of terms in each parenthesis may be infinite).*

The right hand sides of (1) and (4) coincide because $q = q^+ - q^-$. Thus, for a μ-summable model formula (1) has a meaning and

$$w(\mu,\pi) = w^+(\mu,\pi) - w^-(\mu,\pi). \tag{5}$$

Models on a finite interval $[m,n]$ may be considered as special cases of models on the infinite interval $[m,+\infty)$; it suffices to put

$$q(a) = \begin{cases} r(ja) & \text{if } a \in A_{n+1}, \\ 0 & \text{if } a \in A_{n+2} \cup A_{n+3} \cup \cdots. \end{cases}$$

For models on a finite interval the μ-summability above follows from the fact that the functions q and r are bounded above (see condition α of Chapter 1, §12). On an infinite interval this is already not so. Therefore it makes no sense to introduce that condition, and *we will exclude it from our initial premises.* As a consequence, along with the passage to an infinite interval of control we will obtain some strengthened results for the finite interval.

§3. The Fundamental Equation

We shall show that the formulas

$$w(\mu,\pi) = \sum_{x \in X_m} \mu(x)w(x,\pi) \tag{1}$$

and

$$w(x,\pi) = \sum_{a \in A(x)} \pi(a|x)[q(a) + w'(p_a,\pi_a)] \tag{2}$$

(the fundamental equation) established in Chapter 1, §§4,5 and 12 hold now as well. Precisely:

   a) *If the model is μ-summable, then it is x-summable* for all x with $\mu(x) > 0$, and equation (1) is satisfied.*
   b) *If the model is x-summable, then the derived model is p_a-summable for all $a \in A(x)$, and equation (2) is satisfied.*

---

* We will say that the model is *x-summable* when it is summable relative to the μ-distribution concentrated at the point x.

Let us consider first the case of a nonnegative reward function q. Note that if $q = 0$ on all the sets A_t with $t > n$, then the choice of the control after the time n plays no rôle, and the situation reduces to control on the segment $[m,n]$ (for a reward bounded below). Therefore for the reward function q_n defined by

$$q_n = \begin{cases} q & \text{on } A_{m+1} \cup A_{m+2} \cup \cdots \cup A_n \\ 0 & \text{on } A_{n+1} \cup A_{n+2} \cup \cdots \end{cases}$$

relations (1) and (2) follow from the results of §§4,5, and §12 of Chapter 1. As $n \to \infty$ the nonnegative function q_n tends monotonically to q. Under such convergence it is legitimate to pass to the limit both under the expectation sign P and under the summation sign in the series (2.1). This means that for arbitrary μ and π the value $w_n(\mu,\pi)$ with the reward q_n converges nondecreasingly to the value $w(\mu,\pi)$ for the reward q, and the same is true for the derived model. With this convergence, termwise passage to the limit under the summation sign in formulas (1) and (2) is legitimate, and we find that these formulas are valid for any nonnegative reward q.

Now suppose that q can take on values of any sign. We have proved that formula (1) is satisfied for the values w^+ and w^- corresponding to the nonnegative rewards q^+ and q^-:

$$w^+(\mu,\pi) = \sum_{X_m} \mu(x)w^+(x,\pi), \tag{3}$$

$$w^-(\mu,\pi) = \sum_{X_m} \mu(x)w^-(x,\pi). \tag{4}$$

Accordingly, if $w^+(\mu,\pi) < +\infty$, then $w^+(x,\pi) < +\infty$ for all x with $\mu(x) > 0$, and the same is valid for w^-. Taking account of (4) and (3) and employing property S, we obtain equation (1).

Further, suppose that $w^+(x,\pi) < +\infty$ for all strategies π. Fix some $a \in A(x)$ and any strategy π' in the derived model. Let ψ_a be a selector of the correspondence $A(y)$, $y \in X_m$, which assigns to the point x the fixed action a. Applying formula (2) to the strategy $\pi = \psi_a\pi'$ and the nonnegative reward function q^+, we get

$$w^+(x,\pi) = q^+(a) + w^{+\prime}(p_a,\pi_a).$$

But $\pi_a = \pi'$, and hence $w^{+\prime}(p_a,\pi') < \infty$. Applying the analogous arguments to w^-, we obtain the first half of assertion b).

Now suppose that π is any strategy. Applying formula (2) to the reward functions q^+ and q^-, and subtracting one from the other with the aid of property S, we conclude that (2) is satisfied for the function q as well.

* * *

Now that we have extended the fundamental equation to the general case, we may use all the consequences of that equation. In particular, if the model is μ-summable and if ψ_t is a selector of the correspondence $A(x)$, $x \in X_{t-1}$ ($t = m +$

$1, \ldots, n$), and π is any strategy in the derived model of order $n - m$, then

$$w(x, \psi_{m+1} \psi_{m+2} \cdots \psi_n \pi) = T_{\psi_{m+1}} T_{\psi_{m+2}} \cdots T_{\psi_n} w_n(x, \pi). \qquad (5)$$

(cf. formula (1.7.5)). Here the quantity (5) is less than $+\infty$ for either q^+ or q^-.

For a μ-summable model on the finite interval $[m,n]$ with reward functions q and r, formula (5) takes the form

$$w(x, \psi_{m+1} \psi_{m+2} \cdots \psi_n) = T_{\psi_{m+1}} T_{\psi_{m+2}} \cdots T_{\psi_n} r(x)$$

or, if we make use of the formula preceding (2.1),

$$T_{\psi_{m+1}} T_{\psi_{m+2}} \cdots T_{\psi_n} r(x) = \sum_{m+1}^{n} P_x^{\varphi} q(a_t) + P_x^{\varphi} r(x_n), \qquad (6)$$

where φ is the simple strategy given by $\varphi = \psi_{m+1} \psi_{m+2} \cdots \psi_n$. Obviously formula (6) may be applied as well to a model Z on the infinite interval $[m, \infty)$ for any $n > m$ and any function r on X_n, if the "truncated" model $Z^n(r)$ is μ-summable ($Z^n(r)$ is obtained by breaking Z at time n and taking the terminal reward to be r).

Further, for any strategy ρ on the segment $[m,n]$,

$$w(\mu, \rho \pi) = \sum_{m+1}^{n} P_\mu^{\rho} q(a_t) + P_\mu^{\rho} w(x_n, \pi). \qquad (7)$$

For models on a finite interval this formula was proved in Chapter 1, §8. The passage to an infinite time interval in the case $q \geq 0$ is carried out in the same way as in the proof of assertions a) and b). Finally, for a μ-summable model and an arbitrary reward function, formula (7) is obtained by subtracting the corresponding formulas for q^+ and q^-.

Applying formula (7) to the reward q^+, we note that if the model Z is μ-summable then the quantity

$$P_\mu^{\rho} w^+(x_n, \pi) = w^+(v, \pi),$$

is finite, where

$$v(y) = P_\mu^{\rho} \{x_n = y\} \qquad (y \in X_n).$$

Therefore the corresponding derived model is v-summable.

§4. Uniformly ε-Optimal Strategies

For a μ-summable model, the value of the initial distribution μ defined by the formula

$$v(\mu) = \sup_{\pi} w(\mu, \pi)$$

has a meaning, and the definitions of optimal and of ε-optimal strategies for the process Z_μ, given in §§3 and 12 of Chapter 1, need no changes.

A uniformly ε-optimal strategy was defined in Chapter 1 as a strategy π satisfying the condition

$$w(\mu,\pi) \geq v(\mu) - \varepsilon \tag{1}$$

for all initial distributions μ. We proved there that this condition was equivalent to the requirement that

$$w(x,\pi) \geq v(x) - \varepsilon \tag{2}$$

for all $x \in X_m$. Now, these two conditions are no longer equivalent: if $w(x,\pi)$ has a meaning for all $x \in X_m$, this does not mean that $w(\mu,\pi)$ is defined for every μ. Therefore *we shall adopt condition (2) as the definition of ε-optimality, restricting ourselves to the class of those models which are x-summable for all $x \in X_m$.*
   We shall show that:

   a) *For any $\varepsilon > 0$ there exists a uniformly ε-optimal strategy π;*
   b) *If the model is μ-summable, then the function $v(x)$ is also μ-summable* and*

$$v(\mu) = \sum_{X_m} \mu(x)v(x) \qquad (= \mu v). \tag{3}$$

   c) *If the strategy π is uniformly ε-optimal, then $w(\mu,\pi) \geq v(\mu) - \varepsilon$ for all those μ for which the model is μ-summable.*

To prove Result a) it is sufficient to construct, for any $x \in X_m$, a strategy $\pi = \pi_x$ such that

$$w(x,\pi) \geq v(x) - \varepsilon,$$

and then to combine those π_x according to §4 of Chapter 1. Fix some x, and put, for brevity, $\pi_x = \pi$. If $v(x) < +\infty$, the existence of the required strategy π follows from the definition of $v(x)$. If $v(x) = +\infty$, then it follows from the definition of $v(x)$ that for any natural number k there exists a strategy π_k for which

$$w(x,\pi_k) \geq 2^k. \tag{4}$$

Since we may mix strategies (cf. Chapter 1, §3), there exists a strategy π such that

$$Pq(a_t) = \sum_{k=1}^{\infty} 2^{-k} P_k q(a_t) \qquad (t = m+1, m+2, \ldots), \tag{5}$$

where the measure P corresponds to the strategy π and the measures P_k to the strategies π_k; all the processes start at the state x.
   The x-summability of the model, property 2.S and formulas (2.1), (4) and (5), imply that

$$w(x,\pi) = \sum_{k=1}^{\infty} \frac{1}{2^k} w(x,\pi_k) \geq \sum_{k=1}^{\infty} \frac{1}{2^k} \cdot 2^k = +\infty = v(x).$$

---

   * We shall say that the function f is μ-summable above (below) if $\mu f^+ < \infty$ ($\mu f^- < \infty$). For such a function $\mu f = \mu f^+ - \mu f^-$. The words "above" ("below") in the formulation (b) is dropped in accordance with the remarks of §2.

Turning now to the proof of b), we denote by Q and R the subsets of X_m on which $v > 0$ and $v < 0$. Since

$$w(x,\pi) \le v(x),$$

$$-w^-(x,\pi) \le 0 \le w^+(x,\pi), \qquad (x \in X_m),$$

$$-w^-(x,\pi) \le w(x,\pi) \le w^+(x,\pi),$$

we find from formula (3.1) that for any strategy π

$$\mu v^- = -\sum_R \mu(x)v(x) \le -\sum_R \mu(x)w(x,\pi)$$

$$\le \sum_R \mu(x)w^-(x,\pi) \le \sum_{X_m} \mu(x)w^-(x,\pi) = w^-(\mu,\pi),$$

and for a strategy σ of Result a)

$$\mu v^+ = \sum_Q \mu(x)v(x) \le \sum_Q \mu(x)[w(x,\sigma) + \varepsilon]$$

$$\le \sum_Q \mu(x)[w^+(x,\sigma) + \varepsilon] \le \sum_{X_m} \mu(x)[w^+(x,\sigma) + \varepsilon]$$

$$= w^+(\mu,\sigma) + \varepsilon$$

(all the sums have a meaning since the terms are of the same sign).

From these inequalities it is clear that if the model is μ-summable, then the function $v(x)$ is μ-summable as well, so that $\mu v = \mu v^+ - \mu v^-$ has a meaning. The second half of assertion b), and assertion c), are proved just as in Chapter 1, §12. It follows from what has been proved that the two following conditions are equivalent:

1°. *The model Z is μ-summable above.*
2°. $v(\mu) < +\infty$.

Indeed, if $v(\mu) < +\infty$, then $v(\mu)$ has a meaning, which means that the model is μ-summable either above or below. In addition, for any strategy π

$$w(\mu,\pi) = w^+(\mu,\pi) - w^-(\mu,\pi) \le v(\mu) < +\infty$$

so that $w^+(\mu,\pi) < +\infty$. This means that 2° implies 1°. Conversely, if 1° is satisfied, then $v(\mu)$ has a meaning; if $v(x)$ were equal to $+\infty$, then, by a), there would be a strategy π for which $w(\mu,\pi) = w^+(\mu,\pi) - w^-(\mu,\pi) = +\infty$, so that $w^+(\mu,\pi) = +\infty$, which contradicts 1°. Thus 1° implies 2°. We shall show that conditions 1° and 2° are equivalent also to the following:

3°. $\sup_\pi w^+(\mu,\pi) < +\infty$.

To this end we consider the model Z^+ gotten from Z by replacing the reward function q by q^+. Obviously the model Z is μ-summable above if and only if the model Z^+ has the same property. Applying the equivalence of conditions 1° and

$2°$ proved above to the model Z^+, we find that the model Z is summable above if and only if the value of the initial distribution μ in the model Z^+ is finite. But this value is equal to

$$\sup_\pi w^+(\mu,\pi),$$

and this means that $1°$ is equivalent to $3°$.

One establishes the equivalence of the following conditions similarly.

$1a°$. *The model Z is μ-summable below.*

$2a°$. $\inf_\pi w(\mu,\pi) > -\infty$.

$3a°$. $\sup_\pi w^-(\mu,\pi) < +\infty$.

§5. Optimality Equations

For a finite control interval (and for a reward function bounded above) the following results were obtained in §6 and §12 of Chapter 1.

a) *The value v of the model Z is expressed in terms of the value v' of the derived model Z' by the formulas*

$$v = Vu, \qquad u = Uv', \tag{1}$$

where

$$Vg(x) = \sup_{a \in A(x)} g(a) \qquad (x \in X), \tag{2}$$

$$Uf(a) = q(a) + \sum_{y \in X} p(y|a)f(y) \qquad (a \in A); \tag{3}$$

b) *For any $\kappa > 0$ there exists a selector ψ of the correspondence $A(x)(x \in X_m)$, such that*

$$u(\psi(x)) \geq v(x) - \kappa \tag{4}$$

for all $x \in X_m$.

c) *Suppose that ε' and κ are arbitrary nonnegative numbers. If the strategy π' is ε'-optimal for the model Z' and the selector ψ satisfies condition (4), then the strategy $\psi\pi'$ is $(\varepsilon' + \kappa)$-optimal for the model Z.*

In order to generalize these results, we have first of all to show that the functions v and v' have a meaning. The existence of v follows from the x-summability of the model Z for any $x \in X_m$, a condition introduced in the preceding section. Beginning at this point, and up to the end of the chapter, we will assume that *not only Z, but all the models $Z', Z'', \ldots$ derived from Z, are x-summable*. (If this is so we will say that *the model Z is summable*). One may always assure that this additional requirement is satisfied by excluding from X_t all the states in which the condition of x-summability of the corresponding derived model is violated. Such a purge of the state space does not affect the control of the model Z, since in view of 2a)–2b) the excluded states are inaccessible for any strategy. Under the hypotheses made above, finite or infinite values v are defined for the model and all of the models derived from it.

Consider first the case when the model is summable above, so that $v < +\infty$. In this case the validity of the results a)–c) is established in the same way as in §12 of Chapter 1. The possibility of applying the operator U to the function v' and the equation

$$Uv'(a) = q(a) + v'(p_a)$$

follows from 3b) and 4b). The inequality $w'(p_a, \pi') \geq v'(p_a) - \varepsilon$, where π' is an ε'-optimal strategy for Z', follows from 4c).

Now suppose that the model is summable below. The example presented at the end of §13 of Chapter 1 shows that assertion b) can be false for points x at which $v(x) = +\infty$. However the following weakened variant of b) is true.

b') *For any* $\kappa > 0$ *and* $K > 0$ *there exists a selector* ψ *of the correspondence* $A(x)(x \in X_m)$ *such that*

$$u(\psi(x)) \geq \begin{cases} v(x) - \kappa & \text{for } v(x) < +\infty, \\ K & \text{for } v(x) = +\infty. \end{cases}$$

Indeed, if $v(x) < +\infty$ the previous arguments remain valid. If $v(x) = +\infty$, then, by virtue of the fundamental equation, formula (4.3), and the definition of $u(a)$ contained in (1) and (3),

$$+\infty = \sup_{\pi} w(x,\pi) \leq \sup_{\substack{\pi' \\ a \in A(x)}} [q(a) + w'(p_a, \pi')]$$

$$\leq \sup_{a \in A(x)} [q(a) + v'(p_a)] = \sup_{a \in A(x)} u(a).$$

a) follows from b'): if $v(x) < +\infty$ the previous proof remains valid, and if $v(x) = +\infty$ we have $Vu(x) = +\infty = v(x)$ in view of (2).

The result c) and its proof do not depend on whether the model is summable above or below.

In what follows it is convenient to rewrite equations (1) and condition (4) in terms of the operators T_ψ and T defined by the formulas

$$T_\psi f(x) = Uf(\psi(x)), \qquad (x \in X) \tag{5}$$

$$Tf(x) = \sup_\psi T_\psi f(x) = VUf(x) \qquad (x \in X), \tag{6}$$

see the end of §6 of Chapter 1. Moreover, as in §7 of Chapter 1, we suppose that $m = 0$ and denote the model Z and its successive derived models by $Z_0, Z_1, Z_2, \ldots$ and the corresponding model values by $v_0, v_1, v_2, \ldots$. It follows from Result a) that in a summable model the values v_t are connected by the recurrence relations

$$v_t = Tv_{t+1}, \qquad t = 0, 1, 2, \ldots. \tag{7}$$

Result b) implies that, in a model which is summable above, for any sequence of positive numbers $\kappa_1, \kappa_2, \ldots$, it is possible to choose selectors $\psi_1, \psi_2, \ldots$ of the correspondence $A(x)$ $(x \in X_{t-1})$, $t = 1, 2, \ldots$ such that

$$T_{\psi_t} v_t = v_{t-1} - \kappa_t. \tag{8}$$

Finally, it follows from Result c) that for such ψ_t and for any ε'-optimal strategy π in the model Z_n, the product $\psi_1 \psi_2 \cdots \psi_n \pi$ is an ε-optimal strategy for Z with $\varepsilon = \kappa_1 + \kappa_2 + \cdots + \kappa_n + \varepsilon'$ $(n = 1,2,\ldots)$.

It follows from (7) that for any $n > 0$

$$v_0 = T^n v_n. \tag{9}$$

In the next section we shall deduce that under some additional restrictions

$$v_0 = \lim_{n \to \infty} T^n 0.$$

Simply put, this means that the control on a finite, but sufficiently long, interval $[0,n]$ may yield just about the same gain as the control on an infinite time interval.

Now suppose further that $\varphi = \psi_1 \psi_2 \cdots \psi_t \cdots$, where the selectors ψ_t satisfy condition (8), and that $\varepsilon = \kappa_1 + \kappa_2 \cdots + \kappa_t + \cdots$. In §7 we will discuss conditions under which the simple strategy φ is uniformly ε-optimal.

§6. An Expression for the Value of a Model

Consider a summable model Z. Obviously, for any n, the model Z^n, gotten from Z by replacing the reward function q in all the spaces A_t with $t > n$ by 0, is summable. We will denote the values of v and w in the model Z^n by v^n and w^n respectively. It is clear that $v^n = 0$ on X_n, and therefore in view of (4.9) $v^n = T^n 0$ on X_0.

By virtue of (1.1),

$$w(x,\pi) = \lim_{n \to \infty} w^n(x,\pi) \qquad (x \in X_0) \tag{1}$$

so that

$$v(x) = \sup_{\pi} \lim_{n \to \infty} w^n(x,\pi).$$

If exchanging the sup and lim signs were legitimate, we could find from this that

$$v = \lim_{n \to \infty} v^n = \lim_{n \to \infty} T^n 0. \tag{2}$$

But this operation is possible only under quite restrictive conditions, and (2) can be false, as is demonstrated by the following example.

EXAMPLE 1. Consider a homogeneous model, in which all the X_t (and A_t) are identical. The space X_t consists of the point x and of the points y_k, $k = 1, 2, \ldots$ (see

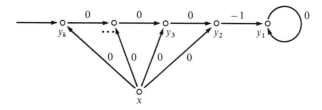

Figure 4.2

figure (4.2)). From y_{k+1} there is a deterministic transition into y_k, and the state y_1 is absorbing. By an appropriate choice of the action, we may pass from x into any of the states y_k with index $k \geq 2$. The reward function q is equal to 0 throughout, aside from the arrow leading from y_2 into y_1; here $q = -1$. Starting from x, we sooner or later pass from y_2 into y_1, so that $v(x) = -1$. However $v^n(x) = 0$ for any integer n, because one may pass from x to y_k with an index k so large that n steps do not suffice to attain y_1.

It follows from equation (1) only that

$$v \leq \varliminf_{n \to \infty} v^n.$$

Indeed, fix x, and then choose an arbitary number c less than $v(x)$. Then $w(x,\pi) > c$ for some strategy π, and, in view of (1), $w^n(x,\pi) > c$ beginning with some $n = n_0$. Thereafter, for $n \geq n_0$, $v^n(x) > c$. Therefore $\varliminf v^n(x) \geq c$.

For v to be equal to the limit v^n, we need to exclude the possibility of substantial losses in an arbitrarily far removed future. Example 1 shows that summability below is not sufficient for this.

In order for equation (2) to be satisfied it suffices that the quantity

$$w_n(x,\pi) = \sum_{t=n+1}^{\infty} P_x^\pi q(a_t) \qquad (4)$$

should satisfy the relation

$$\varliminf_{n \to \infty} \inf_\pi w_n(x,\pi) \geq 0 \qquad (x \in X_0). \qquad (5)$$

Indeed, put $z_n(x) = \inf_\pi w_n(x,\pi)$. For any strategy π

$$w^n(x,\pi) + z_n(x) \leq w^n(x,\pi) + w_n(x,\pi) = w(x,\pi) \leq v(x)$$

so that

$$v^n(x) + z_n(x) \leq v(x).$$

By (5), we have

$$0 \le \varliminf_{n \to \infty} z_n \le \varliminf_{n \to \infty} (v - v^n) = v - \varlimsup_{n \to \infty} v^n.$$

Jointly with inequality (3), this yields (2).

Inequality (5) is evidently satisfied if the reward function q is nonnegative. It is satisfied also if *there exist positive functions* $b_t(x)$, $x \in X_0$, *such that*

$$\sum_1^\infty b_t(x) < +\infty \tag{6}$$

and for sufficiently large t, the estimate

$$P_x^\pi q(a_t) \ge -b_t(x) \tag{7}$$

is valid *for any $x \in X_0$ and any strategy* π. Indeed, it follows from (4) and (7) that for any x and π and sufficiently large n

$$w_n(x,\pi) \ge -\sum_{n+1}^\infty b_t(x) \tag{8}$$

so that (5) follows from (6).

Summable models satisfying the condition in italics above will be said to be *bounded below*. Thus, for any model which is bounded below we have $T^n 0 \to v$.

Any summable model on a finite interval is obviously bounded below on a finite interval. On the other hand, if inequalities (7) are satisfied *for all t*, then the model is x-summable for any initial state x. "Purging" the space X according to the description in §5, we may suppose that the model is summable. Evidently it is bounded below.

§7. Simple ε-Optimal Strategies

In this section we study strategies of the form $\varphi = \psi_1 \psi_2 \cdots \psi_t \cdots$, where the ψ_t are selectors of the correspondences $A(x)$, $(x \in X_{t-1})$ (simple strategies). Our aim is to show that if

$$T_{\psi_t} v_t \ge v_{t-1} - \kappa_t \tag{1}$$

and $\varepsilon = \kappa_1 + \kappa_2 + \cdots$, then

$$w(x,\varphi) \ge v(x) - \varepsilon, \tag{2}$$

i.e. the strategy φ is uniformly ε-optimal. We shall see that this assertion is valid only under certain additional restrictions on the model Z.

Suppose that the model Z is summable. In view of formulas (3.5) and (3.7), for any strategy π and any model Z^n

$$w^n(x,\varphi) + P_x^\varphi w_n(x_n,\pi) = T_{\psi_1} T_{\psi_2} \cdots T_{\psi_n} w_n(x,\pi). \tag{3}$$

It is clear from the definition of the operators T_ψ that they preserve inequalities between functions, and that for any constant c we have $T_\psi(c + f) = c + T_\psi f$; (a constant term may be carried across the operator sign). For each $\varepsilon > 0$ there exists a strategy π_ε for which

$$v_n(x) - \varepsilon \le w_n(x, \pi_\varepsilon) \le v(x) \qquad (x \in X_n).$$

Therefore, since (3) holds for all strategies π_ε it follows that

$$w^n(x,\varphi) + P_x^\varphi v_n(x_n) = T_{\psi_1} T_{\psi_2} \cdots T_{\psi_n} v_n(x). \tag{4}$$

From inequalities (1) and the properties of the operators T_ψ indicated above it follows that

$$
\begin{aligned}
T_{\psi_1} \cdots T_{\psi_{n-1}} T_{\psi_n} v_n &\ge T_{\psi_1} \cdots T_{\psi_{n-1}} v_{n-1} - \kappa_n \\
&\ge T_{\psi_1} \cdots T_{\psi_{n-2}} v_{n-2} - \kappa_{n-1} - \kappa_n \ge \cdots \\
&\ge v - \kappa_1 - \kappa_2 - \cdots - \kappa_n \ge v - \varepsilon.
\end{aligned}
\tag{5}
$$

Since

$$w(x,\varphi) = \lim_{n \to \infty} w^n(x,\varphi) \tag{6}$$

(see (6.1)), then for inequalities (2) to follow from (4) and (5) it suffices that

$$\varlimsup_{n \to \infty} P_x^\varphi v_n(x_n) \le 0. \tag{7}$$

$$* \; * \; *$$

We shall dwell in detail on the case when $v(x)$ is finite. It follows from formula (4) that in this case there exists a limit

$$\delta(x) = \lim_{n \to \infty} P_x^\varphi v_n(x_n). \tag{8}$$

Indeed, of the three terms entering into formula (4), the first has, from (6), a limit $w(x,\varphi)$. The third is monotonically nonincreasing in view of the inequalities

$$T_{\psi_{n+1}} v_{n+1} \le T v_{n+1} = v_n$$

and therefore also has some limit $\lambda(x) \le v_0(x) = v(x)$. Therefore the limit (8) exists and is equal to

$$\delta(x) = \lambda(x) - w(x,\varphi), \tag{9}$$

if at least one of the terms on the right side is finite. The finiteness of λ follows from the inequalities

$$v - \varepsilon \leq \lambda \leq v, \tag{10}$$

the lower estimate being obtained by a passage to the limit from (5).

In view of (9) and (10)

$$v(x) - w(x,\varphi) - \varepsilon \leq \delta(x) \leq v(x) - w(x,\varphi).$$

These inequalities have some interesting consequences:

  1) *Always $\delta \geq -\varepsilon$* (since $w(x,\varphi) \leq v(x)$).
  2) *If $\delta \leq 0$, then the strategy φ is ε-optimal* (This follows also from relations (7) and (8)).
  3) *If the strategy φ is ε-optimal, then $\delta \leq \varepsilon$.*

Applying 2) and 3) to the case $\varepsilon = 0$, we arrive at the following result.

Suppose that the value v is finite and that the selectors ψ_t satisfy the conditions

$$T_{\psi_t} v_t = v_{t-1},$$

$t = 1, 2, \ldots .$ *Put $\varphi = \psi_1 \psi_2 \cdots .$ Then the nonnegative limit*

$$\delta(x) = \lim_{n \to \infty} P_x^\varphi v_n(x_n)$$

exists. For the optimality of the simple strategy φ it is necessary and sufficient that the limit be zero.

$$* \quad * \quad *$$

In analogy to the class of models bounded below (see §6), one may introduce a class of models bounded above. We shall say that a summable model is *bounded above*, if there exist positive functions $c_t(x)$, $x \in X_0$, such that

$$\sum_1^\infty c_t(x) < +\infty, \tag{11}$$

and, for all sufficiently large t, the estimate

$$P_x^\pi q(a_t) \leq c_t(x) \tag{12}$$

is valid for all $x \in X_0$ and all strategies π. We shall show that *for such models the condition (7) is satisfied, so that the strategy φ is ε-optimal.*

First we prove that in a model Z which is bounded above, for sufficiently large n

$$P_x^\varphi w_n(x_n, \pi) \leq \sum_{n+1}^\infty c_t(x) \tag{13}$$

for any $x \in X_0$, any simple strategy φ in the model Z and any strategy π in the derived model Z_n.

Suppose that $\rho = \psi_1 \psi_2 \cdots \psi_n \pi$, where $\psi_1, \psi_2, \ldots \psi_n$ are the first n factors of the strategy φ. Evidently the quantity in the left side of (13) does not depend on the values of the reward function q on the spaces $A_1, \ldots, A_n$. Putting $q = 0$ on these spaces, we find from (3.7) that

$$P_x^{\varphi} w_n(x_n, \pi) = \sum_{t=1}^{n} P_x^{\rho} 0(a_t) + \sum_{t=n+1}^{\infty} P_x^{\rho} q(a_t) = \sum_{n+1}^{\infty} P_x^{\rho} q(a_t).$$

Therefore (13) follows from (12).

Applying (13) to the uniformly ε-optimal strategy π_ε in the model Z_n, we find that

$$P_x^{\varphi} v_n(x_n) \le P_x^{\varphi} w_n(x_n, \pi_\varepsilon) + \varepsilon \le \varepsilon + \sum_{n+1}^{\infty} c_t(x);$$

Because the positive number ε is arbitrary, it therefore follows that

$$P_x^{\varphi} v_n(x_n) \le \sum_{n+1}^{\infty} c_t(x). \tag{14}$$

Equation (7) now obviously follows from (14) and (11).

We note that all summable models on a finite time interval $[m,n]$ are summable both above and below.

For models bounded above, given a fixed initial state one may neglect positive contributions which are introduced into the mean reward at times which are far removed. The following example shows that this condition is at the heart of the matter.

EXAMPLE 1. Consider a homogeneous model with two states, as shown in figure 4.3. In the state x two actions are possible, carrying us into x and y respectively. The state y is absorbing. Obviously $v(x) = 1$ and $v(y) = 0$. The strategy φ, consisting of returning permanently to x, satisfies inequalities (1) for $\kappa_t = 0$, but is not optimal, since $w(x,\varphi) = 0$.

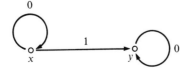

Figure 4.3

We say that the model is *uniformly bounded below* if it is bounded below and all the functions $b_t(x)$ ($t = 1,2, \ldots$) in (6.6) and (6.7) are constants. Models uniformly bounded above are defined analogously with b_t replaced by c_t.

If the model is uniformly bounded both above and below, then in order to obtain a strategy close to an optimal one, a finite number of the conditions (1) suffice. More precisely, we shall show that *if*

$$T_{\psi_t} v_t \geq v_{t-1} - \kappa_t \qquad (t = 1,2, \ldots ,n),$$

then any strategy π coinciding on the first n steps with the product $\varphi = \psi_1 \psi_2 \cdots \psi_n$ is ε-optimal for

$$\varepsilon = \sum_1^n \kappa_t + \sum_{t>n} (b_t + c_t) \tag{15}$$

(Under the assumptions that inequalities (6.7) and (12) are satisfied starting from n.) By taking n sufficiently large and then taking sufficiently small $\kappa_1, \kappa_2, \ldots, \kappa_n$, we may make ε arbitrarily small.

For the proof we observe that $w^n(x,\pi) = w^n(x,\varphi)$, so that, according to formula (4),

$$w(x,\pi) = w^n(x,\pi) + w_n(x,\pi)$$
$$= T_{\psi_1} T_{\psi_2} \cdots T_{\psi_n} v_n(x) - P_x^\varphi v_n(x_n) + w_n(x,\pi).$$

It follows from (5), (14) and (6.8) that the right side is not less than $v(x) - \varepsilon$.

§8. Sufficiency of Markov and Simple Strategies

Recall the main results of §13, Chapter 1.

1. For each initial distribution μ and strategy π there exists a Markov strategy σ such that

$$w(\mu,\sigma) = w(\mu,\pi).$$

2. For any Markov strategy σ there exists a simple strategy φ such that

$$w(\mu,\varphi) \geq w(\mu,\sigma) \quad \text{for all } \mu.$$

To what extent do these results carry over to the present case?

Result 1 remains valid for any μ-summable model. Indeed, we define σ for all $t > 0$ by formula (1.13.3). We have shown that, for any t the probability distributions for a_t relative to the measures P_μ^σ and P_μ^π coincide. Therefore

$$w(\mu,\pi) = \sum_1^\infty P_\mu^\pi q(a_t) = \sum_1^\infty P_\mu^\sigma q(a_t) = w(\mu,\sigma).$$

The situation is different with Result 2. If $v = +\infty$, then this result is false; see example 2 of Chapter 1, §13. It is not known whether this result is valid for any model which is summable above. It is possible to prove it only for models which have a nonnegative reward function q, or, somewhat more generally, for models Z which are summable above and bounded above (see §7).

For any model which is summable above, we may, as in §13 of Chapter 1, choose for each $t = 1, 2, \ldots$ a selector ψ_t of the correspondence $A(x)$, $x \in X_{t-1}$, such that

$$w_{t-1}(x,\psi_t\sigma^t) \geq w_{t-1}(x,\sigma^{t-1}) \tag{1}$$

for all $x \in X_{t-1}$; here we denote by σ^t the restriction of σ to Z^t. From these inequalities we wish to deduce that

$$w(x,\varphi) \geq w(x,\sigma) \qquad (x \in X_0) \tag{2}$$

for the simple strategy $\varphi = \psi_1\psi_2\cdots$.

Using the operators T_ψ, we may write out the inequalities (1) in the form

$$T_{\psi_t}w_t(x,\sigma^t) \geq w_{t-1}(x,\sigma^{t-1}) \qquad (t = 1,2,\ldots).$$

Therefore, using formulas (3.5) and (3.7), we get

$$w(x,\sigma) \leq T_{\psi_1}w_1(x,\sigma^1) \leq T_{\psi_1}T_{\psi_2}w_2(x,\sigma^2)$$
$$\cdots \leq T_{\psi_1}T_{\psi_2}\cdots T_{\psi_n}w_n(x,\sigma^n) = w(x,\psi_1\psi_2\cdots\psi_n\sigma^n)$$
$$= w^n(x,\varphi) + P_x^\varphi w_n(x_n,\sigma^n) \tag{3}$$

for any $n > 0$.

Since $w^n(x,\varphi) \to w(x,\varphi)$, then (2) follows from (3) under the following additional assumption

$$\varlimsup_{n\to\infty} P_x^\varphi w_n(x_n,\sigma^n) \leq 0.$$

This condition is slightly weaker than (7.7), and hence it is satisfied if the model is bounded above.

A close but weaker result may be obtained for models uniformly bounded below (see §7). Indeed, in such models, for any initial distribution μ (for which the model is μ-summable above) any Markov strategy σ and any $\varepsilon > 0$, it is possible to construct a simple strategy φ such that

$$w(\mu,\varphi) > w(\mu,\sigma) - \varepsilon. \tag{4}$$

First of all in view of the results of §3, we may without loss of generality suppose that the model Z is simply summable above. Indeed we are interested only in the process Z_μ, and the model Z is μ-summable above. (cf. the analogous remark in §5).

Since

$$w^n(\mu,\sigma) \to w(\mu,\sigma) < +\infty,$$

then for sufficiently large n

$$w^n(\mu,\sigma) \geq w(\mu,\sigma) - \frac{\varepsilon}{2}. \tag{5}$$

Since the model is uniformly bounded below, for sufficiently large n

$$w(\mu,\pi) = w^n(\mu,\pi) + \sum_{n+1}^{\infty} P_\mu^\pi q(a_t) \geq w^n(\mu,\pi) - \frac{\varepsilon}{2} \tag{6}$$

for any strategy π. Starting out with a number n for which both (5) and (6) are valid, we replace the reward function q on the sets A_t with $t > n$ by zero. Then the value w^n of the previous model turns into the value w of the new model. Since the new model, being summable above, is also bounded above, then there exists in it a simple strategy φ which is uniformly no worse than σ. For such a strategy

$$w^n(\mu,\varphi) \geq w^n(\mu,\sigma), \tag{7}$$

and (4) now follows from (5), (6), and (7).

Thus we have the following results:

a) *In a μ-summable model, for each strategy π there exists a Markov strategy σ such that $w(\mu,\sigma) = w(\mu,\pi)$.*

b) *In a model which is summable above and bounded above, for any Markov strategy σ there exists a simple strategy φ such that*

$$w(x,\varphi) \geq w(x,\sigma)$$

for all $x \in X_0$.

c) *In a model which is uniformly bounded below, for any initial distribution μ for which the model is μ-summable above, any Markov strategy σ and any $\varepsilon > 0$, there exists a simple strategy φ such that $w(\mu,\varphi) \geq w(\mu,\sigma) - \varepsilon$.*

Taking account of the remark with which we began the proof of Result c), and also of formula (3.1), we obtain the following variant of Result b), analogous to Result c):

b′) *In a model which is bounded above, for any initial distribution μ for which the model is μ-summable above and any Markov strategy σ, there exists a simple strategy φ such that $w(\mu,\varphi) \geq w(\mu,\sigma)$.*

The following is a consequence of a), b′), and c).

d) *Suppose that the model is μ-summable above and that π is any strategy. If the model is bounded above, then there exists a simple strategy φ such that $w(\mu,\varphi) \geq w(\mu,\pi)$. If the model is uniformly bounded below, then for any $\varepsilon > 0$ there exists a simple strategy φ such that $w(\mu,\varphi) \geq w(\mu,\pi) - \varepsilon$.*

The question as to the possibility of extending either of the Results b), c), and along with them d), to arbitrary models which are bounded above remains open.

Chapter 5

Borel Models

§1. The Main Results

This chapter stands in the same relation to Chapter 3 as Chapter 4 did to Chapter 1.

As in Chapter 3, we shall study general models with Borel spaces of states and actions, and we will suppose that the set of strategies is nonempty (the nontriviality condition of Chapter 3, §1). In distinction from Chapter 3, the interval of control will be supposed infinite, and the requirement that the reward function q be bounded above will be removed (in semicontinuous models the assumption that q is bounded above at each step is preserved). It follows from the nontriviality of the model, as it did in Chapter 3, §2, that the set of simple strategies is nonempty.

The value $w(\mu,\pi)$ of the strategy π with the initial distribution μ is defined as it was in Chapter 4 by formulas (4.2.1) or (4.2.5) under the hypothesis that the series (4.2.2) or (4.2.3) converges (μ-summability of the model above or below*). Of course, in computing $Pq^+(a_t)$ and $Pq^-(a_t)$ we need now to make use of formula (2.2.4).

In Chapter 3, for Borel models on a finite interval of control, we obtained the following three main results:

    I. *The value v of the model satisfies the optimality equations.*

    IIa. *For each $\varepsilon > 0$ and each initial distribution μ there exists a simple (a.s. μ) ε-optimal strategy.*

    III. *For a fixed initial distribution μ, for each strategy π there exists a simple strategy φ such that $w(\mu,\varphi) \geq w(\mu,\pi)$.*

Are these results valid in the case of a summable model on an infinite interval of control (and for a reward which is not bounded above)?

We shall show that Result I remains valid. We will prove the following weakened variants of results IIa and III:

    II′a. *If a model which is μ-summable above is bounded above, then for any number $\varepsilon > 0$ there exists a simple (a.s. μ) ε-optimal strategy.*

    III′. *Suppose that the model is μ-summable above. If it is bounded above, then for any strategy π there exists a simple strategy φ such that $w(\mu,\varphi) \geq w(\mu,\pi)$. If the model is uniformly bounded below, then for any strategy π and any number $\varepsilon > 0$ there exists a simple strategy φ such that $w(\mu,\varphi) \geq w(\mu,\pi) - \varepsilon$.*

---

* See the remark on terminology on page 104.

If $v(\mu) = +\infty$ Result III is generally speaking false (see Example 1.13.2). The question as to the validity of IIa and III in an arbitrary μ-summable model remain open.

In order to prove I, II'a and III', it is necessary first to extend the results of Chapter 4 to Borel models. We will do this in §2. We will give the proofs of propositions I, II'a and III in §3.

§2. Extension of the Results of Chapter 4 to Borel Models

We summarize the results of Chapter 4, writing out sums over state and action spaces in the form of integrals:

a) If the model is μ-summable, then it is x-summable (a.s. μ) and

$$w(\mu,\pi) = \int_{X_m} w(x,\pi)\mu(dx).$$

b) If the model is x-summable, then the derived model is p_a-summable for every $a \in A(x)$ and

$$w(x,\pi) = \int_{A(x)} \pi(da\,|\,x)[q(a) + w'(p_a,\pi_a)]$$

(the fundamental equation).

c) If the model is x-summable, the ψ_t are measurable* selectors of the correspondences $A(y)$ $(y \in X_{t-1}, t = m+1, \ldots, n)$ and if π is any strategy in the derived model of order $n - m$, then

$$w(x,\psi_{m+1}\psi_{m+2}\cdots\psi_n\pi) = T_{\psi_{m+1}}T_{\psi_{m+2}}\cdots T_{\psi_n}w_n(x,\pi), \tag{1}$$

where the operators T_{ψ_t} are defined by the equation

$$T_{\psi_t}f(x) = q[\psi_t(x)] + \int_{X_t} p(dy\,|\,\psi_t(x))f(y) \qquad (x \in X_{t-1}).$$

Suppose that $Z^n(r)$ is the model obtained by taking the terminal payoff to be r at the time n, and that $\varphi = \psi_{m+1}\psi_{m+2}\cdots\psi_n$ is a strategy on the segment $[m,n]$. If $Z^n(r)$ is x-summable, then

$$T_{\psi_{m+1}}T_{\psi_{m+2}}\cdots T_{\psi_n}r(x) = \sum_{t=m+1}^{n} P_x^\varphi q(a_t) + P_x^\varphi r(x_n). \tag{1'}$$

d) If the model is μ-summable and ρ is any strategy on the segment $[m,n]$, then the derived model of order $n - m$ is v-summable, where v is the distribu-

---

* For discrete models the word "measurable" may be dropped, since all selectors are measurable. However in the general case formula (1) simply would not make sense without the hypothesis of measurability of ψ_t.

tion on X_n given by the formula

$$v(\Gamma) = P_\mu^\rho\{x_n \in \Gamma\} \qquad (\Gamma \in \mathscr{B}(X_n)),$$

while for any strategy π on the interval $[n,\infty)$

$$w(\mu,\rho\pi) = \sum_{m+1}^{n} P_\mu^\rho q(a_t) + P_\mu^\rho w(x_n,\pi).$$

In what follows we will suppose that *the model is x-summable for any* $x \in X_m$, so that the value $v(x)$ is defined for all $x \in X_m$. By definition, the value $v(\mu)$ has a meaning if and only if the model is μ-summable.

e) For any $\varepsilon > 0$ there exists a strategy π such that

$$w(x,\pi) \geq v(x) - \varepsilon \quad \text{for all } x \in X_m$$

(an ε-optimal strategy).

f) If the model is μ-summable, then the value function v is also μ-summable and

$$v(\mu) = \int_{X_m} \mu(dx)v(x) \qquad (=\mu v). \tag{2}$$

g) If the strategy π is uniformly ε-optimal, then

$$w(\mu,\pi) \geq v(\mu) - \varepsilon$$

for all initial distributions μ for which the model is μ-summable.

h) The following three conditions are equivalent:

1°. The model Z is μ-summable above.
2°. $v(\mu) < +\infty$.
3°. $\sup_\pi w^+(\mu,\pi) < +\infty$.

In the formulations i)–p) we suppose that $m = 0$, and that the model Z and all of its derived models are x-summable for any initial state x, i.e. that the model Z is summable.

i) The values $v = v_0, v_1, v_2, \ldots$ of the model $Z = Z_0$ and of its successive derived models $Z_1, Z_2, \ldots$ are connected by the recurrence relations

$$v_{t-1} = Vu_t, \qquad u_t = Uv_t$$

or

$$v_{t-1} = Tv_t,$$

$t = 1, 2, \ldots$, where the operators V, U, and T are defined by the formulas

$$Vg(x) = \sup_{a \in A(x)} g(a) \qquad (x \in X),$$

$$Uf(a) = q(a) + \int_X p(dy\,|\,a)f(y) \qquad (a \in A),$$

$$T = VU.$$

j) In a model which is summable above, for any sequence $\kappa_1, \kappa_2, \ldots$ of positive numbers, it is possible to choose measurable selectors $\psi_1, \psi_2, \ldots$ of the correspondences $A(x)(x \in X_{t-1}, t = 1, 2, \ldots)$ such that

$$T_{\psi_t} v_t \geq v_{t-1} - \kappa_t. \tag{3}$$

k) If the inequalities (3) are satisfied for $t = 1, 2, \ldots, n$, for the measurable selectors $\psi_1, \psi_2, \ldots, \psi_n$ and the nonnegative numbers $\kappa_1, \kappa_2, \ldots, \kappa_n$, then for any strategy π which is ε'-optimal in the model Z_n, the strategy $\psi_1 \psi_2 \ldots \psi_n \pi$ is ε-optimal in the model Z for $\varepsilon = \kappa_1 + \kappa_2 + \cdots + \kappa_n + \varepsilon'$.

l) If the model is bounded below (see Chapter 4, §6), then

$$v = \lim_{n \to \infty} T^n 0.$$

m) If the model is bounded above (see Chapter 4, §7), and if the measurable selectors ψ_t, $t = 1, 2, \ldots$ satisfy (3) with nonnegative κ_t's, then the simple strategy $\varphi = \psi_1 \psi_2 \cdots$ is uniformly ε-optimal for $\varepsilon = \kappa_1 + \kappa_2 + \cdots$; if the model is uniformly bounded both above and below, then for sufficiently large n any strategy π coinciding on the first n steps with the product $\psi_1 \psi_2 \cdots \psi_n$ is ε-optimal with

$$\varepsilon = \sum_1^n \kappa_t + \sum_{t > n} (b_t + c_t),$$

the numbers b_t and c_t satisfying formulas (4.6.7) and (4.7.12).

n) If the model is μ-summable, then for any strategy π there exists a Markov strategy σ such that $w(\mu,\pi) = w(\mu,\sigma)$.

o) If the model is summable above and bounded above, then for any Markov strategy σ there exists a simple strategy φ such that $w(x,\varphi) \geq w(x,\sigma)$ for all $x \in X$.

p) If the model is μ-summable above and uniformly bounded below, then for any Markov strategy σ and any number $\varepsilon > 0$ there exists a simple strategy φ such that $w(\mu,\varphi) \geq w(\mu,\sigma) - \varepsilon$.

Now what changes when the spaces X_t and A_t are uncountable?

Assertions e) *and* j) *are* (generally speaking) *false*, as we saw in Chapter 3, §1 (in Example 2, there were no uniformly ε-optimal strategies for $\varepsilon < 1$). We shall show that *all the remaining assertions hold in the general case*, and that the following weakened variant of proposition e) holds:

e′) *The function $v(x)$ is universally measurable. For any $\varepsilon > 0$ and any initial distribution μ there exists a strategy π such that*

$$w(x,\pi) \geq v(x) - \varepsilon \quad \text{(a.s. } \mu\text{)},$$

i.e. an (a.s. μ) ε-*optimal strategy.*

In the general theory, without j), propositions k) and m) are not of great value. However they will be useful in §6, in the investigation of semicontinuous models, and also in the concrete examples of Chapter 6.

Results a) and b) are established using the same arguments as in the discrete case, except that we have to rely on §3 of Chapter 2 rather than §12 of Chapter 1. In the discrete case we made use of the fact that if the sum of a positive series is finite, then all of its terms were finite as well, and also of property S (see §§3,4 of Chapter 4). In the general case, an analogous rôle is played by the following assertions: 1) If an integral of a nonnegative function is finite, then that function is finite almost everywhere; 2) If the functions f_1 and f_2 are nonnegative and at least one of the integrals $\mu f_1, \mu f_2$ is finite, then the integral $\mu(f_1 - f_2)$ has a meaning and $\mu(f_1 - f_2) = \mu f_1 - \mu f_2$. Both of them follow from the definition of the integral given in §1 of Chapter 2. Moreover, in the proof of b) we need to assure the measurability of the selector ψ_a. It suffices to put

$$\psi_a(y) = \begin{cases} a & \text{for } y = x, \\ \gamma(y) & \text{for } y \neq x, \end{cases}$$

where γ is any measurable selector of the correspondence $A(y)$ ($y \in X_m$).

The derivations of assertions c) and d) do not differ from the discrete case (the Markov property is valid in view of §3 of Chapter 2).

The proof of e') is based on quite different ideas from those employed in the proof of e) in Chapter 4; it requires the general theorems of Chapter 3. We shall go around this proof for the time being; sections 4 and 5 are devoted to it.

The first half of assertion f), the μ-summability of the function v, is proved as in §4 of Chapter 4. The only difference consists in that instead of the function v itself we take a measurable function $\bar{v}$ such that $v = \bar{v}$ (a.s. μ), and instead of a uniformly ε-optimal strategy a strategy $\bar{\pi}$ which is (a.s. μ) ε-optimal; the existence of such a strategy follows from e'). It is clear that the μ-summability of v is equivalent to the μ-summability of $\bar{v}$. Formula (2) is deduced from e') in the same way as in §7 of Chapter 3.

Result g) is an obvious consequence of a) and f).

For the proof of h), as in the discrete case we need to deduce from the equation $v(\mu) = +\infty$ the existence of a strategy π such that $w(\mu,\pi) = +\infty$. This is done using e'), a) and f).

Result i) is deduced from b) and f) as in §7 of Chapter 3. (In the case $v(\bar{x}) = +\infty$ we choose an arbitrarily large K and then, as in §5 of Chapter 4, we find an $\bar{a} \in A(\bar{x})$ such that $u(\bar{a}) > K$.)

Assertion k) follows from formula (1), the inequality $w_n(x,\pi) \geq v_n(x) - \varepsilon'$, the monotonicity of the operators T_ψ and the possibility of carrying a constant term across the signs of these operators.

Results l) and m) are proved as in Chapter 4, except that the derivation of formula (4.7.4) changes; instead of strategies π_ε which are uniformly ε-optimal (relative to the model Z_n), it is necessary to take strategies such that $w_n \geq v_n - \varepsilon$ (a.s. v), where v is the distribution of the state x_n given the initial state x and the strategy φ.

Assertions m), o), and p) are established as in §8 of Chapter 4 (using the results of §8 of Chapter 3 instead of those of §13 of Chapter 1).

§3. Proofs of the Main Results

Result I of §1 coincides with proposition 2i). Result III follows from assertions
2n), 2o), and 2p).

In Chapter 3 the result IIa was obtained from assertion III using Lemma 1 in
§9 of Chapter 3, which says that *for any sequence $\varphi^1, \varphi^2, \ldots$ of simple strategies
and any $\varepsilon > 0$ there exists a simple strategy φ for which*

$$w(x,\varphi) \geq \sup_k w(x,\varphi^k) - \varepsilon \tag{1}$$

for all $x \in X_0$ (we continue here to assume that $m = 0$). The extension of this
lemma to the case of an infinite interval of control will take up the remainder of
this section. Here we have to assume that *the model is summable above and bounded
above.*

The lemma is applied in the same way as it was before, except that we have to
delete from X_0 a set of μ-measure zero which contains all the states x for which
the condition of x-summability above is violated (such a set exists according to
2a)). Models which are bounded below have to be excluded in the passage from
III′ to II′a because it has not been possible to extend the lemma to such models.

* * *

Now we turn to the proof of the lemma for the case at hand. We will denote
the values of strategies in the derived model of order n by w_n, and the restriction
of simple strategies φ to the derived models once again by φ.

Each of the given simple strategies φ^k decomposes into an infinite product

$$\varphi^k = \psi_1^k \psi_2^k \cdots \psi_t^k \cdots$$

of measurable selectors ψ_t^k of the correspondences $A(x)$ $x \in X_{t-1}$. In order to
simplify the notation we write $T_t^k = T_{\psi_t^k}$ and

$$r_t(x) = \sup_k w_t(x,\varphi^k),$$

$x \in X_t, t = 0, 1, 2, \ldots$. Since $r_t \leq v_t$, then

$$T_t^k r_t \leq T_t^k v_t \leq T v_t = v_{t-1},$$

so that

$$\Phi_t(x) = \sup_k T_t^k r_t(x) < +\infty.$$

Therefore, for some k

$$T_t^k r_t(x) \geq \Phi_t(x) - \frac{\varepsilon}{2^t}.$$

Denote the smallest such number k by $k(t,x)$ and put

$$\psi_t(x) = \psi_t^{k(t,x)}(x),$$

$x \in X_{t-1}, t = 1, 2, \ldots$. Since the least upper bound of a countable set of measurable functions is measurable then r_t and Φ_t are measurable functions. Therefore $k(t,x)$ is measurable in x, and ψ_t is a measurable selector of the correspondence $A(x)$ $(x \in X_{t-1})$. We shall prove that the strategy $\varphi = \psi_1 \psi_2 \cdots \psi_k \cdots$ satisfies inequality (1).

We obtain (1) by a passage to the limit from (2.1') for $r = r_n$. In order to make use of that formula, we need, in view of 2c), to verify the x-summability above of the model $Z^n(r_n)$, i.e. that

$$\sum_{t=1}^{n} P_x^\rho q^+(a_t) + P_x^\rho r_n^+(x_n) < +\infty$$

for any strategy ρ on the segment $[0,n]$. The finiteness of each term $P_x^\rho q^+(a_t)$ follows from the summability above of the model Z in question. Since $r_n \leq v_n$, then $r_n^+ \leq v_n^+$ as well. Therefore it suffices to verify that

$$P_x^\rho v_n^+(x_n) < \infty.$$

Note that

$$P_x^\rho v_n^+(x_n) = v v_n^+,$$

where

$$v(\Gamma) = P_x^\rho\{x_n \in \Gamma\}, \qquad \Gamma \in \mathcal{B}(X_n).$$

It follows from proposition 2d) that the derived model is v-summable above. Using proposition 2f), we conclude that the function v_n is also v-summable above, i.e. that $v v_n^+ < +\infty$. So, the x-summability above of the model $Z^n(r_n)$ is proved.

By formula (2.1'), applied to the function r_n and the strategy φ,

$$\sum_{1}^{n} P_x^\varphi q(a_t) + P_x^\varphi r_n(x_n) = T_{\psi_1} T_{\psi_2} \cdots T_{\psi_n} r_n(x). \tag{2}$$

We shall estimate this last quantity from below using the inequalities

$$T_{\psi_t} r_t \geq T_t^k r_t - \frac{\varepsilon}{2^t} \geq T_t^k w_t(\cdot, \varphi^k) - \frac{\varepsilon}{2^t} = w_{t-1}(\cdot, \varphi^k) - \frac{\varepsilon}{2^t} \tag{3}$$

$$\geq r_{t-1} - \frac{\varepsilon}{2^t}$$

following from the definition of ψ_t, the definition of r_t, and the fundamental equation. From (2) and (3) as in §7 of Chapter 4, we get

$$\sum_{1}^{n} P_x^\varphi q(a_t) + P_x^\varphi r_n(x_n) \geq r_0(x) - \varepsilon. \tag{4}$$

The boundedness above of the model implies that

$$\overline{\lim_{n \to \infty}} P_x^\varphi v_n(x_n) \leq 0$$

(see §7 of Chapter 4). Recalling that $r_n \leq v_n$, and passing to the limit in (4), we observe that

$$w(x,\varphi) = \sum_1^\infty P_x^\varphi q(a_t) \geq r_0(x) - \varepsilon$$

for all $x \in X_0$, and we thus have the desired extension of the lemma.

§4. Measures on Infinite Products

In the proof of proposition e′) of §2 we have to operate in the space L of infinite paths, something which we have so far avoided. Now we need results about measures on infinite product spaces, analogous to the results for finite products described in Chapter 3, §4.

Suppose we are given arbitrary sets $E_0, E_1, \ldots, E_t, \ldots$. The points of the space $E = E_0 \times E_1 \times E_t \times \cdots$ are sequences $x = x_0 x_1 \cdots x_t \cdots$, with $x_t \in E_t, t = 0, 1, \ldots$ If each of the E_t is a measurable space, then a measurable structure can be introduced into E as the minimal σ-algebra $\mathscr{B}(E)$ containing for each $t = 0, 1, 2, \ldots$, all the sets

$$C \times E_{t+1} \times E_{t+2} \times \cdots \times E_n \times \cdots, \tag{1}$$

where $C \in \mathscr{B}(E_0 \times E_1 \times \cdots \times E_t)$. One can show that if all the spaces E_t are Borelian, then the space E is also Borelian; see Appendix 1, §4.

We note that any probability measure P on E may be considered also as a probability measure on the finite product $E_0 \times E_1 \times \cdots \times E_t$, by putting $P(C)$ equal to the measure on E of the set (1).

Theorems E and F of Chapter 3, §4, establishing a connection between the transition functions and measures in the product spaces, remain valid for infinite products as well. However Theorem E now turns out to be far from trivial. It was first proved by C. Ionescu Tulcea, and a proof may be found, for example, in Chapter 5, §1 of the book [1] by J. Neveu. Theorem F can be proved in the same way as it is for a finite number of factors; see Appendix 4.

Theorems 1 and 2 of Chapter 3, §4 carry over to infinite product spaces for sets B satisfying the following additional requirement:

The sequence $x_0 x_1 \cdots x_t \cdots \in B$ if the sequence $x_0 x_1 \cdots x_t \in B_t$ for each $t = 0, 1, 2, \ldots$. (The projection B_t of B into the product $E_0 \times E_1 \times \cdots \times E_t$ is defined as in Chapter 3, §4).

The above condition on B implies that

$$B = \bigcap_{t=0}^\infty (B_t \times E_{t+1} \times E_{t+2} \times \cdots).$$

Hence $P(B) = 1$ if and only if $P(B_t) = 1$ for all $t = 0, 1, 2, \ldots$. Therefore, from the fact that Theorems 1 and 2 of Chapter 3, §4 hold for finite products $E_0 \times E_1 \times \cdots \times E_t$, it follows that they hold also for infinite products $E_0 \times E_1 \times \cdots \times E_t \times \cdots$.

§5. Universal Measurability of the Value of a Model and the Existence of a.s. ε-Optimal Strategies

The proof of proposition e′) of §2 is carried out according to the same plan as in Chapter 3, §§2–6.

First of all we introduce the space L of paths of infinite length. This space consists of all possible sequences

$$l = x_m a_{m+1} x_{m+1} \cdots a_n x_n \cdots$$

with $x_n \in X_n$ and $a_n \in A_n$, such that

$$j(a_{n+1}) = x_n, \qquad n = m, m+1, \ldots. \tag{1}$$

Since all the spaces X_t and A_t are Borelian, and each of the countable number of conditions (1) distinguishes a measurable subset of the infinite product $X_m \times A_{m+1} \times \cdots$, then L is also a Borel space.

According to Chapter 3, §3, the class $\mathcal{M}$ of all Borel measures on L also forms a Borel space.

Relying on Theorem 2 of Chapter 3, §4 and on §4 of this Chapter we assign to each initial distribution μ and strategy π a strategic measure P_μ^π in the space L of paths. By formulas (3.4.10)–(3.4.12), for any $n > m$, the distribution P_μ^π for the history at the time n is given by the same formula (2.2.4) as for the distribution for the paths in a model on the segment $[m,n]$, i.e. consistent with the definitions of Chapter 4, §2 and Chapter 5, §1.

The subclass S_0 of the class S of strategic measures and the mapping $S_0 \overset{k}{\to} X_m$ are defined as in Chapter 3, §§2,5. As before, the description of the class S is given by Theorem 1 of Chapter 3, §5, except that this time one writes the relations (3.5.1) for all $t \geq n + 1$. The measurability and convexity of the class S is deduced with the aid of Theorem 3.5.1 as in Chapter 3 §5. The measurability of the class S_0 and of the mapping k are established as in Chapter 3, §6.

In order to extend the proof of universal measurability of the value $v(x)$ carried out in Chapter 3, §2, it remains to represent the value $w(x,\pi)$ of the strategy π as a measurable function $f(P)$ of the measure $P \in S_0$. In distinction from Chapter 3, the function $f(P)$ can take not only finite values, but also the values $+\infty$ or $-\infty$ (depending on whether the model is summable below or above). The concepts of measurability and of universal measurability extend to such functions in an obvious way. Theorem B of Chapter 3, §2 and its proof remain valid for such functions.

To fix ideas, we suppose that the model is summable below. Let P be a measure of the class S_0, corresponding to the initial state x and the strategy π. By the definition given in Chapter 4, §2 and Chapter 5, §1,

$$w(x,\pi) = PI^+ - PI^-,$$

where

$$I^- = \sum_m^\infty q^-(a_t), \qquad I^+ = \sum_m^\infty q^+(a_t)$$

and

$$PI^- < +\infty. \tag{2}$$

Each of the functions

$$f_1(P) = PI^+, \qquad f_2(P) = PI^- \qquad (P \in S_0)$$

is measurable in view of Chapter 3, §2. It therefore follows from (2) that their difference $w(x,\pi) = f_1(P) - f_2(P)$ is measurable as well.

* * *

In order to extend to an infinite interval the proof of the existence of an (a.s. μ) ε-optimal strategy given in Chapter 3, §§2,6, we need the following two additions to that proof.

First, in Chapter 3, §2 we started from the fact that for any $x \in X_m$ and any $\varepsilon > 0$ there exists a strategy $\pi = \pi_x$ such that $w(x,\pi) \geq v(x) - \varepsilon$. This assertion, trivial for $v(x) < +\infty$, has to be extended to the case when $v(x) = +\infty$. It follows from the convexity of the class of strategic measures established above that these measures may be mixed in the same way that we mixed the strategies in Chapter 1, §3. Hence the construction presented in Chapter 4, §4 carries over to the general case, yielding for $v(x) = +\infty$ a strategic measure $P \in S(x)$ for which the mathematical expectation of the reward is also equal to $+\infty$.

Second, we have to carry over to the case of infinite paths the formula

$$P_\mu^\pi = \int_{X_m} P_x^\pi \mu(dx), \tag{3}$$

which we used at the end of §6 of Chapter 3 (see (3.6.6)). This can be done using Theorem E of Chapter 3, §4 and of this chapter, §4, according to which two measures P and Q coincide if the corresponding distributions of histories of any length coincide.

§6. Semicontinuous Models

Now we turn to the investigation of the semicontinuous models with which we were concerned, for a finite interval of control, in Chapter 2. We shall study under which conditions the fundamental result of Chapter 2, the existence of a uniformly optimal strategy and the possibility of finding it from the optimality equations, carry over to the infinite interval of control.

In the definition of a semicontinuous model (see Chapter 2, §4) all the spaces X_t and A_t were supposed to be separable metric spaces. In order to be able to use the results on general models proved in the preceding sections, *we suppose in addition that all the spaces X_t and A_t are Borelian**.

The main result of this section is the following: *If the semicontinuous model Z and all derived models Z', Z'', ... are uniformly bounded both above and below, then there exists a simple strategy $\varphi = \psi_1\psi_2 \cdots \psi_t \cdots$ such that*

$$v_{t-1} = T_{\psi_t} v_t, \qquad t = 1, 2, \ldots,$$

and this strategy is uniformly optimal.

From Theorem 2.5.A, applied to the spaces $E = X_{t-1}$, $E' = A$ and to the function $f = 0$, it follows that the correspondence $A(x)$ $(x \in X_{t-1})$ admits a measurable selection for each $t = 1, 2, \ldots$. Therefore the model is nontrivial, and the results of the preceding sections are applicable to it.

The boundedness above of the reward function q at each step (see condition 2.4.D) and the boundedness above of the model $Z^{(t)}$ imply that the model $Z^{(t)}$ is summable above. Hence all the values v_t are bounded above and satisfy the optimality equations $v_{t-1} = Vu_t$, $u_t = Uv_t$.

We have to prove the existence of selectors ψ_t satisfying equations (1). To apply Theorem 2.5.A, we need to know that the function v_t is semicontinuous. In Chapter 2, §5 this was established by induction from t to $t - 1$. Now we have no beginning for the induction. Instead of induction we appeal to the formula $v_t = \lim T^n 0$ proved under the hypothesis that the model is bounded below (see §2, proposition *l*). The semicontinuity of $T^n 0$ follows from the results of Chapter 2. In order to deduce the semicontinuity of v from this fact, we have to suppose that the model is uniformly bounded both below and above. Then $T^n 0$ converges to v uniformly on all of X_0, and semicontinuity is preserved under such convergence.

In a summable model bounded both above and below, for any $\varepsilon > 0$ there is an N such that for $n \geq N$

$$|w(x,\pi) - w^n(x,\pi)| \leq \varepsilon$$

for all $x \in X_0$ and any strategy π. Hence

$$w(x,\pi) \leq w^n(x,\pi) + \varepsilon \leq v^n(x) + \varepsilon = T^n 0(x) + \varepsilon,$$

so that

$$v(x) \leq T^n 0(x) + \varepsilon.$$

Conversely,

$$w^n(x,\pi) \leq w(x,\pi) + \varepsilon \leq v(x) + \varepsilon,$$

so that

$$T^n 0(x) = v^n(x) \leq v(x) + \varepsilon.$$

According to condition 2.4.C, it follows from the semicontinuity and boundedness above of the function v_t that the function

$$u_t(a) = (Uv_t)(a) = q(a) + \int_{X_t} v_t(y)p(da|a), \qquad a \in A_t,$$

---

* For this it is sufficient for example that they be complete, or be Borel subsets in their completions.

has the same properties. Applying Theorem 2.5.A to the spaces $E = X_{t-1}$, $E' = A_t$ and to the function $f = u_t$, we conclude that there exists a measurable selector ψ_t of the correspondence $A(x)$ $(x \in X_{t-1})$ such that

$$u_t(\psi_t(x)) = \sup_{a \in A(x)} u_t(a) = (Vu_t)(x) = v_{t-1}(x),$$

i.e. equation (1) is satisfied. It follows from proposition 2m) that the simple strategy $\varphi = \psi_1 \psi_2 \cdots \psi_t \cdots$ is uniformly optimal.

Finally we note one case in which *one may* omit the requirement that the models Z^t are bounded below in the main result of this section; if the spaces X_t are discrete (e.g. finite), then all functions on X_t are continuous and no special proof of semi-continuity of v_t is needed.

Chapter 6

Homogeneous Models

§1. Introduction

A *homogeneous model* is given by a mapping j of the action space A onto the state space X, a transition function p of A into X, and a running reward function q on A (see Chapter 1, §2). In order to apply the theory constructed in the preceding chapters to homogeneous models, it suffices to consider an infinite number of copies $X_0, X_1, \ldots, X_t, \ldots$ of the space X, an infinite number of copies $A_1, \ldots,$ $A_t, \ldots$ of the space A, and carry j, p, and q over to them, regarding j as mapping A_t onto X_{t-1} and p as the transition function from A_t into X_t. This construction may be called an *evolute* of the homogeneous model into time. Making use of the evolute, we may apply all the concepts previously introduced (history, strategy, summability, etc.) to homogenous models.

In the theory of homogeneous models a special role is played by simple strategies $\psi_1 \psi_2 \cdots \psi_t \cdots$ for which $\psi_1 = \psi_2 = \cdots = \psi_t = \cdots$ (*stationary strategies*). The central result of this chapter asserts that under certain conditions we may without any loss consider only stationary strategies.

The rôle of the stationary strategies persists for a somewhat wider class of models in which the reward at the instant t is equal to $\beta^{t-1}q(a_t)$, where β is an arbitrary positive number. The value of the strategy π for the initial distribution μ is defined by the formula

$$w(\mu,\pi) = \sum_{t=1}^{\infty} \beta^{t-1} P_\mu^\pi q(a_t). \tag{1}$$

Such a situation arises, for example, if the utility of the reward decreases as the time to which it relates moves further into the future. If one deposits today x rubles in the savings bank, then after t years one will receive $\lambda^t x$ rubles (in the Soviet Union, $\lambda = 1.03$ for time deposits and $\lambda = 1.02$ for demand deposits). It is therefore natural to value q rubles t years from now as having a present value of $x = \lambda^{-t}q$ rubles. The number $\beta = \lambda^{-t}$ is called the *discount coefficient*. We agree to call such models also *homogeneous*.

Thus, every homogeneous model is defined by elements X, A, j, p, q, β (the case described at the beginning of this section corresponding to $\beta = 1$). Its evolute is constructed just as it is for $\beta = 1$, with the exception of the running reward $\tilde{q}$, which is defined by the formula

$$\tilde{q}(a) = \beta^{t-1}q(a), \qquad a \in A_t. \tag{2}$$

At first glance it may appear that if $\beta \geq 1$ the series (1) will diverge except in trivial cases. Indeed this is not so: For the convergence of (1) for any $\beta > 0$ it suffices for example that for any strategy the system, after a finite number of steps, should fall into a state after which one can obtain only a zero payoff.

§2. Application of the Results of Chapter 4

As in the preceding chapters, we begin with the case of discrete, i.e. finite or countable, spaces X and A.

Let $\tilde{Z}$ be the evolute of the homogeneous model Z. Its derived model $\tilde{Z}_n$ of order n differs from $\tilde{Z}$ only in that the running reward is multiplied by the number β^n. Symbolically, $\tilde{Z}_n = \beta^n\tilde{Z}$. Hence it follows that

$$w_n(\mu,\pi) = \beta^n w(\mu,\pi), \qquad v_n(x) = \beta^n v(x). \tag{1}$$

(Here we have made the natural identification of a strategy in the model $\tilde{Z}_n$ to a strategy in the model $\tilde{Z}$, w_n and v_n are values in the model $\tilde{Z}_n$, and w and v values in the model $\tilde{Z}$.)

We will not rewrite all the results of Chapter 4, rather noting only the changes in the formulas which appear on the transition from $\tilde{Z}$ to Z.

The *fundamental equation* of Chapter 4, §3 can be written in the form

$$w(x,\pi) = \sum_{a \in A(x)} \pi(a|x)[q(a) + \beta w(p_a,\pi_a)]. \tag{2}$$

Here we are assuming that the model is x-summable, and asserting that it is p_a-summable for all $a \in A(x)$.

Formula (4.3.5) remains unchanged if one puts

$$T_\psi f(x) = q(\psi(x)) + \beta \sum_X f(y)p(y|\psi(x)). \tag{3}$$

Formulas (4.3.6) and (4.3.7) then take on the form

$$T_{\psi_{m+1}}T_{\psi_{m+2}}\cdots T_{\psi_n}r(x) = \sum_{t=m+1}^n \beta^{t-1}P_x^\varphi q(a_t) + \beta^n P_x^\varphi r(x_n) \tag{4}$$

and

$$w(\mu,\rho\pi) = \sum_{t=m+1}^n \beta^{t-1}P_\mu^\rho q(a_t) + \beta^n P_\mu^\rho w(x_n,\pi). \tag{5}$$

(If the model Z is μ-summable, it is also v-summable, where $v(y) = P_\mu^\rho\{x = y\}$, $y \in X$.)

The *summability* of the homogeneous model in the sense of Chapter 4, §5 reduces to the x-summability of Z for each $x \in X$. The value v of a summable model satifies the *optimality equations*

$$v = Vu, \qquad u = Uv \quad [\text{or } v = Tv], \tag{6}$$

where V has the preceding sense and the operator U is defined by the formula

$$Uf(a) = q(a) + \beta \sum_X f(y)p(y|a), \qquad a \in A \tag{7}$$

(see Chapter 4, §5). The relationships between the operators V, U, T, and T_ψ remain the same as before.

Let the selectors $\psi_1, \psi_2, \ldots, \psi_n$ of the correspondence $A(x)(x \in X)$ in a summable model satisfy the inequalities

$$T_{\psi_t} v \geq v - \kappa_t, \tag{8}$$

where $\kappa_1, \kappa_2, \ldots, \kappa_n$ are nonnegative numbers. If the strategy π is ε'-optimal, then the strategy $\psi_1 \psi_2 \cdots \psi_n \pi$ is ε-optimal for

$$\varepsilon = \sum_{t=1}^{n} \beta^{t-1} \kappa_t + \beta^n \varepsilon'. \tag{9}$$

Indeed, by (1.2), (1) and (4) the inequalities (8) are equivalent to the inequalities

$$\tilde{T}_{\psi_t} v_t \geq v_{t-1} - \beta^{t-1} \kappa_t, \tag{10}$$

where $\tilde{T}_{\psi_t}$ is the operator in the model $\tilde{Z}$ corresponding to the selector ψ_t of the correspondence $A(x)$ from the space X_{t-1} into A_t. In view of (1), the condition of ε'-optimality of the strategy π may be written in the form

$$w_n(x,\pi) \geq v_n(x) - \beta^n \varepsilon', \qquad x \in X_n = X. \tag{11}$$

Thus the strategy π is $\beta^n \varepsilon'$-optimal in the derived model $\tilde{Z}_n$. Hence our assertion reduces to what was proved in Chapter 4, §5.

In the definitions of the *models bounded above and below*, inequalities (4.6.7) are replaced by

$$\beta^{t-1} P_x^\pi q(a_t) \geq -b_t(x) \tag{12}$$

and inequalities (4.7.12) by

$$\beta^{t-1} P_x^\pi q(a_t) \leq c_t(x). \tag{13}$$

If $\beta < 1$, for the model to be uniformly bounded below (above), it is sufficient that the reward q should be bounded below (above). (In particular, if q is bounded, then the model is uniformly bounded both above and below.)

For models bounded above, it follows from inequalities (8) for $t = 1, 2, \ldots$ that the simple strategy $\varphi = \psi_1 \psi_2 \cdots \psi_t \cdots$ is ε-optimal, where

$$\varepsilon = \sum_{t=1}^{\infty} \beta^{t-1} \kappa_t. \tag{14}$$

For models which are uniformly bounded both above and below, it follows from inequalities (8) for $t = 1, \ldots, n$, with n sufficiently large, that any strategy π coinciding with $\psi_1 \psi_2 \cdots \psi_n$ on the first n steps is ε-optimal for

$$\varepsilon = \sum_{t=1}^{\infty} \beta^{t-1} \kappa_t + \sum_{t=n+1}^{\infty} (b_t + c_t). \tag{15}$$

In order to reduce these results to what was proved in Chapter 4, §7, we need only to rewrite inequalities (8) in the form (10).

* * *

In distinction from nonhomogeneous models, where we had an infinite system of optimality equations, relating to one another the values of the derived models of various orders, here we have only one optimality equation $v = Tv$ for the single unknown function v. The question as to the uniqueness of the solution of this equation now naturally arises.

It is not hard to show that *if the reward function q is bounded and the discount coefficient β is less than 1, then the equation $v = Tv$ has a unique bounded solution.* The proof of this fact is based on the following contraction inequality

$$\|Tf - Tg\| \le \beta \|f - g\|, \tag{16}$$

where

$$\|f\| = \sup_{x \in X} |f(x)|.$$

In order to deduce (16), we note that for any two bounded functions f and g on X

$$f \le g + \|f - g\|,$$

so that

$$Tf \le Tg + \beta \|f - g\|. \tag{17}$$

Inequality (16) follows from (17) and the relation gotten from (17) by exchanging f and g. If v and $\tilde{v}$ are two solutions of the optimality equation, then, by (16),

$$\|v - \tilde{v}\| = \|Tv - T\tilde{v}\| \le \beta \|v - \tilde{v}\|,$$

so that $\|v - \tilde{v}\| = 0$ and $v = \tilde{v}$.

§3. Stationary Optimal Strategies

We now take up the problem of the existence of stationary optimal strategies. We recall that the simple strategy $\varphi = \psi_1 \psi_2 \cdots \psi_t \cdots$ is said to be stationary if $\psi_1 = \psi_2 = \cdots = \psi_t = \cdots = \psi$; we agree to use the abbreviation $\varphi = \psi^\infty$. First we show that *if ψ^∞ is a stationary optimal strategy in a summable homogeneous model, then ψ satisfies the equation*

$$T_\psi v = v. \tag{1}$$

Indeed, for the strategy $\varphi = \psi^\infty$ the strategy φ_a coincides with φ for any $a \in A$, so that the fundamental equation (2.2) takes the form

$$w(x,\varphi) = q(\psi(x)) + \beta w(p_{\psi(x)}, \varphi).$$

Since

$$w(p_{\psi(x)}, \varphi) = \sum_X p(y \mid \psi(x)) w(y, \psi),$$

then

$$w(x,\varphi) = T_\psi w(x,\varphi) \tag{2}$$

(see (2.3)). If φ is optimal, then $w(x,\varphi) = v(x)$ for all x, and (2) reduces to (1).

Thus, all the stationary strategies are generated by solutions of equation (1), and the question as to the existence of such strategies may be broken into two questions:

1. *Is there a solution of equation (1)?*
2. *Does every solution of equation (1) generate a stationary optimal strategy?*

In order to answer the first question, it is convenient to rewrite equation (1) for the selector ψ in the form

$$u(\psi(x)) = \sup_{a \in A(x)} u(a), \qquad x \in X, \tag{3}$$

where

$$u(a) = q(a) + \beta \sum_X v(y) p(y \mid a), \qquad a \in A.$$

(The equivalence of (1) and (3) follows from the equation $T = VU$ and formulas (2.3) and (2.7).) It is clear from (3) that *that equation, and thus equation (1), has a solution if and only if the function $u(a)$ attains a maximum on each fibre $A(x)$*. It suffices for example that all the fibres $A(x)$ should be finite.

The answer to the second question can be negative, as one sees from the following simple example.

EXAMPLE 1. Consider the homogeneous model of Example 1 of Chapter 4, §7 (figure 4.3). Obviously there are only two distinct selectors of the correspondence $A(x)$. One of them, ψ_1, prescribes remaining at x, and the second, ψ_2, prescribes going over from x to y. At y there is only one way of behaving—remain at y. The value v of the model is equal to 1 in state x and equal to 0 in state y, and both selectors ψ_1 and ψ_2 satisfy equation (1). However here the strategy ψ_2^∞ is optimal, but the strategy ψ_1^∞ is not.

Example 1 motivates us to alter the statement of the second question and to ask:

> 2a. *Do there exist solutions of equation* (1) *which generate stationary optimal strategies?*

It turns out that the answer can be negative.

EXAMPLE 2. Suppose that the states x are enumerated by the numbers $0, 1, 2, \dots$ (see figure 6.1). The state 0 is absorbing. In each of the states k with $k = 1, 2, \dots$, two actions are possible. The first carries us from k to $k + 1$ and carries a zero reward, and the second takes us from k to 0 and carries the reward $(k - 1)/k$. Obviously $v = 1$ at all states other than $k = 0$, and the unique solution of equation (1) is the selector ψ which prescribes passing from k to $k + 1, k = 1, 2, \dots$. However the corresponding stationary strategy ψ^∞ never leads us from a state $k \geq 1$ into the state 0. Hence $w(k, \psi^\infty) = 0$ for all $k \geq 1$.

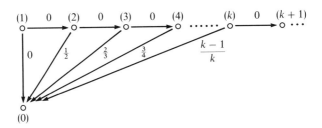

Figure 6.1

We note that in both examples we are dealing with models which are unbounded above (since we may guarantee ourselves a certain positive reward at a time arbitrarily far off in the future).

If the model is bounded above, question 2 can be given a positive answer. In this case *any solution of equation* (1) *generates a stationary optimal strategy.* One sees this directly from formulas (2.8) and (2.14) by putting $\psi_t = \psi$, $\kappa_t = 0$ for all t.

* * *

Now let us drop the requirement of boundedness above, and suppose only that *the model is summable above,* i.e. $v(x) < +\infty$ for all $x \in X$. We shall show that *if the model Z is finite, then there exists a solution of equation* (1) *generating a stationary optimal strategy.* Thus in this case both questions 1 and 2a may be answered in the affirmative.

The idea of the proof consists in the following. For each $\gamma > 0$, consider a homogeneous model $Z(\gamma)$, gotten from Z by replacing the discount coefficient β by γ. Because the model $Z = Z(\beta)$ is summable above, the models $Z(\gamma)$ with $\gamma < \beta$ are bounded above. By what has been proved, in the model $Z(\gamma)$ there exists a stationary optimal strategy $\varphi(\gamma) = \psi(\gamma)^\infty$. But in view of the finiteness of the spaces X and $A(x)$ there are only finitely many distinct selectors of the correspondence $A(x)(x \in X)$. Hence there exists a selector ψ such that $\psi(\gamma_n) = \psi$ for some sequence $\gamma_n \uparrow \beta$. Obviously for $\gamma = \gamma_n$

$$w_\gamma(x,\psi^\infty) \geq w_\gamma(x,\pi) \quad \text{for all } \pi \tag{4}$$

(the index γ here denotes that the value is taken in the model $Z(\gamma)$). We show below that for any strategy π

$$w_\beta(x,\pi) = \lim_{\gamma \uparrow \beta} w_\gamma(x,\pi). \tag{5}$$

From this and (4) it will then follow that

$$w_\beta(x,\psi^\infty) \geq w_\beta(x,\pi) \quad \text{for all } \pi, \tag{6}$$

i.e. that the stationary strategy ψ^∞ is optimal for the model $Z(\beta) = Z$.

It remains to prove that the models $Z(\gamma)$ with $\gamma < \beta$ are bounded above, and to deduce formula (5). We note that according to what was proved at the end of Chapter 4, §4, the summability above of the model $Z(\beta)$ is equivalent to the boundedness above, for each x, of the quantity

$$w_\beta^+(x,\pi) = \sum_{t=1}^\infty \beta^{t-1} P_x^\pi q^+(a_t).$$

In view of the finiteness of the space X this quantity does not exceed some finite constant K for all $x \in X$ and all strategies π. Hence

$$\gamma^{t-1} P_x^\pi q(a_t) \leq \left(\frac{\gamma}{\beta}\right)^{t-1} \beta^{t-1} P_x^\pi q^+(a_t) \leq K\left(\frac{\gamma}{\beta}\right)^{t-1}$$

and the series of numbers on the right converges.

In order to prove (5) we recall that

$$w_\gamma = w_\gamma^+ - w_\gamma^-, \tag{7}$$

where

$$w_\gamma^+ = \sum_{t=1}^\infty \gamma^{t-1} P_x^\pi q^+(a_t)$$

and w_γ^- is expressed analogously in terms of q^- (see §1 and Chapter 4, §2). In both series one may pass to the limit as $\gamma \uparrow \beta$ term by term, because all the terms of these series are nonnegative and are nondecreasing functions of γ. Since $w_\beta^+ < +\infty$, (5) follows from (7) and from the formula $w_\beta = w_\beta^+ - w_\beta^-$.

$$* * *$$

By a slight variation of the argument given at the beginning of this section it is not hard to show that *if equation* (1) *does not have a solution in a homogeneous model which is summable above, then there is not only no stationary strategy for that model, but no optimal strategy at all.*

Indeed, if π is an optimal strategy, it follows from the fundamental equation (2.2) and the optimality equations (2.6) that

$$v(x) = w(x,\pi) = \sum_{a \in A(x)} \pi(a|x)[q(a) + \beta w(p_a,\pi_a)]$$

$$\leq \sum_{a \in A(x)} \pi(a|x)[q(a) + \beta v(p_a)]$$

$$\leq \sup_{a \in A(x)} [q(a) + \beta v(p_a)] = Tv(x) = v(x)$$

for each $x \in X$. Therefore

$$\sum_{a \in A(x)} \pi(a|x)[q(a) + \beta v(p_a)] = v(x) < +\infty.$$

Applying Lemma 1.13.1 to the probability distribution $\pi(\cdot|x)$ and the function $f(a) = q(a) + \beta v(p_a)$ on the space $A(x)$, we conclude that there exists an action $\psi(x)$ in the fibre $A(x)$ for which $f[\psi(x)] \geq v(x)$. This last inequality shows that for the selector ψ

$$q(\psi(a)) + \beta v(p_{\psi(x)}) \geq v(x) \qquad (x \in X),$$

i.e. $T_\psi v \geq v$. Since $T_\psi v \leq Tv = v$, then ψ satisfies equation (1).

Combining this result with the answer to question 2, we come to the following conclusion: *If in a homogeneous model Z which is bounded above there is any optimal strategy π at all, then there is in that model a stationary optimal strategy.*

§4. The Bus, Streetcar, or Walk Problem

Now we turn to the problem of the selection of transport of Chapter 1, §10, this time with an unbounded number of stages. We have a finite homogeneous model with a nonpositive reward function q and discount coefficient $\beta = 1$ (see figure 1.10). This model is bounded above, its value v satisfies the optimality equations (2.6), and equation (3.1) defines a selector ψ which generates an optimal strategy ψ^∞.

Denoting as in Chapter 1, §10 by α the action "Walk" by β the action "Take the Bus", by γ the action "Take the Streetcar", by δ' "Wait" in state 0, by δ'' "Wait" in state D, and finally by ε a fictitious action in state B, we may write out the equations $v = Vu$, $u = Uv$ in the form

$$v(0) = \max\{u(\alpha), u(\delta')\},$$
$$v(C) = u(\beta),$$
$$v(D) = \max\{u(\gamma), u(\delta')\},$$
$$v(B) = u(\varepsilon),$$
$$u(\alpha) = -20 + v(B),$$
$$u(\beta) = -3 + v(B),$$
$$u(\gamma) = -10 + v(B),$$
$$u(\delta') = -\frac{cd}{c+d} + \frac{d}{c+d}v(C) + \frac{c}{c+d}v(D),$$
$$u(\delta'') = -\frac{cd}{c+d} + \frac{d}{c+d}v(C) + \frac{c}{c+d}v(D),$$
$$u(\varepsilon) = v(B).$$

Obviously

$$u(\varepsilon) = v(B) = 0,$$

so that from the system (1) we find immediately that

$$u(\alpha) = -20, \qquad u(\beta) = -3, \qquad u(\gamma) = -10,$$
$$v(C) = -3, \qquad u(\delta') = u(\delta''). \tag{2}$$

Inasmuch as $u(\delta')$ and $u(\delta'')$ coincide, we may denote their common value by $u(\delta)$, understanding by δ waiting in either state 0 or D. After doing this, there remain for $v(0)$, $v(D)$ and $u(\delta)$ the equations

$$v(0) = \max\{-20, u(\delta)\},$$
$$v(D) = \max\{-10, u(\delta)\},$$
$$u(\delta) = \frac{-cd - 3d + cv(D)}{c+d}. \tag{3}$$

The system we have just found has two maximum signs. We shall investigate it under various assumptions as to which actions yield these maxima.

Obviously there are three possibilities: 1) $u(\delta) \in (-\infty, -20]$, 2) $u(\delta) \in [-20, -10]$; 3) $u(\delta) \in [-10, \infty)$. We shall consider each of them separately.

In case 1) we have

$$\max\{u(\alpha), u(\delta)\} = \max\{-20, u(\delta)\} = -20 = u(\alpha),$$
$$\max\{u(\gamma), u(\delta)\} = \max\{-10, u(\delta)\} = -10 = u(\gamma),$$

so that the selector ψ with values $\psi(0) = \alpha$, $\psi(C) = \beta$, $\psi(B) = \varepsilon$, $\psi(D) = \gamma$ satisfies the conditions (3.3). Thus the optimal strategy is to walk, not waiting for transport, taking the bus or the streetcar only if one happens to be there*. In this case the system (3) becomes

$$v(0) = -20,$$

$$v(D) = -10, \tag{4}$$

$$u(\delta) = \frac{-cd - 3d - 10c}{c + d}$$

so that the inequality $u(\delta) \le -20$ defining case 1) reduces to the inequality

$$20 \le \frac{cd + 3d + 10c}{c + d}. \tag{5}$$

In case 2)

$$\max\{u(\alpha), u(\delta)\} = \max\{-20, u(\delta)\} = u(\delta),$$

$$\max\{u(\gamma), u(\delta)\} = \max\{-10, u(\delta)\} = -10 = u(\gamma).$$

Therefore the selector with values $\psi(0) = \delta$, $\psi(C) = \beta$, $\psi(D) = \gamma$, $\psi(B) = \varepsilon$ satisfies conditions (3.3), and the optimal stationary strategy is the one recommending that one wait for the arrival of the first streetcar or bus and then to go on it. In this case the system (3) takes the form

$$v(0) = u(\delta),$$

$$v(D) = -10,$$

$$u(\delta) = \frac{-cd - 3d - 10c}{c + d}$$

and the inequalities defining the case at hand reduce to

$$10 \le \frac{cd + 3d + 10c}{c + d} \le 20. \tag{6}$$

In case 3)

$$\max\{u(\alpha), u(\delta)\} = \max\{-20, u(\delta)\} = u(\delta),$$

$$\max\{u(\gamma), u(\delta)\} = \max\{-10, u(\delta)\} = u(\delta)$$

and it is the selector ψ given by $\psi(0) = \delta$, $\psi(D) = \delta$, $\psi(C) = \beta$, $\psi(B) = \varepsilon$ that satisfies conditions (3.3). Here the optimal stationary strategy prescribes that one should under all circumstances wait for the bus. The system (3) is now written

---

* In our model the probabilities of the states C and D at the initial time are equal to zero, but in real life this is not so, because the bus and streetcar have to stop for a few seconds at the waiting point.

out as

$$v(0) = u(\delta),$$

$$v(D) = u(\delta),$$

$$u(\delta) = \frac{-cd - 3d + cu(\delta)}{c + d},$$

so that

$$u(\delta) = -c - 3,$$

and the inequality yielding the third case reduces to

$$c + 3 \le 10,$$

or, equivalently,

$$\frac{cd + 3d + 10c}{c + d} \le 10. \tag{7}$$

Comparing (5), (6), and (7), we see that the three possible cases are determined by the values of the number

$$\kappa = \frac{cd + 3d + 10c}{c + d}.$$

If $\kappa \le 10$ one should take the bus, if $10 \le \kappa \le 20$ one should take the first transportation that comes along, and if $20 \le \kappa$ one should walk. These conditions have a simple intuitive meaning. According to figure (1.10), the mathematical expectation of the time spent on the trip if one goes by the first streetcar or bus that comes along is equal to

$$\frac{cd}{c + d} + \frac{d}{c + d} \cdot 3 + \frac{c}{c + d} \cdot 10 = \kappa.$$

If this time is larger than the time of going by foot, then one should go on foot; otherwise one should wait. One should wait for the bus in the case when κ is less than the time of going by streetcar.

§5. The Replacement Problem

Now we turn to the replacement problem (see Chapter 1, §§3,11). Passing to an infinite interval of control $[0,\infty)$, we introduce a discounting which guarantees the finiteness of the total reward. Thus, we consider a homogeneous countable model with states $0, 1, 2, \ldots, x, \ldots$, two actions c and d in each state, the transition function

$$p(x + 1 \mid xc) = p_x, \qquad p(0 \mid xc) = 1 - p_x = q_x,$$

$$p(0 \mid xd) = 1 \qquad (x = 0,1,2,\ldots), \tag{1}$$

the running reward

$$q(xc) = p_x h_x, \qquad q(xd) = \alpha \qquad (x = 0,1,2,\ldots) \tag{2}$$

and the discount coefficient $\beta < 1$; here

$$h_0 \geq h_1 \geq \cdots \geq h_x \geq \cdots \geq \alpha > 0, \tag{3}$$

$$1 \geq p_0 \geq p_1 \geq \cdots \geq p_x \geq \cdots \geq 0. \tag{4}$$

(Here, in accordance with the remark in Chapter 1, §11, we take the parameter γ to be equal to 0; if $\beta < 1$ this does not decrease the generality, since the change of all the rewards by a constant does not affect the convergence of the total outcome. In Chapter 1, in order to stay within the realm of finite models, we supposed that the probability q_x of breakdown was equal to 1 for sufficiently large x. Now we have no need for this restriction.)

The model is bounded above and all the fibres $A(x)$ are finite. Therefore the value of the model satisfies the optimality equation $v = Tv$, there exists a selector ψ with $T_\psi v = v$, and the stationary strategy $\varphi = \psi^\infty$ is optimal. Since the reward q is bounded and $\beta < 1$, v is the unique bounded solution of the equation $v = Tv$.

Since the action space consists of the two elements c and d, then the function ψ is defined by the set C on which it takes the value c, i.e. the set of those states where one continues to make use of the equipment; on the complementary set $D = X \backslash C$ one replaces the equipment. The operator T_ψ operates according to the formula

$$T_\psi f(x) = \begin{cases} p_x h_x + \beta[p_x f(x+1) + q(x)f(0)] & \text{if } x \in C, \\ \alpha + \beta f(0) & \text{if } x \in D. \end{cases} \tag{5}$$

The operator T is defined by the formula

$$Tf(x) = \max\{p_x h_x + \beta[p_x f(x+1) + q_x f(0)], \alpha + \beta f(0)\}. \tag{6}$$

We introduce the values of the actions c and d in any state x:

$$u(xc) = p_x h_x + \beta[p_x v(x+1) + q_x v(0)],$$
$$u(xd) = \alpha + \beta v(0). \tag{7}$$

Since $T_\psi v = v = Tv$, then it follows from (5)–(7) that

$$C = \{x : u(xc) \geq u(xd)\}, \qquad D = \{x : u(xc) < u(xd)\} \tag{8}$$

(As in Chapter 1, §11, we assign the state x to the set C if it is indifferent as to which action should be used in that state.)

The direct solution of the system $v = Tv$ is difficult. First we use the result of Chapter 4, §6 that $v = \lim T^n 0$; the function $v^n = T^n 0$ here is the value of our model on the control interval $[0,n]$ with a zero terminal reward. We can apply to that interval and that reward the arguments of Chapter 1, §11 which show that $v^n(x)$ is a nonincreasing function of x. (We have only to replace v_t by βv_t in the formulas of Chapter 1, §11; the finiteness of the space X was not used in the proof.) It follows from $v^n \to v$ that $v(x)$ is also a nonincreasing function of x. Knowing that v is monotone, we establish just as we did in Chapter 1, §11 that if some state x belongs to the set C, then $u(x - 1,c) \geq u(x,c) \geq u(x,d) = u(x - 1,d)$, so that $(x - 1) \in C$. Thus, analogously to the case of a finite control interval, the sets C and D have the form

$$C = \{0,1, \ldots ,k - 1\}, \qquad D = \{k,k + 1, \ldots\} \qquad (9)$$

(one of the sets C, D can be empty, in which case $k = 0$ or ∞). Thus the construction of a stationary optimal strategy reduces to finding the number k.

Thus, we get the optimal strategy if we choose the best among the selectors defined by the formula

$$\psi_m(x) = \begin{cases} c & \text{for } x < m, \\ d & \text{for } x \geq m, \end{cases} \qquad (10)$$

$(m = 0,1,2, \ldots ,\infty)$. Fix an $m < \infty$. The value $w = w_m$ of the strategy $\varphi = \psi_m^\infty$ satisfies the fundamental equation $w = T_{\psi_m} w$. Using (5), we can write this equation in the following expanded form

$$w(0) = p_0 h_0 + \beta p_0 w(1) + \beta(1 - p_0)w(0),$$
$$w(1) = p_1 h_1 + \beta p_1 w(2) + \beta(1 - p_1)w(0),$$
$$\vdots$$
$$w(m - 1) = p_{m-1}h_{m-1} + \beta p_{m-1}w(m) + \beta(1 - p_{m-1})w(0), \qquad (11)$$
$$w(m) = \alpha + \beta w(0),$$
$$w(m + 1) = \alpha + \beta w(0),$$
$$\vdots$$

where we have dropped the argument φ in the value $w(x,\varphi)$. Using the abbreviation

$$L_x = \beta^x p_0 p_1 \cdots p_x, \qquad (12)$$

$x = 0, 1, 2, \ldots$, multiplying the equation for $w(1)$ by βL_0, the equation for $w(2)$ by $\beta L_1, \ldots$, the equation for $w(x - 1)$ by βL_{x-2}, and adding them, we get

$$w_m(x) = \frac{1}{\beta L_{x-1}}\{w_m(0)[(1 - \beta) + \beta(1 - \beta)(L_0 + L_1 + \cdots + L_{x-2}) + \beta L_{x-1}]$$
$$- (L_0 h_0 + L_1 h_1 + \cdots + L_{x-1}h_{x-1})\}], \qquad (13)$$

$x = 1, 2, \ldots, m$. Equating the expressions for $w(m) = w_m(m)$ in (11) and (13), we get

$$w_m(0) = \frac{L_0 h_0 + L_1 h_1 + \cdots + L_{m-1} h_{m-1} + \beta L_{m-1} \alpha}{(1 - \beta)[1 + \beta(L_0 + L_1 + \cdots + L_{m-1})]}. \tag{14}$$

It is clear from (11) and (13) that if

$$w_k(0) = \sup_m w_m(0), \tag{15}$$

then at the same time

$$w_k(x) = \sup_m w_m(x)$$

for all x, so that the corresponding stationary strategy $\varphi = \psi_k^\infty$ is uniformly optimal. Thus, the number k is found from (14) and (15) (according to our convention to use the action c when $u(xc) = u(xd)$, the largest of the maximum points is taken). By an easy contradiction argument one verifies that if the supremum in (15) is not attained, then $k = \infty$ and the set D is empty, i.e. one should continue to use the equipment no matter how old.

$$* * *$$

Let us prove that if

$$w_0(0) \leq w_1(0) \leq \cdots \leq w_m(0) \tag{16}$$

and

$$w_m(0) \geq w_{m+1}(0), \tag{17}$$

then the strategy $\varphi = \psi_m^\infty$ is optimal (hence, for finite k, in order to find the optimal strategy it is not necessary to compare infinitely many numbers).

By virtue of the uniqueness of the solution of the optimality equations it suffices to verify that $w_m = Tw_m$. Since $w_m = T_{\psi_m} w_m$, this equation may be written in the form $T_{\psi_m} w_m = Tw_m$. According to (5) and (6), this last relation is equivalent to the system

$$p_x h_x + \beta p_x w_m(x+1) + \beta q_x w_m(0) \geq \alpha + \beta w_m(0) \qquad (x = 0,1,\ldots,m-1), \tag{18}$$

$$p_x h_x + \beta p_x w_m(x+1) + \beta q_x w_m(0) \leq \alpha + \beta w_m(0) \qquad (x = m, m+1,\ldots). \tag{19}$$

Putting the value of $w_m(x+1)$ given by formula (13) into (18) and using (14) we find after simple calculations, that (18) is equivalent to the condition $w_m(0) \geq w_x(0)$ which follows from (16). Taking account of formula (11), we can rewrite (19) in the form

$$p_x h_x + \beta q_x w_m(0) \leq [\alpha + \beta w_m(0)](1 - \beta p_x).$$

This last relation follows easily from (17), (14) and the equation $L_x = \beta p_x L_{x-1}$.

§6. Stationary ε-Optimal Strategies

When equation (3.1) (or (3.3)) does not have solutions, there are no optimal strategies. It is then natural to seek stationary ε-optimal strategies for $\varepsilon > 0$. Suppose that $\varphi = \psi^\infty$ is an ε-optimal strategy. In view of the fundamental equation

$$v(x) - \varepsilon \leq w(x,\varphi) = q(\psi(x)) + \beta \sum_{y \in X} w(y,\varphi)p(y\,|\,\psi(x))$$

$$= T_\psi w(x,\varphi) \leq T_\psi v(x).$$

Thus, all the stationary ε-optimal strategies ψ^∞ are generated by selectors ψ satisfying the inequality

$$T_\psi v \geq v - \varepsilon. \tag{1}$$

Inequality (1) is equivalent to the inequality

$$u(\psi(x)) \geq \sup_{a \in A(x)} u(a) - \varepsilon \tag{2}$$

(cf. formulas (3.1) and (3.3)). If the model is summable above, so that

$$v(x) = \sup_{a \in A(x)} u(a) < +\infty,$$

then inequality (2) (in contrast to equation (3.3)) has a solution for any $\varepsilon > 0$. Therefore, of the two questions investigated in §3, there remains only the second. It is natural to pose it in the following form:
   Is there, for each $\varepsilon > 0$, a $\kappa > 0$ such that the inequality

$$T_\psi v \geq v - \kappa \tag{3}$$

implies the ε-optimality of the stationary strategy ψ^∞?
   Example 1 of §3 shows that the answer to this question may be negative for models which are unbounded above. Indeed, the selector ψ_1 prescribing that one remain in the state x satisfies these equations for all $\kappa \geq 0$, while

$$w(x,\psi_1^\infty) = v(x) - 1,$$

so that the stationary strategy ψ_1^∞ is not ε-optimal for any $\varepsilon < 1$.
   Now let a model be bounded above. Then we may use formulas (2.8) and (2.14), putting $\psi_t = \psi$ and $\kappa_t = \kappa$. For $\beta < 1$, the series (2.14) converges to $\kappa/(1 - \beta)$, and we arrive at the following result: *If the model is bounded above and $\beta < 1$, then every selector ψ satisfying (1) for $\kappa = \varepsilon(1 - \beta)$ generates an ε-optimal strategy.* Thus, in this case the answer to our question is positive.
   The answer remains positive for $\beta \geq 1$ as well, if the model is uniformly bounded both above and below. In fact, by the concluding result of §2, the strategy ψ^∞ will

be ε-optimal for

$$\varepsilon = \kappa \sum_{t=1}^{n} \beta^{t-1} + \sum_{t=n+1}^{\infty} (b_t + c_t).$$

By first choosing n large enough, and then κ small enough, we may make ε arbitrarily small.

* * *

The preceding considerations leave open the question as to the existence of stationary ε-optimal strategies in models which are bounded only above (with $\beta \geq 1$) or only below. The following two examples show that in both cases stationary ε-optimal strategies may fail to exist.

EXAMPLE 1. (A model bounded below). Imagine a gambling house with the following rules. Before the first play the gambler lays a stake s of at least one dollar on the table. This constitutes his initial "fortune". Before each play the gambler must decide whether to wager all of his fortune, or to leave the game; in the latter case the house levies on him a fee of one dollar. If he stays, the game is played, using some random device. The gambler's fortune now either doubles, or is lost entirely, each with probability $\frac{1}{2}$.

The corresponding model can be described by figure 6.2. The state 0, "Out of the game", is absorbing. In the state B_m, "With fortune m", the gambler must make his decision as to whether to stake the m dollars or to go; the corresponding actions are depicted in the figure by arrows. In the first case he passes, with a zero reward, to the state C_m, and in the second case, with a reward of $m - 1$, into the state 0.

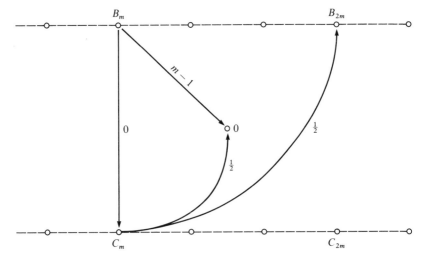

Figure 6.2

The gambler does nothing at state C_m; the random device is now employed, and he passes, with zero reward, into state 0 or into state B_{2m}, each with probability $\frac{1}{2}$.

Consider the strategy π_k—"Play k times, if not wiped out earlier". Beginning with the state B_m, we receive $m2^k - 1$ if all k plays are successful, and nothing otherwise. Therefore

$$w(B_m,\pi_k) = 2^{-k}(m2^k - 1) = m - 2^{-k}.$$

Since k is arbitrary,

$$v(B_m) \geq m.$$

The strategy π_k is not stationary. Now suppose that φ is any stationary strategy. There either exists a state B_m at which φ prescribes leaving, so that

$$w(B_m,\varphi) = m - 1,$$

or else φ prescribes playing in each state B_m, so that

$$w(B_m,\varphi) = 0.$$

In both cases the strategy is not uniformly ε-optimal for any $\varepsilon < 1$.

EXAMPLE 2. (A model bounded above, $\beta = 1$.) A criminal is hiding from the authorities. Each day he selects his lodging from a countable set of possibilities. The probability that he will be arrested at the mth lodging is equal to 2^{-m}. How shall he act so as to minimize the probability of arrest?

We introduce three states: A is "Freedom", B is "Arrest", and C is "Jail" (see figure 6.3). In state A there are a countable number of actions, i.e. of choosing lodging $1, 2, 3, \ldots$. If he chooses the action m he passes into state B with probability 2^{-m}, and remains in state A with probability $1 - 2^{-m}$. The corresponding rewards are equal to zero. From state B only one transition is possible, and that is into state C, with a reward equal to -1. The value $w(A,\pi)$ of the strategy π is equal to the minus probability of arrest, so that the problem consists of maximizing $w(A,\pi)$.

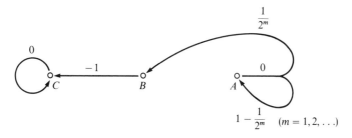

Figure 6.3

Consider the following strategy π_m: pass the tth night in lodging $t + m$. Clearly

$$w(A,\pi_m) = -1 + \prod_{t=1}^{\infty} \left(1 - \frac{1}{2^{m+t}}\right)$$

a quantity which tends to zero as $m \to \infty$. Therefore $v(A) = 0$. At the same time, any stationary strategy φ would have to specify spending the whole time in one and the same lodging m, which would lead the criminal to disaster, because obviously $w(A,\varphi) = -1$.

§7. Extension of the Results to Borel Models

Now we shall drop the assumption that the state space X and the action space A are finite or countable; as in Chapters 3 and 5, we will suppose them to be arbitrary Borel spaces.

Using the same evolute of the homogeneous model as in §2, but now using the results of Chapter 5, §2 instead of Chapter 4, we may extend the results of §2 to the general case. The fundamental equation takes the form

$$w(x,\pi) = \int_{A(x)} \pi(da|x)[q(a) + \beta w(p_a,\pi_a)] \tag{1}$$

(we are assuming the model Z to be x-summable). The operators T_ψ, T, U, and V operate according to the formulas

$$T_\psi f(x) = q(\psi(x)) + \beta \int_X f(y)p(dy|\psi(x)), \tag{2}$$

$$Tf(x) = \sup_{a \in A(x)} \left[q(a) + \beta \int_X f(y)p(dy|a)\right], \tag{3}$$

$$Uf(a) = q(a) + \beta \int_X f(y)p(dy|a), \tag{4}$$

$$Vg(x) = \sup_{a \in A(x)} g(a) \tag{5}$$

where in the formula for T_ψ the ψ is a measurable selector of the correspondence $A(x)$ of X into A. As before,

$$T = VU, \qquad T_\psi f(x) = Uf(\psi(x)), \qquad Tf(x) = \sup_\psi T_\psi f(x). \tag{6}$$

The only formula in (6) which requires a separate verification is the last one. It asserts that

$$\sup_{a \in A(x)} u(a) = \qquad u(\psi(x)), \tag{7}$$

where $u = Uf$. Inasmuch as $\psi(x) \in A(x)$, the left side of (7) is not less than the right side. The reverse inequality is:

$$u(a) \leq \sup_{\psi} u(\psi(x)) \quad \text{for all } a \in A(x). \tag{8}$$

Since the model is nontrivial, there exists some measurable selector ψ_1 of the correspondence $A(y)$ $(y \in X)$. Inasmuch as in a Borel space one-point sets are measurable, the selector

$$\psi(y) = \begin{cases} \psi_1(y) & \text{for } y \neq x, \\ a & \text{for } y = x, \end{cases} \tag{9}$$

is also measurable. For such a selector $u(a) = u(\psi(x))$ so that (8), and hence the last equality in (6), are proved.

Formulas (2.4) and (2.5) and the optimality equation (2.6) remain unchanged. The formulations of §2 concerning ε-optimal strategies, models bounded above and below, and the uniqueness of the solution of the optimality equations, also remain valid.

* * *

With the obvious replacement of sums by integrals the derivation of the equation

$$T_\psi v = v \tag{10}$$

carries over to Borel models as a necessary condition for the stationary strategy ψ^∞ to be optimal (§3). But in the general case there is no reason to expect that equation (10) has a solution. Therefore it is natural to search for ε-optimal stationary strategies rather than optimal ones.

It is true however that the results of §3 on models which are bounded above, asserting that every (measurable) solution of equation (10) generates a stationary optimal strategy, and that if there is any optimal strategy at all there exists a stationary optimal strategy, remain valid. The proof of the first of these statements does not change, and the proof of the second makes use of Theorem 3.2.A instead of Lemma 1.13.1.

* * *

The *semicontinuous models* (in which, as in Chapter 5, §6, we suppose that the spaces X and A are Borelian) constitute an exception. Applying the results of Chapter 5, §6 to the evolute of such a model, we conclude that, in the following two cases, *there exists a measurable selector of the correspondence $A(x)$ which satisfies* (10) *and hence generates a stationary optimal strategy*:

   a) *If the model is uniformly bounded both above and below* (in particular if q is bounded and $\beta < 1$);
      b) *If the space X is discrete and $\beta < 1$.*

In case (b) the model is uniformly bounded above because $\beta < 1$ and $q \leq C < \infty$ (see the definition of a homogeneous semicontinuous model).

* * *

All the conclusions of §6 on stationary ε-optimal strategies also remain in force, with one qualification: since we can only consider measurable selectors ψ, the question as to the existence of a ψ satisfying the inequality

$$T_\psi v \geq v - \varepsilon \tag{11}$$

for a given $\varepsilon > 0$ is no longer trivial; in fact the answer to that question is in general *negative*.

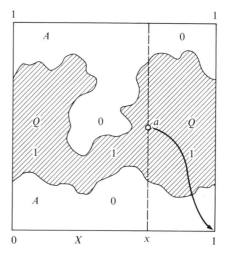

Figure 6.4

EXAMPLE 1. Suppose that A is the unit square, X its base, j the orthogonal projection of A onto X. Suppose that the measure $p(\cdot|a)$ is for any a concentrated at the point $x = 1$ (see figure 6.4). The reward q is equal to 1 on Q and equal to 0 on $A \backslash Q$, where Q is a Borel subset of the square A such that Q projects onto all of X, and not one measurable selector ψ of the correspondence $A(x)$ satisfies the condition that $\psi(x) \in Q$ for all $x \in X$ (see Example 3.1.1). Suppose that the discount coefficient $\beta = \frac{1}{2}$. Obviously, for any initial state x we can get a reward 1 at the first step, and then, falling into the state $x = 1$, collect a reward $\frac{1}{2} + (\frac{1}{2})^2 + \cdots = 1$; it is not possible to obtain any more. This means that $v(x) = 2$ for all $x \in X$. At the same time, for any selector ψ

$$T_\psi v(x) = q(\psi(x)) + \beta v(1) = q(\psi(x)) + 1.$$

Now if the selector ψ is measurable, then for at least one $y \in X$ we have $\psi(y) \notin Q$, so that $q(\psi(y)) = 0$. But then $T_\psi v(y) = v(y) - 1$, so that for $\varepsilon < 1$ the inequality

$$T_\psi v \geq v - \varepsilon$$

is not satisfied for any measurable selector ψ.

Inasmuch as inequality (11) is a necessary condition for ε-optimality of the strategy ψ^∞, it follows from the example just presented that generally speaking ε-optimal strategies may fail to exist. In analogy with the nonhomogeneous case, the question naturally arises as to whether there exist stationary (a.s. μ) ε-optimal strategies, i.e. strategies ψ^∞ such that

$$w(x, \psi^\infty) \geq v(x) - \varepsilon \quad \text{(a.s. } \mu)$$

(cf. Chapter 3, §1). The next section is devoted to this question.

§8. Stationary a.s. ε-Optimal Strategies

In distinction from the discrete case, where we directly constructed the desired stationary strategy here we have to resort to a roundabout procedure. First we construct a nonstationary simple (a.s. μ) ε-optimal strategy φ, and then, starting from φ, we construct a stationary strategy ψ^∞ which is not worse than φ.

The first step was carried out in Chapter 5. There we showed the existence of an (a.s. μ) ε-optimal strategy for any $\varepsilon > 0$ under the assumption that the model was μ-summable above and bounded above.

In this section we show that *if in a homogeneous model Z the reward function q is bounded and the discount coefficient β is less than 1, then for any simple strategy φ and any number $\varepsilon > 0$ there exists a stationary strategy ψ^∞ such that*

$$w(x, \psi^\infty) \geq w(x, \varphi) - \varepsilon \tag{1}$$

for all $x \in X$. Since under the above hypotheses the model is bounded above and μ-summable above for any μ, we have in sum established that *under the same hypotheses, for any $\varepsilon > 0$ and any initial distribution μ there exists a stationary (a.s. μ) ε-optimal strategy.*

$$* \quad * \quad *$$

So, suppose that in a homogeneous Borel model

$$\sup_{a \in A} |q(a)| < +\infty, \qquad \beta < 1.$$

Suppose that $\varphi = \psi_1 \psi_2 \cdots \psi_t \cdots$ is an arbitrary simple strategy and μ an arbitrary initial distribution. We assign to the strategy φ an operator S, defined by the formula

$$Sf(x) = \sup_t T_{\psi_t} f(x) \qquad (x \in X). \tag{2}$$

The existence of a measurable selector ψ satisfying (1) follows obviously from the following three propositions

    1°. The equation

$$Sh = h$$

has a unique bounded measurable solution h.

    2°. For any $\kappa > 0$ there exists a measurable selector ψ of the correspondence $A(x)$ such that

$$T_\psi h \geq h - \kappa. \tag{3}$$

    3°. If the measurable selector ψ satisfies inequality (3), then (1) holds for $\varepsilon = \kappa/(1 - \beta)$.

We make the abbreviation

$$T_t = T_{\psi_t}, \qquad t = 1, 2, \ldots .$$

The proof of 1° is based on the contraction inequality

$$\|Sf - Sg\| \leq \beta\|f - g\|, \tag{4}$$

which is deduced in the same way as was the analogous estimate (2.16) for the operator T. It follows from (4) that $S^n f$ converges uniformly to a solution h of the equation $Sh = h$, moreover the solution is unique in the class of bounded measurable functions (the fixed-point theorem for a contraction operator*).

Now we prove 2°. Inasmuch as $h = Sh = \sup_t T_t h$, then for each x

$$T_t h(x) \geq h(x) - \kappa \tag{5}$$

for some t. Denote by $t(x)$ the smallest of these numbers and put

$$\psi(x) = \psi_{t(x)}(x). \tag{6}$$

Then

$$T_\psi h(x) = T_{t(x)} h(x) \geq h(x) - \kappa.$$

In order to verify the measurability of ψ, we denote by Y_t the set of all x satisfying inequality (5). This set is measurable, and for any $\Gamma \in \mathscr{B}(A)$

$$\{x : \psi(x) \in \Gamma\} = \bigcup_{t=1}^{\infty} \left[\{x : \psi_t(x) \in \Gamma\} \cap (X \backslash Y_1) \cap (X \backslash Y_2) \cap \cdots \cap (X - Y_{t-1}) \cap Y_t \right] \in \mathscr{B}.$$

It remains to prove 3°. By formula (2.4), applied to $r = h$,

$$T_1 T_2 \cdots T_n h(x) = \sum_{t=1}^{n} \beta^{t-1} P_x^\varphi q(a_t) + \beta^n P_x^\varphi h(x_n).$$

---

* See e.g. Kolmogorov and Fomin [1].

Since the function h is bounded and $\beta < 1$, the last term tends to 0 as $n \to \infty$. The sum by definition tends to $w(x,\varphi)$. Accordingly,

$$T_1 T_2 \cdots T_n h(x) \to w(x,\varphi). \tag{7}$$

But $T_1 T_2 \cdots T_n h \leq S^n h = h$, so that

$$w(x,\varphi) \leq h(x). \tag{8}$$

On the other hand, since $T_\psi(f + c) = T_\psi f + \beta c$ for any constant c (recall the definition of the operator T_ψ by formula (2.3)), it follows from inequality (3) that

$$
\begin{aligned}
T_\psi^n h &\geq T_\psi^{n-1}(h - \kappa) = T_\psi^{n-1} h - \beta^{n-1}\kappa \geq T_\psi^{n-2}(h - \kappa) - \beta^{n-1}\kappa \\
&= T_\psi^{n-2} h - (\beta^{n-2} + \beta^{n-1})\kappa \\
\cdots &\geq h - \kappa(1 + \beta + \beta^2 + \cdots + \beta^{n-1}).
\end{aligned}
$$

Thus for any n

$$T_\psi h \geq h - \kappa \sum_0^\infty \beta^t = h - \frac{\kappa}{1 - \beta}.$$

But when it is applied to the strategy ψ^∞, formula (7) yields

$$T_\psi^n h(x) \to w(x, \psi^\infty)$$

so that

$$w(x, \psi^\infty) \geq h(x) - \frac{\kappa}{1 - \beta}. \tag{9}$$

3° follows on comparing (8) and (9).

§9. Allocation of a Resource Between Production and Consumption

We continue the study of the examples considered for a finite control interval in Chapter 2, §§7, 9–11.

In order to apply the results of the present chapter to the problem of Chapter 2, §7, we will suppose that $q_t = \beta^{t-1} q$, where the discount coefficient β lies between 0 and 1. The value of the strategy π for the initial state x is given by the formula

$$w(x,\pi) = \sum_{t=1}^\infty \beta^{t-1} P_x^\pi q(x_{t-1} - a_t). \tag{1}$$

Here x_{t-1} is the resource at the beginning of the period t, and a_t is the investment into production over that period. The transition from x_{t-1} to x_t is given by the recurrence equation

$$x_t = F(a_t, s_t), \tag{2}$$

where s_t are independent identically distributed random variables.

If the function q is bounded above, then our model is summable above, and according to §2 the value v satisfies the optimality equation $v = Tv$ or

$$v(x) = \sup_{0 \le a \le x} [q(x - a) + \beta E v(F(a, s_t))] \qquad (0 \le x < \infty). \tag{3}$$

Under our current assumption the model is bounded above as well. If the supremum in (3) is attained at the points $a = \psi(x)$, and the function ψ is measurable, then $T_\psi v = v$, and in accordance with §3 the stationary strategy $\varphi = \psi^\infty$ is an optimal strategy. If the function q is bounded below, then the model is bounded below and $v = \lim_{n \to \infty} T^n 0$ (Chapter 4, §6).

The function q is usually assumed to be concave. This does not imply that q is bounded above but it does imply that it is majorized by some linear function:

$$q(c) \le Kc + L, \tag{4}$$

where K and L are positive constants. In this case, for summability above and boundedness above of the model it is sufficient that for some positive number γ satisfying the condition

$$\beta\gamma < 1, \tag{5}$$

the inequality

$$EF(a, s_t) \le \gamma a + N \tag{6}$$

should be satisfied for all $a \in [0, \infty)$, where N is any positive constant.

Indeed, since $0 \le a_t \le x_{t-1}$, it follows from (2) and (6) that the conditional mathematical expectation of x_t for the given history $x_0 a_1 x_1 \cdots a_{t-1} x_{t-1}$ does not exceed $\gamma(x_{t-1} + N)$. Hence by induction one easily finds that for any strategy π

$$P_x^\pi x_t \le \gamma^t x + N(1 + \gamma + \gamma^2 + \cdots + \gamma^{t-1}) \le \gamma^t(x + Nt). \tag{7}$$

Here we are assuming, without loss of generality, that $\gamma > 1$. It follows from (4), (7), and the inequalities $0 \le x_t - a_{t+1} \le x_r$ that

$$\beta^t P_x^\pi q(x_t - a_{t-1}) \le \beta^t [P_x^\pi (Kx_t + L)] \le K(\beta\gamma)^t(x + Nt) + L\beta^t, \tag{8}$$

so that the series (1) is majorized by a convergent series with positive terms not depending on the strategy π (but depending, in general, on the initial state x).

* * *

Now we continue the study of the special case in which

$$q(c) = c^\alpha, \qquad \alpha \in (0,1)$$

(see Chapter 2, §7). Here the model is uniformly bounded below. The function q satisfies the condition (4), and the conditions (5)–(6) for the function F take the form

$$Es_t < 1/\beta. \tag{9}$$

We have

$$Tf(x) = \sup_{0 \le a \le x} \left[(x - a)^\alpha + \beta Ef(as_t) \right]. \tag{10}$$

In Chapter 2, §7, we investigated this operator for $\beta = 1$ and established that it leaves invariant the set of functions of the form

$$f(x) = bx^\alpha \qquad (b \ge 0). \tag{11}$$

This result is valid in the general case as well, except that formulas (2.7.7) are replaced by

$$\xi(b) = \frac{(\beta b \lambda)^{\frac{1}{1-\alpha}}}{1 + (\beta b \lambda)^{\frac{1}{1-\alpha}}}, \tag{12}$$

$$\chi(b) = \left[1 + (\beta b \lambda)^{\frac{1}{1-\alpha}} \right]^{1-\alpha}$$

where, as before,

$$\lambda = Es_t^\alpha.$$

For the function (11) to satisfy the optimality equation $v = Tv$, it is necessary and sufficient that $b = \chi(b)$. This last equation has the unique root

$$b^* = \frac{1}{\left[1 - (\beta \lambda)^{\frac{1}{1-\alpha}} \right]^{1-\alpha}}. \tag{13}$$

We note that

$$\xi(b^*) = (\beta \lambda)^{\frac{1}{1-\alpha}}$$

Accordingly, the equation $v = Tv$ has the solution

$$v(x) = b^* x^\alpha, \tag{14}$$

and for such a function v the selector

$$\psi(x) = (\beta \lambda)^{\frac{1}{1-\alpha}} x \tag{15}$$

satisfies the equation $T_\psi v = v$. The resulting formulas have a meaning (and yield a solution of the corresponding equations) for $\beta \lambda < 1$, or, in view of the equation following (12), under the condition

$$Es_t^\alpha < 1/\beta, \tag{16}$$

which is more general than condition (9)*.

Since we do not know whether the solution of the equation $v = Tv$ is unique or not, we may for the time being only presuppose that the function (14) is the value of the model, and consequently the strategy $\varphi = \psi^\infty$ is an optimal strategy. We will prove this, passing to the limit in the formulas obtained in Chapter 2, §7 for a finite control interval.

In view of the boundedness below of the model,

$$v = \lim_n T^n 0.$$

According to formulas (2.7.8) and (2.7.12)–(2.7.14),

$$T^n 0(x) = \left[\sum_{t=0}^{n-1} (\beta\lambda)^{\frac{t}{1-\alpha}} \right]^{1-\alpha} x^\alpha$$

(discounting requires us to replace λ by $\beta\lambda$ in these formulas). Passing to the limit, we find that the inequality $\beta\lambda < 1$ (or the condition (16)) is necessary and sufficient for the finiteness of the value v, (i.e. the summability of the model above) and that if this condition is fulfilled, the value v is indeed given by formula (14).

If condition (9) is satisfied, then the optimality of the strategy $\varphi = \psi^\infty$ follows from the general result formulated in the first part of this section. Under the more general condition (16) it is necessary to calculate $T_\psi^n 0(x) = w^n(x,\varphi)$ directly and to verify that the limit of this quantity coincides with the function (14). The operator T_ψ acts according to the formula

$$T_\psi f(x) = [x - \psi(x)]^\alpha + \beta Ef(\psi(x)s_t)$$
$$= (1 - c)^\alpha x^\alpha + \beta Ef(cxs_t),$$

where

$$c = (\beta\lambda)^{\frac{1}{1-\alpha}}, \tag{17}$$

---

* This follows from the well-known moment inequality

$$(Es_t)^\alpha \geq E(s_t^\alpha), \qquad \alpha \in (0,1),$$

for which see e.g. B. V. Gnedenko, [1], §28. It is a special case of Jensen's inequality $Ef(s_t) \leq f(Es_t)$ for concave functions f; here $f(s) = s^\alpha$. Jensen's inequality is to be found in many places, for example Hardy, Littlewood, and Polya's book *Inequalities* [1], Theorem 86. It follows from (9) and the above moment inequality that

$$Es_t^\alpha \leq \beta^{-\alpha} \leq \beta^{-1}.$$

and successive calculation shows that

$$T_\psi^n 0(x) = (1 - c)^\alpha \sum_{t=0}^{n-1} (\beta \lambda c^\alpha)^t x^\alpha.$$

Taking into account (17) and the condition $\beta\lambda < 1$, one finds that in the limit the right side of this last equation coincides with the right side of formula (14)

§10. The Problem of Allocation of Stakes

In the problem of the allocation of stakes (see Chapter 2, §9), there was only a terminal reward and therefore this problem loses its meaning when applied to an infinite control interval. However the problem remains meaningful in the special case when starting from an initial state x, we have to reach a state at least equal to 1, with the maximal probability. We will consider here the same version of this problem as we did in Chapter 2, §9: the gambler loses his stake with probability q, and with probability $p < q$ receives twice his stake; $p + q = 1$.

The formalization of this problem employed in Chapter 2 does not work for an infinite interval of control since we cannot use the concept of a terminal reward. We construct another model, introducing an additional state Δ, and supposing that from the states $x \geq 1$ and Δ one must necessarily pass into Δ. The reward is equal to 1 when one passes from $x \geq 1$ into Δ, and equal to 0 for all the remaining transitions (including the transition $\Delta \to \Delta$). Each path contains no more than one transition $x \to \Delta$, and the total payoff I corresponding to this path is equal to 1 if and only if there is such a transition; otherwise the gain is equal to 0. Originally we were interested in the event

$$C = \{x_t \geq 1 \text{ for some } t\}. \tag{1}$$

Obviously C coincides with the event $\{I = 1\}$. Therefore the probability of C is equal to the probability of the event $\{I = 1\}$, which in turn is equal to the mathematical expectation of I. Therefore our problem reduces to the standard problem of maximization of the expected total reward.

Suppose that I_n is the gain over the interval $[0,n]$. Obviously the event $\{I_n = 1\}$ coincides with the event

$$C_n = \{x_t \geq 1 \text{ for some } t \leq n - 1\}. \tag{2}$$

Therefore the value v^n of our model on the interval $[0,n]$ is given by the function f_{n-1} calculated in Chapter 2, §9 (see the remark at the end of that section). Obviously our model is nonnegative and summable. Therefore, in view of the general result 5.2.1, the value v of the model on an infinite control interval is equal to

$$v = \lim_{n \to \infty} v^n = \lim_{n \to \infty} f_n = f_\infty.$$

In Chapter 2, §9, we described a bold strategy φ, consisting in placing the maximum possible stake x when $x \leq \frac{1}{2}$, the stake $1 - x$ when $x \in [\frac{1}{2},1]$, and no stake when $x \geq 1$. In the new model there corresponds to it a stationary strategy $\bar{\varphi}$ which prescribes the same behavior when $x < 1$ (no choice being available when $x \geq 1$ or in the state Δ). The value of the strategy $\bar{\varphi}$ on the segment $[0,n + 1]$ coincides with the value f_n of the strategy φ on the segment $[0,n]$. Therefore, the value $\bar{\varphi}$ on an infinite control interval is equal to $\lim f_n = v$, so that $\bar{\varphi}$ is an optimal strategy.

$$* \quad * \quad *$$

We note that our model is not bounded above.

Indeed, for any $x \in (0,1)$ there exists a strategy π such that $w(x,\pi) > 0$, which implies that $P_x^\pi q(a_t) > 0$ for some t. Suppose that π_n is a strategy consisting in making a zero stake at the first n steps, and then applying the strategy π. It is clear that

$$P_x^{\pi_n} q(a_{n+t}) = P_x^\pi q(a_t) > 0.$$

However, if the model were bounded above, then the left side would not exceed $c_{n+t}(x)$, where $\sum_t c_t(x) < \infty$.

Therefore we may not conclude from the general results of this chapter that the selector ψ generates a stationary optimal strategy if $T_\psi v = v$. For instance, in the above example the selector ψ prescribing a zero stake whenever $x < 1$ satisfies the equation $T_\psi v = v$, but leads to a zero reward when $x < 1$.

Further the function equal to 0 at the points 0 and Δ, and equal to 1 at the remaining points, satisfies the equation $v = Tv$, although it is not the value of the model (this in spite of the fact that it satisfies the requirement of being equal to 0 at the absorbing states). Therefore the equation $v = Tv$ is insufficient for calculating v (its solution is nonunique).

§11. Allocation of a Resource Among Consumption and Several Productive Sectors

In this problem, taking account of discounting, the operator T operates according to the formula

$$Tf(x) = \sup_{0 \leq i \leq x} \left\{ q(x - i) + \beta \sup_{0 \leq \gamma \leq 1} Ef[i\gamma\sigma_t + i(1 - \gamma)\tau_t] \right\} \tag{1}$$

(see formula (2.10.3)). In the homogeneous case all the pairs σ_t, τ_t are identically distributed.

For summability and boundedness above it suffices that the function a should be bounded above and that the coefficient β should be less than 1. Instead of the boundedness above of q we may assume that q is majorized by a linear function and that

$$E\sigma_t < \frac{1}{\beta}, \qquad E\tau_t < \frac{1}{\beta}. \tag{2}$$

Indeed, by modifying the arguments at (9.7)–(9.8), we easily find from conditions (2) that for any triple t, x, π

$$P_{x^-x_t}^\pi \leq \delta^t x,$$

where δ is the larger of the numbers $E\sigma_t$, $E\tau_t$. Since $\delta\beta < 1$, it follows as in §9 that the model is bounded and summable above.

$$* \quad * \quad *$$

Let us consider in detail the special case

$$q(c) = c^\alpha, \qquad \alpha \in (0,1);$$

(see Chapter 2, §9). With a discount coefficient β the operator T leaves, as it did before, invariant the set $\mathscr{L}$ of functions of the form

$$f(x) = bx^\alpha \qquad (b \geq 0),$$

but now the coefficients $\xi(b)$ and $\chi(b)$ in formulas (2.10.8)–(2.10.9) are defined by relations (9.12) rather than (2.7.7). Using the solution of the equation $b = \chi(b)$ found in §9, we obtain that the optimality equation $v = Tv$ has the same solution (9.13)–(9.14) as in §9, but

$$\lambda = \sup_{0 \leq \gamma \leq 1} E[\gamma\sigma_t + (1-\gamma)\tau_t]^\alpha. \tag{3}$$

As before, this solution has a meaning if

$$\beta\lambda < 1. \tag{4}$$

The supremum (3) was investigated in Chapter 2, §9. If it is attained for the value $\gamma = \gamma^*$, then the equality $v = T_\psi(v)$ is fulfilled for the function v at hand and the selector

$$\psi(x) = \{i(x), \gamma(x)\} = \left\{ (\beta\lambda)^{\frac{1}{1-\alpha}} x, \gamma^* \right\}. \tag{5}$$

Since we do not know whether the solution of the equation $v = Tv$ is unique, and we do not as yet know whether the model is summable (or bounded) above under condition (4), the optimality of the stationary strategy (5) is still to be proved. This can be done exactly as in §9. In addition the argument shows that condition (4) is necessary and sufficient for the summability above.

§12. The Stabilization Problem

Finally we consider, on the infinite time interval, the stabilization problem (see §2 of Chapter 1 and §11 of Chapter 2). From the nature of the problem, the value w^n over the control interval $[0,n]$ tends to $-\infty$ as $n \to \infty$. We therefore introduce a discount coefficient $\beta < 1$ (In Chapter 7, §12 we will study another, perhaps more natural statement of the problem).

The operator T here operates according to the formula

$$Tf(x) = \sup_{-\infty < a < +\infty} [-b(x-a)^2 - ca^2 + \beta Ef(x - a + s_t)]. \tag{1}$$

The running reward is negative, so that the model is summable and bounded above and the value v satisfies the equation $v = Tv$. By redoing the calculations of Chapter 2, §11, taking account of the coefficient β, we find that for a nonnegative l

$$T(-lx^2 - m) = -l'x^2 - m', \tag{2}$$

where

$$l' = \frac{\beta cl + bc}{\beta l + b + c}, \qquad m' = \beta \sigma^2 l + \beta m, \tag{3}$$

while the maximum in (1) is attained for

$$a = \psi(x) = \frac{\beta l + b}{\beta l + b + c} x. \tag{4}$$

Equating l' to l and m' to m, and retaining only the positive root l, we find that

$$l = \frac{[b^2 + c^2(1-\beta)^2 + 2bc(3-\beta)]^{\frac{1}{2}} - b - (1-\beta)c}{2\beta}, \tag{5}$$

$$m = \frac{\beta \sigma^2 l}{1 - \beta}. \tag{6}$$

For these values the function

$$v(x) = -lx^2 - m \tag{7}$$

is invariant relative to the operator T. The remainder of this section will be devoted to the verification that v is in fact the value of the model and that the stationary strategy $\varphi = \psi^\infty$ is optimal.

In view of Chapter 4, §6, the value of the model does not exceed the function

$$v^\infty = \lim_{n \to \infty} T^n 0;$$

we cannot say more than this, since our model is unbounded below. On the other hand, this value is not less than

$$w(x,\varphi) = \lim_{n \to \infty} T_\psi^n 0(x).$$

Therefore it suffices to verify that

$$v^\infty = -lx^2 - m = w(x,\varphi). \tag{8}$$

By an obvious modification of the calculations of Chapter 2, §11, we find from formulas (2) and (3) that

$$T^n 0(x) = -l_n x^2 - m_n,$$

where

$$l_n = \frac{l(1 - \lambda^n)}{1 - (\lambda^n l/\tilde{l})}, \qquad m_n = \sigma^2(\beta l_{n-1} + \beta^2 l_{n-2} + \cdots + \beta^n l_0),$$

$$\lambda = \frac{c - l}{c - \tilde{l}}, \tag{9}$$

where l and $\tilde{l}$, with $\tilde{l} < l$, are the roots of the quadratic equation

$$\beta l^2 + (b + c - \beta c)l - bc = 0, \tag{10}$$

coinciding with the equation from which the number l was found in formula (5). Since $|\lambda| < 1$ and $\beta \in (0,1)$ it follows from (9) that $l_n \to l$, $m_n \to m$, and the left equation in (8) is proved.

Further,

$$T_\psi f(x) = -b[x - \psi(x)]^2 - c\psi(x)^2 + \beta Ef[x - \psi(x) + s_t],$$

so that we may easily deduce from (4) that

$$T_\psi(-Lx^2 - M) = -L'x^2 - M', \tag{11}$$

where

$$L' = \frac{(\beta L + b)c^2 + (\beta l + b)^2 c}{(\beta l + b + c)^2}, \qquad M' = \beta\sigma^2 L + \beta M. \tag{12}$$

It follows from (11) that

$$w(x,\varphi) = \lim_{n \to \infty} T^n 0(x) = -Lx^2 - M,$$

where L and M can be obtained by equating L' to L, M' to M. From (12) we find that

$$L = \frac{c(\beta l + b)^2 + bc^2}{(\beta l + b + c)^2 - \beta c^2},$$ (13)

$$M = \frac{\beta \sigma^2 L}{1 - \beta}.$$ (14)

For the proof of the second of the equations (8) we need to show that $L = l$ and $M = m$. We use the following identity

$$\frac{c(\beta l + b)^2 + bc^2}{(\beta l + b + c)^2 - \beta c^2} = l + \frac{[\beta l^2 + (b + c - \beta c)l - bc](l + b + c)}{(\beta l + b + c)^2 - \beta c^2}.$$

In view of (10) and (13) it therefore follows that $L = l$. It follows from (6) and (14) that $M = m$.

Chapter 7

Maximization of the Average Reward
Per Unit Time

§1. Introduction. Canonical Strategies

In the preceding chapters we estimated the value of a strategy in terms of the total gain over the entire time of control. If the gain over the time interval $[0,n]$ increases unboundedly as $n \to \infty$, then it is natural to prefer those strategies for which the growth of the gain is faster. In §§1–7 we construct strategies which are optimal from this point of view for homogeneous finite models (without discount).

In fact, what we shall construct will be a stationary strategy φ such that for any initial distribution μ and any strategy π

$$w^n(\mu,\pi) \le w^n(\mu,\varphi) + C, \tag{1}$$

where C is a constant independent of n, μ, and π, and

$$w^n(\mu,\pi) = P_\mu^\pi \sum_{t=1}^n q(a_t) \tag{2}$$

is the value of the strategy π on the interval $[0,n]$. The strategy φ which we shall construct will have other remarkable properties as well. Introduce a terminal reward r at the time n. Then the value $w^n(\mu,\pi)$ is replaced by

$$w_r^n(\mu,\pi) = P_\mu^\pi \left[\sum_{t=1}^n q(a_t) + r(x_n) \right]$$
$$= w^n(\mu,\pi) + P_\mu^\pi r(x_n). \tag{3}$$

It turns out that with an appropriate choice of r the strategy φ is optimal for the control problem on all finite intervals $[0,n]$; in other words, for any n, μ, and π

$$w_r^n(\mu,\pi) \le w_r^n(\mu,\varphi). \tag{4}$$

It is clear that (1) follows from (4).

Further, we shall show that

$$w_r^n(x,\varphi) = r(x) + nv(x) \qquad (x \in X), \tag{5}$$

where v is some function on X. It is clear from (5), (3) and (1) that for any terminal reward function f

$$\mu v = \lim_{n \to \infty} \frac{w_f^n(\mu,\varphi)}{n} \geq \overline{\lim_{n \to \infty}} \frac{w_f^n(\mu,\pi)}{n}, \tag{6}$$

so that the strategy φ maximizes in the limit the average expected reward per unit time. (Note that by virtue of (4), for $f = r$ the strategy φ maximizes the average expected reward for each fixed n).

A triple (v,φ,r) satisfying the conditions (4) and (5) will be called a *canonical triple* for the model Z. (Here $\varphi = \psi^\infty$ is a stationary strategy, ψ being a selector of the correspondence $A(x)$, and v and r are functions on X.) We shall say that the stationary strategy φ itself is *canonical* if it appears in some canonical triple.

By (6), the canonical strategy φ is *asymptotically optimal* in the sense that for any initial distribution μ

$$\overline{\lim} \frac{w^n(\mu,\pi)}{n} \leq \underline{\lim} \frac{w^n(\mu,\varphi)}{n} \tag{7}$$

for any strategy π. The function v appearing in the canonical triple is uniquely determined by the model Z: in view of (6) we have

$$v(x) = \sup_\pi \overline{\lim} \frac{w^n(x,\pi)}{n} = \sup_\pi \underline{\lim} \frac{w^n(x,\pi)}{n}. \tag{8}$$

These equations justify us in calling the function $v(x)$ the *asymptotic value of the model*. Replacing the state x in (8) by the initial distribution μ, we obtain the *asymptotic value* $v(\mu)$ *of the initial distribution* μ. It follows from (6) that $v(\mu) = \mu v$. By (7) and (8), the strategy π is asymptotically optimal if and only if for any initial distribution μ the limit

$$w(\mu,\pi) = \lim_{n \to \infty} \frac{w^n(\mu,\pi)}{n}$$

exists and is equal to $v(\mu)$*.

Our objective is to prove the existence of canonical strategies and find methods of constructing them.

---

* Formula (8) and all the succeeding formulas remain valid also in the case when the upper limit in (7) is replaced by a lower limit. It is clear that in this case the right side of (8) does not exceed $v(x)$. On the other hand, it is not less than $v(x)$, since in view of (6)

$$\underline{\lim} \frac{w^n(x,\varphi)}{n} = \overline{\lim} \frac{w^n(x,\varphi)}{n} = v(x).$$

§2. Canonical Equations

In this section we prove that *the triple* (v, ψ^∞, r) *is canonical if and only if the following equations are satisfied*:

$$v(x) = \sum_{y \in X} v(y)p(y|\psi(x)) = \sup_{a \in A(x)} \sum_{y \in X} v(y)p(y|a),$$

$$r(x) + v(x) = q(\psi(x)) + \sum_{y \in X} r(y)p(y|\psi(x))$$

$$= \sup_{a \in A(x)} \left[q(a) + \sum_{y \in X} r(y)p(y|a) \right] (x \in X)$$

(the *canonical equations*). They may be written in more compact form as follows:

$$v = P_\psi v = V\Pi v, \tag{1}$$

$$r + v = T_\psi r = Tr, \tag{2}$$

where the operators are given by the formulas

$$\Pi f(a) = \sum_{y \in X} f(y)p(y|a) \qquad (a \in A), \tag{3}$$

$$P_\psi f(x) = \sum_{y \in X} f(y)p(y|\psi(x)) = \Pi f(\psi(x)) \qquad (x \in X), \tag{4}$$

$$T_\psi f(x) = q(\psi(x)) + P_\psi f(x) \qquad (x \in X), \tag{5}$$

$$Vg(x) = \sup_{a \in A(x)} g(a) \qquad (x \in X), \tag{6}$$

$$Tf(x) = V(q + \Pi f)(x) = \sup_\psi T_\psi f(x) \qquad (x \in X) \tag{7}$$

cf. formulas (1.1.5), (1.6.10), (1.6.11).

In order to arrive at the canonical equations, we make use of the results of Chapter 1, §7. According to formula (1.7.5)

$$w_r^n(x, \varphi) = T_r^n r(x).$$

It follows from (1.7.9)–(1.7.10) that

$$\sup_\pi w_r^n(x, \pi) = T^n r(x).$$

Therefore conditions (1.4) and (1.5) are equivalent to the requirement

$$r + nv = T_\psi^n r = T^n r \qquad (n = 1, 2, \ldots). \tag{8}$$

On putting $n = 1$ in (8), we get (2). Further, replacing n by $n + 1$ in (8), we have

$$r + (n + 1)v = T_\psi^{n+1}r = T^{n+1}r.$$

Substituting the values of $T_\psi^n r$ and $T^n r$ from (8) into this last expression, we arrive at the equations

$$r + v + nv = T_\psi(r + nv) = T(r + nv). \tag{9}$$

From (5) and (2) we find that

$$T_\psi(r + nv) = T_\psi r + nP_\psi v = r + v + nP_\psi v. \tag{10}$$

Comparing (9) and (10), we obtain the equation $P_\psi v = v$. By (7)

$$T(r + nv) = V(q + \Pi r + n\Pi v).$$

It therefore follows from (9) that

$$\frac{r + v}{n} + v = \frac{1}{n} V(q + \Pi r + n\Pi v) = V\left(\frac{q + \Pi r}{n} + \Pi v\right).$$

Suppose c is the largest value of the function $|q + \Pi r|$. The quantity in parentheses differs from Πv by no more than c/n. Therefore as $n \to \infty$ the right side tends to $V\Pi v$, and we obtain the equation $v = V\Pi v$.

It remains to be proved that (8) follows from (1) and (2). For $n = 1$ equation (8) coincides with (2). Now suppose that (8) is true for some n; we shall prove that then it is true for $n + 1$ as well. Applying the operator T_ψ to both sides of the equation $r + nv = T_\psi^n r$ and using equation (1), we find that

$$T_\psi^{n+1}r = T_\psi(r + nv) = T_\psi r + P_\psi(nv) = r + v + nP_\psi v$$
$$= r + (n + 1)v. \tag{11}$$

If we apply the operator T to both sides of the equation $r + nv = T^n r$, we get

$$T^{n+1}r = T(r + nv) = V(q + \Pi r + n\Pi v) \le V(q + \Pi r)$$
$$+ nV\Pi v = Tr + nv = r + (n + 1)v \tag{12}$$

(see (1) and (2); it is obvious that $V(f + g) \le Vf + Vg$ for any functions f and g on A). Since $Tr \ge T_\psi r$ and therefore $T^{n+1}r \ge T_\psi^{n+1}r$, it follows from (11) and (12) that indeed

$$T^{n+1}r = r + (n + 1)v. \tag{13}$$

Formulas (11) and (13) show that equation (8) is valid for $n + 1$ as well.

Thus, in order to construct a canonical triple, it suffices to solve equations (1)–(2). First we shall investigate the simpler system

$$w = P_\psi w,$$
$$f + w = T_\psi f, \tag{14}$$

where ψ is any selector of the correspondence $A(x)$. (The canonical equations reduce to equations (14) in the case when there is only one action $a = \psi(x)$ at each state x).

Taking account of (5) and dropping the subscript ψ, we may rewrite the system (14) in the form

$$w = Pw,$$
$$f + w = q + Pf, \tag{15}$$

where q is the function on X given by the formula

$$q(x) = q(\psi(x)). \tag{16}$$

We shall call equations (15) (and also (14)) *Howard's equations*. It is convenient to consider equations (15) as matrix equations: w, f, and g are interpreted as column vectors and P as a square matrix with the elements

$$P(x,y) = P(y|\psi(x)). \tag{17}$$

The elements of P are nonnegative and the sum of the elements of any row is equal to 1. A matrix with such properties is said to be *stochastic*. Our initial objective is to prove that *for any stochastic matrix P and any vector q, the Howard system (15) has a solution.*

§3. Solution of the Howard Equations

So, suppose that P is any finite stochastic matrix and q any vector. We wish to construct a pair (w, f) of vectors satisfying the relations

$$w = Pw, \tag{1}$$

$$f + w = q + Pf. \tag{2}$$

We note first of all that if these relations are satisfied, then

$$w = \lim_{n \to \infty} \frac{1}{n} \sum_{t=0}^{n-1} P^t q. \tag{3}$$

Indeed, multiplying (2) by P^t and taking account of (1), we have

$$P^t f + w = P^t q + P^{t+1} f.$$

Summing these equations on t from 0 to $t - 1$, we get

$$f + nw = \sum_{t=0}^{n-1} P^t q + P^n f. \tag{4}$$

It is easy to verify that the product of stochastic matrices is also a stochastic matrix. This means that all the elements of the vectors $P^n f$, $n = 0, 1, 2, \ldots$, are bounded, and (3) follows from (4).

Formula (3) is the starting point for the construction of the solution of the system (1)–(2). In order to make use of it, we must first verify the existence of a limit in the right side of (3). All the elements of the matrix

$$A_n = \frac{1}{n} \sum_{t=0}^{n-1} P^t$$

lie between 0 and 1. Therefore, for some subsequence $n_1 < n_2 < \cdots$ the limit

$$M = \lim_{k \to \infty} A_{n_k}$$

exists. It remains to verify that the sequence A_n cannot have other limit points. Suppose that for another subsequence $m_1 < m_2 < \cdots$

$$\lim_{k \to \infty} A_{m_k} = M'.$$

We have

$$PA_{n_k} = A_{n_k} P = A_{n_k} + \frac{P^{n_k} - E}{n_k},$$

where E is the identity matrix, and in the limit $PM = MP = M$. Hence it follows that $A_{m_k} M = M A_{m_k} = M$, and in the limit $M'M = MM' = M$. Analogously one finds that $MM' = M'M = M'$, and therefore $M = M'$.

Thus we have proved that the limit

$$M = \lim_{n \to \infty} \frac{1}{n} (E + P + P^2 + \cdots + P^{n-1}) \tag{5}$$

exists, and

$$MP = PM = M, \tag{6}$$

$$M^2 = M. \tag{7}$$

In view of (6) the formula

$$w = Mq, \tag{8}$$

(equivalent to (3)) yields a solution of equation (1). Taking account of (8), we may bring equation (2) into the form

$$(E - P)f = (E - M)q. \tag{9}$$

It remains to find the solution f of this equation.

We note that

$$(E - M)q = \lim_{n \to \infty} \frac{1}{n} \sum_{t=0}^{n-1} (E - P^t)q = \lim_{n \to \infty} (E - P)f_n, \tag{10}$$

where

$$f_n = \frac{1}{n} \sum_{t=0}^{n-1} (E + P + P^2 + \cdots + P^{t-1})q. \tag{11}$$

Suppose for the moment that f_n has a limit f. Then $\lim(E - P)f_n = (E - P)f$, equation (10) reduces to (9), and the function f that we need is constructed. In fact the situation is somewhat more complicated. It is clear from (5) that the expression in parentheses in (11) behaves like tM as $t \to \infty$. Therefore, instead of the sequence f_n, which, generally speaking, is unbounded, it is better to consider

$$g_n = \frac{1}{n} \sum_{t=0}^{n-1} (E + P + P^2 + \cdots + P^{t-1} - tM)q. \tag{12}$$

In view of (6), the difference $f_n - g_n$ is carried by the operator $E - P$ into zero, so that (10) is equivalent to the equation

$$(E - M)q = \lim_{n \to \infty} (E - P)g_n. \tag{13}$$

Denote by $\|g\|$ the maximum of the absolute values of the coordinates of the vector g. We shall show that the sequence $\|g_n\|$ cannot tend to infinity. Indeed, suppose that this were the case, and put $h_n = g_n/\|g_n\|$. Then it follows from (13) that

$$\lim_{n \to \infty} (E - P)h_n = 0. \tag{14}$$

Since the sequence h_n is bounded, one can find a convergent subsequence; in view of (14) the limit h of this subsequence satisfies the equation $h = Ph$. Hence, using (5), it follows that $h = Mh$. On the other hand, it follows from (6) and (12) that $Mg_n = 0$, so that $Mh = 0$. Hence $h = 0$, which is impossible since $\|h_n\| = 1$.

Since $\|g_n\|$ does not tend to infinity, one can select from the sequence $\{g_n\}$ a convergent subsequence $\{g_{n_k}\}$. Its limit f satisfies equation (9).

We have proved the existence of a solution of equation (9), and hence of the system (1)–(2). We note that the vector f which we have constructed satisfies the additional condition

$$Mf = 0, \tag{15}$$

since $Mg_n = 0$ for each n.

We have already seen that *the vector w is defined uniquely by equations* (1)–(2) (it is given by formula (3) or (8)). Now we shall show that f *is also defined uniquely under the additional condition* (15). Indeed, if some pair (w, f') satisfies these equations, then $f - f' = P(f - f')$. Hence, in view of (5) and (15),

$$f - f' = M(f - f') = Mf - Mf' = 0.$$

§4. Modification of the Canonical Equations

Now we have to pass from Howard's equations to the canonical equations (2.1)–(2.2). These may be written in the form

$$v(x) = \Pi v(\psi(x)) = \max_{a \in A(x)} \Pi v(a), \tag{1}$$

$$r(x) + v(x) = q(\psi(x)) + \Pi r(\psi(x))$$
$$= \max_{a \in A(x)} \left[q(a) + \Pi r(a) \right] \qquad (x \in X). \tag{2}$$

It is more convenient to solve the modified system

$$v(x) = \Pi v(\psi(x)) = \max_{a \in A(x)} \Pi v(a), \tag{3}$$

$$r(x) + v(x) = q(\psi(x)) + \Pi r(\psi(x))$$
$$= \max_{a \in A_v(x)} \left[q(a) + \Pi r(a) \right] \qquad (x \in X), \tag{4}$$

where

$$A_v(x) = \{a : a \in A(x), \Pi v(a) = v(x)\}. \tag{5}$$

In the following section we shall indicate a recurrence procedure for the solution of the system (3)–(4), inapplicable to the system (1)–(2).

Now let us show that if (v, ψ, r) is a solution of the system (3)–(4), then the triple (v, ψ, r') with $r' = r + cv$ satisfies the system (1)–(2) for any sufficiently large constant c.

Equation (1) coincides with (3). Since

$$\Pi r' = \Pi r + c \Pi v,$$

it follows from (3) and (4) that

$$r'(x) + v(x) = r(x) + v(x) + cv(x)$$
$$= q(\psi(x)) + \Pi r(\psi(x)) + c \Pi v(\psi(x))$$
$$= q(\psi(x)) + \Pi r'(\psi(x))$$

and we have the first equation of (2). It remains to prove that for each x

$$q(a) + \Pi r'(a) \leq r'(x) + v(x) \qquad (a \in A(x))$$

or, what is the same thing,

$$q(a) + \Pi r(a) + c\Pi v(a) \leq r(x) + v(x) + cv(x) \qquad (a \in A(x)). \tag{6}$$

If $a \in A_v(x)$, then in view of (4) and (5) we have

$$q(a) + \Pi r(a) \leq r(x) + v(x), \qquad \Pi v(a) = v(x),$$

and for such an a equation (6) is satisfied for any c. If $a \in A(x) \backslash A_v(x)$, then $\Pi v(a) \neq v(x)$, and in view of (3) $\Pi v(a) < v(x)$. Equation (6) is satisfied for sufficiently large c. Since the set of all pairs (x,a) is finite, then for large c equation (6) will be satisfied simultaneously for all pairs (x,a) with $x \in X$ and $a \in A(x)$.

From the result just proved it follows that if (v,ψ,r) is any solution of the modified canonical system (3)–(4), then v is the asymptotic value and $\varphi = \psi^\infty$ is the canonical strategy.

§5. Howard's Strategy Improvement Procedure

Choose any selector ψ and calculate the corresponding solutions w and f of Howard's equations

$$w(x) = \Pi w(\psi(x)) \qquad (x \in X), \tag{1}$$

$$f(x) + w(x) = q(\psi(x)) + \Pi f(\psi(x)) \qquad (x \in X), \tag{2}$$

satisfying the condition (3.15). For the triple (w,ψ,f) to be a solution of the modified canonical system (4.3)–(4.4), it is necessary and sufficient that

$$w(x) = \max_{a \in A(x)} \Pi w(a) \qquad (x \in X), \tag{3}$$

$$f(x) + w(x) = \max_{a \in A_w(x)} [q(a) + \Pi f(a)] \qquad (x \in X), \tag{4}$$

where

$$A_w(x) = \{a : a \in A(x), w(x) = \Pi w(a)\}. \tag{5}$$

Since $\psi(x) \in A(x)$, $\Pi w(\psi(x))$ does not exceed the maximum $\Pi w(a)$ on the fibre $A(x)$, and, from (1),

$$w(x) \leq \max_{a \in A(x)} \Pi w(a). \tag{6}$$

In view of (1), $\psi(x) \in A_w(x)$. Therefore it follows from (2) that the inequality

$$f(x) + w(x) \leq \max_{a \in A_w(x)} \left[q(a) + \Pi f(a) \right] \tag{7}$$

holds.

If equation (3) does not hold, then there exist x_0, and an $a_0 \in A(x_0)$, such that

$$w(x_0) < \Pi w(a_0).$$

If equation (3) does hold, but equation (4) is not satisfied, then we consider an x_0 and an $a_0 \in A(x_0)$ for which

$$f(x_0) + w(x_0) < q(a_0) + \Pi f(a_0).$$

We define a new selector χ by the formula

$$\chi(x) = \begin{cases} \psi(x) & \text{if } x \neq x_0, \\ a_0 & \text{if } x = x_0. \end{cases}$$

The passage from ψ to χ is called a *Howard strategy improvement procedure.*

Repeating this procedure, we either, after a finite number of steps, arrive at a triple (w, ψ, f) satisfying the modified canonical system, or else we construct an infinite sequence of selectors ψ_n, in which each successive selector is an improvement of its predecessor. In the first case we obtain a canonical strategy $\varphi = \psi^\infty$.

In order to prove the impossibility of the second case, we assign to each strategy π a function

$$w_\beta(x, \pi) = \sum_{t=1}^\infty \beta^{t-1} P_x^\pi q(a_t) \qquad (0 < \beta < 1) \tag{8}$$

(the value of the strategy π for control over an infinite interval of time, with discount coefficient β). In the following two sections we will prove that *if χ is an improvement of ψ, then, for β sufficiently close to 1, $w_\beta(x, \chi) \geq w_\beta(x, \psi)$ for all $x \in X$ and $w_\beta(x_0, \chi) > w_\beta(x_0, \psi)$ for some $x_0 \in X$;* this is one of the justifications for the term "improvement". Therefore no two functions $w_\beta(x, \psi_n)$ can coincide with one another. Since the set of all selectors is finite, the improvement procedure necessarily terminates.

§6. Asymptotics of the Discounted Reward

We obtain the following expression for the discounted reward for the stationary strategy $\varphi = \psi^\infty$, when $\beta \uparrow 1$:

$$w_\beta(x, \varphi) = \frac{w(x)}{1 - \beta} + f(x) + o(1), \tag{1}$$

w and f being solutions of the Howard system

$$w = Pw, \qquad (2)$$

$$f + w = q + Pf, \qquad (3)$$

$$Mf = 0 \qquad (4)$$

(the matrix P and the vector q corresponding to the selector ψ are defined by formulas (2.16) and (2.17), and the matrix M by formula (3.5)).

We want to transform formula (5.8) for $w_\beta(x,\varphi)$. The probabilities

$$P(t,x,y) = P_x^\varphi \{x_t = y\}$$

obviously satisfy the relation

$$P(t + 1,x,y) = \sum_{z \in X} P(t,x,z)P(1,z,y),$$

Therefore they form the matrix $P^{t}*$. We have

$$P_x^\varphi q(a_t) = \sum_{y \in X} q(\psi(y))P_x^\varphi \{x_{t-1} = y\} = \sum_{y \in X} P(t-1,x,y)q(y),$$

so that, for the stationary strategy φ, equation (5.8) in matrix form takes on the form

$$w_\beta(\varphi) = \sum_{t=0}^{\infty} \beta^t P^t q, \qquad (5)$$

$w_\beta(\varphi)$ being a column vector with components $w_\beta(x,\varphi)$.

We express q from equation (3) and substitute into (5). Recalling that $Pw = w$, after some simple calculations we have

$$w_\beta(\varphi) = \frac{w}{1 - \beta} + f - g,$$

where

$$g = (1 - \beta) \sum_{t=1}^{\infty} \beta^{t-1} P^t f. \qquad (6)$$

In order to arrive at the expression (1), it remains to verify that $g \to 0$ as $\beta \uparrow 1$.

Equation (4) means that

$$\lim_{t \to \infty} \frac{S_t}{t} = 0, \qquad (7)$$

---

* The random sequence of states $x_0 x_1 x_2 \ldots$ is a homogeneous Markov chain with transition matrix P.

where

$$s_t = \sum_{k=1}^{t} P^k f; \tag{8}$$

(see the definition (3.5) of the matrix M). It follows from (7) that for any $\varepsilon > 0$ there exists a number T such that

$$\|s_t\| \le \varepsilon t \quad \text{for } t > T, \tag{9}$$

and from (8) that

$$\|s_t\| \le \|f\| t \quad \text{for all } t, \tag{10}$$

Here, as in §3, the norm of a vector means the maximum of the absolute values of its components.

In order to use (9) and (10) for the estimation of g, we express the coefficients of the power series (6) in terms of their sums s_t:

$$g = (1 - \beta)[s_1 + \beta(s_2 - s_1) + \beta^2(s_3 - s_2) + \cdots]$$

$$= (1 - \beta)^2 \sum_{t=1}^{\infty} \beta^{t-1} s_t. \tag{11}$$

Here we recall that, in view of (10), the series $s_1 + \beta s_2 - \beta s_1 + \cdots$ converges absolutely, so that the transformation we have just made is legitimate. It follows from (9), (10), and (11) that

$$\|g\| \le (1 - \beta)^2 \left(\|f\| \sum_{1}^{T} t\beta^{t-1} + \varepsilon \sum_{T+1}^{\infty} t\beta^{t-1} \right)$$

$$\le (1 - \beta)^2 \left(\|f\| \sum_{1}^{T} t + \varepsilon \sum_{1}^{\infty} t\beta^{t-1} \right)$$

$$= (1 - \beta)^2 \left[\frac{T(T + 1)}{2} \|f\| + \frac{\varepsilon}{(1 - \beta)^2} \right],$$

a quantity not exceeding 2ε for β sufficiently close to 1.

§7. Increase of the Discounted Reward with the Howard Improvement

In order to prove the existence of canonical strategies, it remains to verify that *the Howard improvement does not decrease the value* $w_\beta(x,\psi^\infty)$, *and increases this value in at least one state* (if β is sufficiently close to 1).

In order to compare the values $w_\beta(x,\psi^\infty)$ and $w_\beta(x,\chi^\infty)$, it is convenient to introduce a sequence of nonstationary strategies $\pi_0, \pi_1, \pi_2, \ldots$. The strategy π_n consists in using the selector χ at the first n steps, and then using the selector ψ. The values

$$g_n(x) = w_\beta(x,\pi_n)$$

are connected by the simple recurrence relation

$$g_{n+1} = T_\chi(\beta g_n) \tag{1}$$

(see the fundamental equation (6.2.2)). We note that

$$w_\beta(x,\psi^\infty) = g_0(x)$$

and

$$w_\beta(x,\chi^\infty) = \lim_{n \to \infty} g_n(x).$$

Indeed, in view of (5.8)

$$w_\beta(x,\chi^\infty) - w_\beta(x,\pi_n) = \sum_{t=n+1}^{\infty} \beta^{t-1}[P_x^{\chi^\infty} q(a_t) - P_x^{\pi_n} q(a_t)],$$

and the right side is majorized by the sum

$$2\|q\| \sum_{t=n+1}^{\infty} \beta^{t-1},$$

which tends to 0 as $n \to \infty$. The result which we need follows from the following two propositions:

a) *If the selector χ is an improvement of the selector ψ, then*

$$g_0(x) = g_1(x) \quad \text{if } \psi(x) = \chi(x), \tag{2}$$

$$g_0(x) < g_1(x) \quad \text{if } \psi(x) \neq \chi(x) \text{ and } \beta \text{ is close to 1.} \tag{3}$$

b) *If $g_0 \leq g_1$, then $g_n \leq g_{n+1}$ for all $n \geq 1$.*

Since the operator T_χ preserves the inequality, assertion b) follows directly from formula (1). Equation (2) is also obvious, since if the initial state x satisfies the condition $\psi(x) = \chi(x)$ then the strategies π_0 and π_1 prescribe at each stage the same action.

If now $\chi(x_0) = a_0 \neq \psi(x_0)$ for some $x_0 \in X$, then, in view of §5 either

$$\Pi w(a_0) > w(x_0), \tag{4}$$

or

$$\Pi w(a_0) = w(x_0), \qquad q(a_0) + \Pi f(a_0) > f(x_0) + w(x_0), \tag{5}$$

where (w,f) is the solution of the Howard system (6.2)–(6.4) for the selector ψ. By the asymptotic expression (6.1)

$$g_0(x_0) = \frac{w(x_0)}{1 - \beta} + f(x_0) + o(1). \tag{6}$$

By virtue of (1) and the same asymptotic expression

$$g_1(x_0) = T_\chi \left[\frac{\beta w}{1 - \beta} + \beta f + o(1) \right](x_0).$$

Since $\chi(x_0) = a_0$, for any function h we have $T_\chi h(x_0) = q(a_0) + \Pi h(a_0)$ (see the definitions of these operators in §2). Hence

$$g_1(x_0) = q(a_0) + \Pi \left[\frac{\beta w}{1 - \beta} + \beta f + o(1) \right](a_0)$$

$$= \frac{\Pi w(a_0)}{1 - \beta} + [q(a_0) + \Pi f(a_0) - \Pi w(a_0)] + o(1). \tag{7}$$

If case (4) holds, then inequality (3), for the state x_0, follows from the comparison of the first terms of the expansions (6) and (7). In case (5) the leading terms in these formulas coincide, and the inequality which we need is gotten from the comparison of the second terms.

§8. Extension to Infinite Models

We shall compare two problems: (A) Maximize the total reward across n steps, and (B) Maximize the average per unit time reward.

Of course, problem (B) does not differ from problem (A) if the average is taken over n steps, so that problem (B) has an interest in itself only for an infinite control interval. In the case of finite models problem (B) yielded as good a solution as problem (A) (though more delicate considerations were required). The rôle of the simple strategies is played here by the stationary strategies. One might hope that an analogy between the two problems holds for infinite models as well. However this expectation is dashed by examples.

We begin with semicontinuous models, for which problem (A) is solved as well as it is for finite models. The following example shows that for the problem (B) the situation is quite different: there do not exist strategies maximizing the average reward (even if one admits nonstationary strategies).

EXAMPLE 1. The space X consists of three states x, y, and z. The states y and z are absorbing, i.e. in each of them there exists only one action, which does not change the state (see figure 7.1). The set of actions in the state x is a segment Δ of the positive half-line containing the point 0. The transition function for each action δ of Δ is given by the formulas

$$p(y|\delta) = \delta,$$
$$p(z|\delta) = \delta^2,$$
$$p(x|\delta) = 1 - \delta - \delta^2;$$

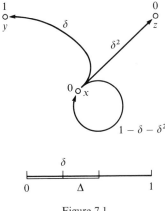

Figure 7.1

(for these formulas to have a meaning we have to assume that $1 - \delta - \delta^2 \geq 0$ for all $\delta \in \Delta$, i.e., $\Delta \subset [0, (\sqrt{5} - 1)/2]$. The running reward function q is equal to 1 in the state y and 0 otherwise.

It is not hard to verify that this homogeneous model is semicontinuous (where one uses the usual R^1 metric on Δ, and the actions in the states y and z are two isolated points of the action space A).

We will regard x as the initial state. We fix some strategy π, and denote by P the corresponding measure P_x^π. Consider the time τ of the first exit from the state x: if $\tau < \infty$, then $x_0 = x_1 = \cdots = x_{\tau-1} = x$ and $x_\tau = x_{\tau+1} = \cdots$ equals y or z; if $\tau = \infty$, then $x_t = x$ for all t. Since the reward function is equal to 1 in the state y and otherwise 0, then

$$w^n(x,\pi) = p_0 + p_1 + \cdots + p_{n-1}, \tag{1}$$

where $p_t = P\{x_t = y\}$. But

$$\{x_t = y\} = \{\tau \leq t, x_\tau = y\}.$$

Therefore, as $t \to \infty$, p_t tends to

$$p = P\{\tau < \infty, x_\tau = y\},$$

i.e. to the probability of at least one arrival into y. In view of (1) the limit

$$w(x,\pi) = \lim_{n \to \infty} \frac{w^n(x,\pi)}{n} = p.$$

exists.

At the time of exit from x we arrive in z with a positive probability. Therefore either $P\{\tau < \infty\} = 0$, or $P\{\tau < \infty, x_\tau = z\} > 0$. In both cases

$$p = P\{\tau < \infty, x_\tau = y\} < 1.$$

On the other hand, for the stationary strategy δ^∞

$$P\{\tau = t,\, x_\tau = y\} = (1 - \delta - \delta^2)^{t-1}\delta.$$

Therefore

$$p = P\{\tau < \infty,\, x_\tau = y\} = \sum_{t=1}^{\infty} (1 - \delta - \delta^2)^{t-1}\delta$$

$$= \begin{cases} 0 & \text{if } \delta = 0, \\[2ex] \dfrac{1}{1 + \delta} & \text{if } \delta > 0. \end{cases}$$

For a sufficiently small positive δ the average reward $w(x,\delta^\infty) = p$ is arbitrarily close to 1, but there is no strategy π for which it is equal to 1.

* * *

For countable models, problem (A) was considered in §§12,13 of Chapter 1, where we proved the existence of uniformly ε-optimal strategies for arbitrary $\varepsilon > 0$ (A weakened version of this result, for general models with Borelian action and state spaces was deduced in Chapter 3.). As is clear from the following example, for problem (B) and stationary strategies the analogous result is not valid.

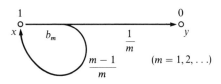

Figure 7.2

EXAMPLE 2. X consists of the two states x and y, y being absorbing (see figure 7.2). In x there is a countable collection of controls $b_1, b_2, \ldots, b_m, \ldots$. The transition function is given by the formulas

$$p(x\,|\,b_m) = \frac{m-1}{m},$$

$$p(y\,|\,b_m) = \frac{1}{m}.$$

The reward function is equal to 1 at state x and 0 at state y.

Considerations analogous to those given in Example 1 show that for any strategy π the asymptotic value $w(x,\pi)$ exists and is equal to the probability p of never leaving x.

Each stationary strategy φ is given by the choice of some action b_m. For such a strategy the probability p_t of remaining in state x through the first t steps is equal to $(1 - (1/m))^t$, so that

$$w(x,\varphi) - \lim_{t \to \infty} p_t = 0.$$

At the same time, nonstationary strategies make it possible to bring the mean reward p arbitrarily close to 1. Indeed, if the strategy π consists of choosing the action $b_{m(t)}$ at step t, then

$$w(x,\pi) = \prod_{t=1}^{\infty} \left[1 - \frac{1}{m(t)} \right] \geq 1 - \sum \frac{1}{m(t)} \quad {}^*$$

For $m(t) = 2^{k+t}$ we will have $w(x,\pi) \geq 1 - 2^{-k}$ for any integer k.

Thus, using stationary strategies we cannot get an outcome less than one unit away from the asymptotic value $v(x)$.

* * *

Examples 1 and 2 destroyed the analogy between problems (A) and (B). Other complications arising in infinite models are connected with the definition itself of the asymptotic value. In the general case we have to deal with the lower values

$$\underline{w}(\mu,\pi) = \underline{\lim} \frac{w^n(\mu,\pi)}{n}, \qquad \underline{v}(\mu) = \sup_{\pi} \underline{w}(\mu,\pi) \tag{2}$$

and the upper values

$$\overline{w}(\mu,\pi) = \overline{\lim} \frac{w^n(\mu,\pi)}{n}, \qquad \overline{v}(\mu) = \sup_{\pi} \overline{w}(\mu,\pi). \tag{3}$$

(In the case of the finite models, the values $\underline{v}$ and $\overline{v}$ coincided, and it was possible to consider only the strategies for which $\underline{w} = \overline{w}$.) The following two examples show that in the general case the values $\underline{v}$ and $\overline{v}$ may well be distinct, and in addition that some properties of values and strategies to which we have become accustomed are violated.

---

* The inequality $(1 - \alpha_1)(1 - \alpha_2) \dots (1 - \alpha_n) \geq 1 - \alpha_1 - \alpha_2 - \dots - \alpha_n$ for $\alpha_t \in [0,1]$ is easily proved by induction.

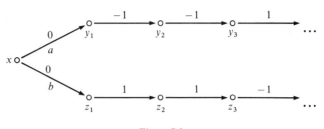

Figure 7.3

EXAMPLE 3. Suppose that X consists of the state x and two sequences $\{y_1, y_2, \ldots\}$, $\{z_1, z_2, \ldots\}$ of states (see figure 7.3). At x there are two actions a and b, leading to y_1 and z_1. From y_m we pass deterministically into y_{m+1}, and from z_m into z_{m+1}, $m = 1, 2, \ldots$; denote the action in each of these states by the same letter as the state. The reward function is given by

$$q(a) = q(b) = 0,$$

$$q(y_m) = \begin{cases} -1 & \text{if } 1 \leq m < m_1, \\ 1 & \text{if } m_1 \leq m < m_2, \\ -1 & \text{if } m_2 \leq m < m_3, \\ 1 & \text{if } m_3 \leq m < m_4, \\ \vdots \end{cases}$$

the sequence of integers $1 < m_1 < m_2 < \cdots$ having been chosen as to increase so rapidly that

$$m_1 + m_2 + \cdots + m_n = o(m_{n+1}) \quad \text{as } n \to \infty,$$

Clearly here

$$\underline{v}(y_m) = \underline{v}(z_m) = -1, \qquad \overline{v}(y_m) = \overline{v}(z_m) = 1, \tag{4}$$

so that $\underline{v} \neq \overline{v}$.

For the initial distribution μ with $\mu(y_1) = \mu(z_1) = \frac{1}{2}$ the rewards $q(y_m)$ and $q(z_m)$ will cancel each other out, and we get $w^n(\mu, \pi) = 0$ for each n. Accordingly

$$\underline{v}(\mu) = \overline{v}(\mu) = 0,$$

while in view of (4)

$$\mu\underline{v} = -1, \qquad \mu\overline{v} = 1.$$

Thus, here $\mu\underline{v} \neq \underline{v}(\mu)$, $\mu\overline{v} \neq \overline{v}(\mu)$.

For the initial state x both the simple strategies a and b have lower values equal to -1. However, the Markov strategy σ, consisting of the choice of a or b, each with probability $\frac{1}{2}$, leads to the same result $w(x, \sigma) = 0$ as the distribution μ.

Hence a Markov strategy may be essentially better than the simple strategies if we use the lower values*.

EXAMPLE 4. We alter example 3, excluding the state x and introducing at each state y_m and z_m one further action, leading as before to y_{m+1} and z_{m+1} respectively, but yielding for each m a payoff of -1. Then the strategy of always receiving the payoff -1 will, from the point of view of the lower values, be asymptotically optimal for each initial state, but it will not be optimal for the initial distribution $\mu(y_1) = \mu(y_2) = \frac{1}{2}$. If one replaces q here by $-q$, then $\bar{v}$ becomes equal to 1 for all initial distributions. Using a strategy φ which prescribes that one should take -1 wherever possible, we will have $\bar{w}(s,\varphi) = 1 = \bar{v}(s)$ for all initial states s, but $\bar{w}(\mu,\varphi) = 0 < 1 = \bar{v}(\mu)$ for that same initial distribution μ. Thus, also when one is dealing with the upper values, asymptotic optimality for all initial states does not guarantee asymptotic optimality for an arbitrary initial distribution.

$$* \quad * \quad *$$

What positive results for the problem (B) are nevertheless preserved in the infinite models?

In the investigation of finite models the basic tool was the canonical system

$$v = P_\psi v = V \Pi v,$$
$$r + v = T_\psi r = Tr.$$

(5)

We have proved that

a) *The triple (v,ψ,r) satisfies these equations if and only if for any strategy π and any n*

$$w_r^n(x,\pi) \le r(x) + nv(x) \le w_r^n(x,\psi^\infty).$$

(6)

The corresponding stationary strategy $\varphi = \psi^\infty$ is asymptotically optimal, and the function v is the asymptotic value of the model:

$$\bar{w}(\mu,\pi) \le \mu v = w(\mu,\psi)$$

(7)

for any initial distribution μ and any strategy π.

b) *The canonical equations have a solution, which can be found by Howard's improvement procedure.*

In §9 it will be proved that under certain conditions of boundedness and measurability result a) is valid for general models as well. That result will be extended also to the case of asymptotic ε-optimality.

In order to save result b) in some or other form, we need to impose conditions of quite another character on the model. One of these conditions will be considered in §10.

---

* These considerations are not applicable to the upper values. The upper values play the same rôle in the problem of minimizing w as the lower values do in the maximization problem.

§9. Canonical and ε-Canonical Triples and Systems for General Models

In order to pass to general models we need to introduce certain measurability and boundedness conditions, which in the finite case were fulfilled automatically. We shall suppose that *the one-point sets in the space X are measurable*, that there exists at least one measurable selector of the correspondence A(x) of X into A, that the running reward function q is bounded, and we will consider only triples* (v,φ,r) *with bounded measurable v and r, and* $\varphi = \psi^\infty$, *where* ψ *is a measurable selector.* With these restrictions, result a) of the preceding section is valid for general models. In order to avoid repetitions, we shall prove this result in a somewhat more general form.

Let ε be any positive number. The strategy σ will be called *asymptotically ε-optimal* if for any initial distribution μ and any strategy π

$$\overline{w}(\mu,\pi) \le \underline{w}(\mu,\sigma) + \varepsilon \tag{1}$$

We shall say that *the triple* (v,φ,r) *is ε-canonical* if for any $x \in X$ and all $n = 1, 2, \ldots$

$$w_r^n(x,\pi) \le r(x) + nv(x) \le w_r^n(x,\varphi) + n\varepsilon^- \tag{2}$$

for any strategy π. The system of relations

$$v = V\Pi v = P_\psi v, \tag{3}$$

$$Tr \le r + v \le T_\psi r + \varepsilon \tag{4}$$

will be called an *ε-canonical system*. (The operators entering into this system are given by the same formulas as in §2, except that the sums are replaced by integrals.[†] If $\varepsilon = 0$ we come back to the concepts introduced in §§1,2.) We shall prove the following result:

  a′) *If the triple* (v,φ,r) *satisfies the ε-canonical system, then that triple is ε-canonical; if* $\varepsilon = 0$ *the converse holds as well. If* (v,φ,r) *is ε-canonical, then the strategy* φ *is asymptotically ε-optimal and the function v, for any initial distribution* μ, *satisfies the inequalities*

$$\sup_\pi \overline{w}(\mu,\pi) \le \mu v \le \sup_\pi \underline{w}(\mu,\pi) + \varepsilon. \tag{5}$$

  Integrating all the terms of the inequality (2) with respect to the initial distribution μ, dividing by n and passing to the limit, we find that

$$\overline{w}(\mu,\pi) \le \mu v \le \underline{w}(\mu,\varphi) + \varepsilon;$$

---

* For this it suffices for example that the space X be Borelian (see Appendix 1).

[†] The equality of the two expressions in (2.7) for the operator T was proved at (6.7.7) under the hypothesis that the one-point sets were measurable.

in passing to the limit the bounded function r can here be replaced by zero. Hence we get both (1) and (5). It remains to prove that (2) follows from (3)–(4), and that for $\varepsilon = 0$ (3) and (4) follow conversely from (2).

Suppose that the system (3)–(4) is satisfied. We shall prove (2) by induction on n. For $n = 0$ conditions (2) are valid, since $w_r^0(x,\pi) = w_r^0(x,\varphi) = r(x)$. Suppose that (2) is satisfied for some n. In view of the fundamental equation and the induction hypothesis, for any strategy π

$$w_r^{n+1}(x,\pi) = \int_{A(x)} \pi(da|x)\left[q(a) + \int_X p(dy|a)w_r^n(y,\pi_a) \right]$$

$$\leq \int_{A(x)} \pi(da|x)\left[q(a) + \int_X p(dy|a)(r(y) + nv(y)) \right]$$

$$= \int_{A(x)} \pi(da|x)[q(a) + \Pi r(a) + n\Pi v(a)]$$

$$\leq V(q + \Pi r + n\Pi v)(x). \tag{6}$$

Taking into account the fact that the supremum of a sum does not exceed the sum of the suprema, and also the first parts of formulas (3)–(4), we have

$$V(q + \Pi r + n\Pi v) \leq V(q + \Pi r) + nV\Pi v = Tr + nv \leq r + (n+1)v.$$

Hence the left inequality in (2) is satisfied also for the value $n + 1$. Further, by the fundamental equation and the induction hypothesis, for the stationary strategy φ

$$w_r^{n+1}(x,\varphi) = q(\psi(x)) + \int_X p(dy|\psi(x))w_r^n(y,\varphi)$$

$$\geq q(\psi(x)) + \int p(dy|\psi(x))[r(y) + nv(y) - n\varepsilon]$$

$$= T_\psi r(x) + nP_\psi v(x) - n\varepsilon.$$

Using the second parts of formulas (3)–(4), we get

$$T_\psi r + nP_\psi v - n\varepsilon \geq r + v - \varepsilon + nv - n\varepsilon = r + (n+1)(v - \varepsilon),$$

so that the first inequality in (2) is valid for the value $n + 1$.

Finally, suppose that (2) is satisfied for $\varepsilon = 0$, i.e. for any strategy π,

$$w_r^n(x,\pi) \leq r(x) + nv(x) \leq w_r^n(x,\varphi),$$

$n = 1, 2, \ldots$, or, equivalently,

$$\sup_\pi w_r^n(x,\pi) = r(x) + nv(x) = w_r^n(x,\varphi), \tag{7}$$

$n = 1, 2, \ldots$. As in the finite case we wish to rewrite (7) in the form

$$T^n r = r + nv = T_\psi^n(r), \tag{8}$$

$n = 1, 2, \ldots$. The canonical system is derived from (8) in the same way as in §2. The equation

$$w_r^n(x,\varphi) = T_\psi^n r(x)$$

is the n-fold iteration of the fundamental equation for the stationary strategy φ, and is valid for general models as well. It remains to verify that (7) implies the formula

$$\sup_\pi w_r^n(x,\pi) = T^n r(x), \tag{9}$$

equivalent to the optimality equation (see Chapter 1, §9; for arbitrary general models we have not established the optimality equation).

Formula (9) is deduced from (7) by induction on n. For $n = 0$ it becomes the identity $r = r$. Suppose (9) is true for some n. Taking account of the left equation in (7), we may repeat the calculation (6), which shows that

$$w_r^{n+1}(x,\pi) \le V(q + \Pi r + n\Pi v)(x) = T(r + nv)(x).$$

By the induction hypothesis and by (7), $r + nv = T^n r$, so that

$$w_r^{n+1}(x,\pi) \le T^{n+1} r(x). \tag{10}$$

On the other hand, in view of the induction hypothesis and formula (7)

$$T^{n+1} r(x) = T w_r^n(x,\varphi) = \sup_\chi T_\chi w_r^n(x,\varphi).$$

But by the fundamental equation

$$T_\chi w_r^n(x,\varphi) = w_r^{n+1}(x,\sigma),$$

where σ is the simple strategy consisting in using the selector χ at the first step and thereupon the selector ψ. Therefore

$$T^{n+1} r(x) \le \sup_\pi w_r^{n+1}(x,\pi),$$

from which, along with (10), it follows that formula (9) is valid for the value $n + 1$ as well.

Result a') is completely proved.

§10. Models with Minorants

We consider the simplest of the conditions reestablishing an analogy between the problems (A) and (B) in infinite models*. We will suppose that *the spaces X and A in the model Z are Borelian.*

---

* For other conditions see for example Ross [2] and Gubenko and Štatland [1].

We shall say that the transition function p of the model Z *has a minorant* v if v is a measure on X such that $0 < v(X) < 1$ and $v(\Gamma) \leq p(\Gamma \mid a)$ for any action a and any measurable set Γ in X.

Put $\beta = 1 - v(x)$ and define a new transition function $\tilde{p}$ by the formula

$$\tilde{p}(\Gamma \mid a) = \frac{1}{\beta}[p(\Gamma \mid a) - v(\Gamma)].$$

Consider the model $\tilde{Z}$ gotten from Z by the replacement of p by $\tilde{p}$ and the introduction of the discount coefficient β (see Chapter 6, §1). Since the reward function q is bounded and $\beta < 1$, the model $\tilde{Z}$ is bounded and its value $\tilde{v}$ is a bounded function.

Now we shall prove that *if the value $\tilde{v}$ of the model $\tilde{Z}$ is measurable and the stationary strategy $\varphi = \psi^{\infty}$ is ε-optimal in the model $\tilde{Z}$, then the number $v\tilde{v}$, the strategy φ, and the function $\tilde{v}$ form an ε-optimal triple in the model Z.*

It suffices to verify that the triple $(v\tilde{v}, \varphi, \tilde{v})$ satisfies the ε-canonical system (9.3)–(9.4). Relation (9.3) is satisfied for any constant v and, in particular, for $v = v\tilde{v}$. In order to verify (9.4) we note that the corresponding operators in the models Z and $\tilde{Z}$ are connected by the formulas

$$\tilde{\Pi}f(a) = \int_X \tilde{p}(dx \mid a)f(x) = \frac{1}{\beta}\int_X [p(dx \mid a) - v(dx)]f(x)$$

$$= \frac{1}{\beta}\Pi f(a) - \frac{vf}{\beta}, \tag{1}$$

$$\tilde{T}_{\psi}f(x) = q(\psi(x)) + \beta\tilde{\Pi}f(\psi(x)) = q(\psi(x)) + \Pi f(\psi(x)) - vf$$

$$= T_{\psi}f(x) - vf, \tag{2}$$

$$\tilde{T}f(x) = V[q(a) + \beta\tilde{\Pi}f(a)] = V[q(a) + \Pi f(a) - vf]$$

$$= Tf(x) - vf; \tag{3}$$

(the definition of operators for models with discount are taken from (6.7.2)–(6.7.5)).

By the results of Chapter 6, §§6,7, the value $\tilde{v}$ of the model $\tilde{Z}$ and the ε-optimal strategy φ satisfy the conditions

$$\tilde{T}\tilde{v} = \tilde{v} \leq \tilde{T}_{\psi}\tilde{v} + \varepsilon.$$

Taking into account (2) and (3), we reduce them to the form

$$T\tilde{v} = \tilde{v} + v\tilde{v} \leq T_{\psi}\tilde{v} + \varepsilon,$$

i.e. relations (9.4) are satisfied for the triple $(v\tilde{v}, \varphi, \tilde{v})$.

Note that if in the model $\tilde{Z}$ a stationary ε-optimal strategy φ exists for any $\varepsilon > 0$, then the number $v = v\tilde{v}$ is the asymptotic value of the model Z (indeed (1.8) follows from the fact that (9.5) is satisfied for all $\varepsilon > 0$).

* * *

It is clear from (1) that if the model Z is semicontinuous, then the model $\tilde{Z}$ is semicontinuous as well. In a homogeneous semicontinuous model $\tilde{Z}$ with a bounded reward function and a discount coefficient $\beta < 1$ the value $\tilde{v}$ is measurable, and there exists a stationary optimal strategy. By what has been proved, this strategy will be canonical in the model Z. Thus, *if in a semicontinuous model the transition function has a minorant, then there exists a canonical (and therefore stationary, asymptotically optimal) strategy.*

* * *

If the model Z is countable, then the model $\tilde{Z}$ is countable as well. In a countable model $\tilde{Z}$ with a bounded reward function q and discount coefficient $\beta < 1$, for any $\varepsilon > 0$ there exists an ε-optimal stationary strategy. By what has been proved, this strategy is asymptotically ε-optimal for the model Z. Accordingly, *if in a countable model the transition function has a minorant, then for any $\varepsilon > 0$ there exists a stationary asymptotically ε-optimal strategy.* In addition the number $v = v\tilde{v}$ is equal to the asymptotic value of the model Z.

Note that in the countable case the existence of a minorant is equivalent to the following requirement *for some state y and some number $c > 0$ we have $p(y|a) \geq c$ for all $a \in A$.*

In the special case *when all the fibres $A(x)$ are finite,* one can choose a selector ψ such that $\tilde{T}_\psi \tilde{v} = \tilde{v}$. Hence under this additional assumption, *in a countable model with a minorant there exists a canonical strategy.*

§11. The Replacement Problem

As an example we consider the replacement problem (see Chapter 1, §§2,11, and Chapter 6, §5). In this countable model one can easily find a canonical triple, and consequently a stationary asymptotically optimal strategy. We note two special cases in which the existence of a canonical triple follows from the general results of this chapter. The first is the case of bounded service time (where $p_K = 0$ for some K). In this case we have a finite model with a state space $\{0,1,2,\ldots,K\}$. The second case is when the failure probability q_0 for new equipment is positive. Since

$$p(0|xd) \geq p(0|xc) = q_x \geq q_0 > 0,$$

the transition function in this case has a minorant, and we may make use of the concluding remark of §10.

The results of Chapter 1, §11 and Chapter 6, §5, compel us to suppose that the canonical strategy is given for some m by the selector

$$\psi(x) = \psi_m(x) = \begin{cases} c & \text{for } x < m, \\ d & \text{for } x \geq m. \end{cases} \tag{1}$$

Let us calculate the asymptotic average reward for the strategy $\varphi = \psi^\infty$. Under this strategy we hit the state 0 in no more than m steps, and from then on we remain in the finite set $\{0,1,2,\ldots,m\}$. This implies first that, $w(x) = w(0)$, so that $w(x)$ is a constant, and, secondly, that the results of §3 are applicable, and that the number w may be determined from the Howard system

$$w = P_\psi w,$$
$$f + w = T_\psi f \tag{2}$$

(see formulas (3.1)–(3.3)). Note that

$$T_\psi f(x) = \begin{cases} p_x h_x + p_x f(x+1) + q_x f(0) & \text{if } 0 \leq x < m, \\ \alpha + f(0) & \text{if } m \leq x, \end{cases} \tag{3}$$

so that the second of the equations (2) is equivalent to the system

$$\left.\begin{aligned} f(0) + w &= p_0 h_0 + p_0 f(1) + q_0 f(0), \\ f(1) + w &= p_1 h_1 + p_1 f(2) + q_2 f(0), \\ &\vdots \\ f(m-1) + w &= p_{m-1} h_{m-1} + p_{m-1} f(m) + q_{m-1} f(0), \\ f(m) + w &= \alpha + f(0), \end{aligned}\right\} \tag{4}$$

$$\left.\begin{aligned} f(m+1) + w &= \alpha + f(0). \\ &\vdots \end{aligned}\right. \tag{5}$$

(The first of equations (2) is satisfied for any constant w). If a function f satisfies the system (4)–(5), then so does any function f + constant.

Reflecting the dependence of ψ, w, and f on the index m (see (1)), we will now write ψ_m, w_m, and f_m. We may suppose that

$$f_m(0) = 0. \tag{6}$$

It follows from (5) that

$$f_m(m) = f_m(m+1) = f_m(m+2) = \cdots = \alpha - w_m. \tag{7}$$

Putting

$$L_x = p_0 p_1 \cdots p_x \qquad (L_{-1} = 1), \tag{8}$$

multiplying equations (4) in turn by $1, L_0, \ldots, L_{m-2}$ and summing them from 1 to $x - 1$, we get*

$$f_m(x) = \frac{(1 + L_0 + \cdots + L_{x-2})w_m - (L_0 h_0 + L_1 h_1 + \cdots + L_{x-1} h_{x-1})}{L_{x-1}} \qquad (9)$$

$x = 0, 1, \ldots, m$. Comparing the two values of $f(m)$ obtained in (7) and (9), we have

$$w_m = \frac{L_0 h_0 + L_1 h_1 + \cdots + L_{m-1} h_{m-1} + L_{m-1}\alpha}{1 + L_0 + L_1 + \cdots + L_{m-1}}. \qquad (10)$$

The system (2) of equations for the selector ψ_m is solved.

We suppose first that there exists a largest number among the numbers w_0, $w_1, \ldots, w_n, \ldots$. Suppose this is w_k. We shall show that $(v, \varphi, r) = (w_k, \psi_k^\infty, f_k)$ *is a canonical triple, and therefore that the stationary strategy* $\varphi = \psi_k^\infty$ *is asymptotically optimal.*

It is clear from (7) that the function f_k is bounded. In view of the results of §§2,9, it suffices to verify that the canonical equations (2.1)–(2.2) are satisfied. Equation (2.1) is valid for any constant v, in particular for $v = w_k$. The left equation in (2.2) coincides with the second equation in the Howard system (2), which means that it is also satisfied. It remains to verify the right equation in (2.2):

$$T_{\psi_k} f_k = T f_k. \qquad (11)$$

The operator T acts on functions satisfying (6) according to the formula

$$Tf(x) = \max[p_x h_x + p_x f(x + 1), \alpha].$$

Comparing this expression with (3) (where we now have $m = k$), we see that equation (11) reduces to the system

$$p_x h_x + p_x f_k(x + 1) \geq \alpha \qquad (0 \leq x < k), \qquad (12)$$

$$p_x h_x + p_x f_k(x + 1) \leq \alpha \qquad (k \leq x). \qquad (13)$$

of inequalities. It is not hard to deduce from (9) and (10) that the inequalities (12) are equivalent to the inequalities

$$w_k \geq w_x, \qquad x = 0, \ldots, k - 1. \qquad (14)$$

---

* We are supposing that all the probabilities p_x are nonzero, so that $L_{x-1} \neq 0$. If this were not so, the model would be in fact finite, and we would only need to consider numbers x and k not exceeding the first index K for which $p_K = 0$. In this case the final results are formulated in just the same way as for probabilities p_x which are not equal to zero.

It follows from (7) that (13) is equivalent to the system

$$w_k \geq h_x - \alpha \frac{1 - p_x}{p_x}, \qquad k = 1, \ldots, x. \tag{15}$$

Since $\alpha > 0$ and the functions h_x and p_x are nonincreasing, the system (15) reduces to the single inequality

$$w_k \geq h_k - \alpha \frac{1 - p_k}{p_k}.$$

An elementary calculation using formulas (8) and (10) shows that this inequality is equivalent to the condition

$$w_k \geq w_{k+1}. \tag{16}$$

But inequalities (14) and (16) are satisfied in view of the choice of the index k. Hence relations (12)–(13) hold as well, and our assertion is proved.

Moreover, we see that *the optimal number k may be defined as the first number m for which $w_{m+1} \leq w_m$.*

There remains the case when

$$w_m < w_{m+1} \quad \text{for all } m. \tag{17}$$

We shall show that in this case a canonical triple is given by the formulas

$$\begin{aligned} w_\infty &= \lim_{k \to \infty} w_k, \\ \psi_\infty(x) &= c \quad \text{for all } x, \end{aligned} \tag{18}$$

$$f_\infty(x) = \begin{cases} 0 & \text{if } x = 0, \\[2ex] \dfrac{(1 + L_0 + \cdots + L_{x-2})w_\infty - (L_0 h_0 + L_1 h_1 + \cdots + L_{x-1}h_{x-1})}{L_{x-1}} \\[2ex] & \text{if } x \geq 1^*. \end{cases} \tag{19}$$

For this it suffices to verify that

$$f_\infty + w_\infty = T_{\psi_\infty} f_\infty = T f_\infty \tag{20}$$

and that the function f_∞ is bounded.

---

* Under the hypothesis (17) all the L_x are nonzero, since, if $L_{x-1} > L_x = 0$, then, from (10),

$$w_x - w_{x+1} = \frac{L_{x-1}\alpha}{1 + L_0 + \cdots + L_{x-1}} > 0.$$

The first of the equations in (20) is obtained by passing to the limit in the corresponding equation for w_m, ψ_m, and f_m (see (2), or, in expanded form, (4)–(5)). The second equation in (20) is equivalent to the system of inequalities

$$p_x h_x + p_x f_\infty(x + 1) \geq \alpha \qquad (x = 0,1,2, \ldots);$$

(see the passage from (11) to (12)–(13)). In view of (10) and (19) these inequalities are equivalent to the inequalities

$$w_\infty \geq w_x \qquad (x = 0,1,2, \ldots).$$

The latter are true in view of (17) and (18).

In order to prove the boundedness of the function f_∞ we have to consider two cases: 1) among the numbers p_x, at least one is less than unity, and 2) $p_x = 1$ for all x.

In the first case, in view of the monotonicity of the p_x, the numbers L_x tend to 0 no more slowly than the terms of a convergent geometric series, and it follows from (10) that

$$w_\infty = \kappa/\lambda, \tag{21}$$

where

$$\kappa = \sum_0^\infty L_x h_x, \qquad \lambda = 1 + \sum_0^\infty L_x. \tag{22}$$

Putting these values into (20), we have, after some simplifications,

$$f_\infty(x) = \frac{1}{\lambda L_{x-1}}[(1 + L_0 + L_1 + \cdots + L_{x-2})(L_x h_x + L_{x+1} h_{x+1} + \cdots)$$
$$- (L_0 h_0 + L_1 h_1 + \cdots + L_{x-1} h_{x-1})(L_{x-1} + L_x + \cdots)]. \tag{23}$$

It follows from (8) and the monotonicity of p_x that

$$L_{x+y} \leq L_y L_{x-1} \qquad (y \geq 0).$$

Moreover, by the hypotheses of the problem,

$$h_{x+y} \leq h_x \qquad (y \geq 0).$$

Therefore

$$L_x h_x + L_{x+1} h_{x+1} + \cdots \leq L_{x-1} \kappa,$$
$$L_{x-1} + L_x + \cdots \leq \lambda L_{x-1},$$

and it follows from (23) that

$$|f_\infty(x)| \leq 2\kappa.$$

In the second case $L_x = 1$ for all x and formula (10) reduces to

$$w_m = \frac{h_0 + h_1 + \cdots + h_{m-1} + \alpha}{m}. \tag{24}$$

Since h_x is a nonincreasing function of x, the limit

$$h_\infty = \lim_{x \to \infty} h_x$$

exists. It follows from (24) that

$$w_\infty = h_\infty.$$

In addition, we get from (20) the expression

$$f_\infty(x) = (x - 1)h_\infty - (h_0 + h_1 + \cdots + h_{x-1})$$
$$= -[(h_0 - h_\infty) + (h_1 - h_\infty) + \cdots + (h_{x-1} - h_\infty)]. \tag{25}$$

The condition (18) for the values of w_m defined by formula (24) reduces to

$$(h_0 - h_m) + (h_1 - h_m) + \cdots (h_{m-1} - h_m) < h_m - \alpha.$$

Passing to the limit as $m \to \infty$, we get*

$$\sum_{x=0}^{\infty} (h_x - h_\infty) \le h_\infty - \alpha.$$

Thus the series at hand converges and the partial sums appearing in (25) are finite.

§12. The Stabilization Problem

In this problem (see Chapter 2, §11 and Chapter 6, §12), the reward function q is unbounded below, and therefore the results of §9 are inapplicable. However in Chapter 2, §11 we saw that the optimal strategy over the time interval $[0,n]$ passed as $n \to \infty$ into a stationary strategy $\varphi = \psi^\infty$ generated by the selector

$$a = \psi(x) = \frac{1 + b}{1 + b + c} x \qquad (-\infty < x < \infty), \tag{(1)}$$

---

* The legitimacy of this operation follows from a general lemma on the passage to the limit under the sign of an infinite sum: if $c_x(m) \ge 0$ and $c_x(m) \uparrow c_x(\infty)$ as $m \to \infty$, then $\sum_{x=0}^{\infty} c_x(m) \uparrow \sum_{x=0}^{\infty} c_x(\infty)$. This lemma is analogous to the well-known theorem on the monotone passage to the limit under the integral sign. In order to apply this lemma to the series of interest to us, we need to put $c_x(m) = h_x - h_m$ for $x < m$ and $c_x(m) = 0$ for $x \ge m$.

where l is the positive root of the quadratic equation

$$l^2 + bl - bc = 0. \tag{2}$$

We get the same strategy if we pass to the limit as $\beta \uparrow 1$ in formulas (6.12.4)–(6.12.5). It is reasonable to expect that this strategy is asymptotically optimal. We shall show that this is the case.

Fix an initial distribution μ. We will prove that

$$\lim_{n \to \infty} \frac{w^n(\mu,\varphi)}{n} = \lim_{n \to \infty} \frac{v^n(\mu)}{n}, \tag{3}$$

where $v^n(\mu)$ is the value of the initial distribution μ, on the control interval $[0,n]$. The asymptotic optimality of the strategy φ follows from equation (3): for any strategy π we have $w^n(\mu,\pi) \le v^n(\mu)$, so that inequality (1.7) is satisfied.

By Chapter 2, §11,

$$v^n(x) = -l_n x^2 - m_n, \tag{4}$$

while

$$\lim l_n = l, \tag{5}$$

$$m_n = \sigma^2(l_0 + l_1 + \cdots + l_n). \tag{6}$$

In view of (4)

$$v^n(\mu) = -l_n \alpha - m_n,$$

where

$$\alpha = \int_{-\infty}^{+\infty} x^2 \mu(dx),$$

and we have

$$\lim_{n \to \infty} \frac{v^n(\mu)}{n} = \begin{cases} -\sigma^2 l, & \text{if } \alpha < \infty, \\ -\infty, & \text{if } \alpha = \infty. \end{cases} \tag{7}$$

We will now calculate $w^n(x,\varphi) = T_\psi^n 0(x)$. From formulas (6.12.11)–(6.12.12) for $\beta = 1$, we have

$$w^n(x,\varphi) = -L_n x^2 - M_n, \tag{8}$$

where

$$L_0 = 0, \tag{9}$$

$$L_{n+1} = \frac{(L_n + b)c^2 + (l + b)^2 c}{(l + b + c)^2}, \tag{10}$$

$$M_0 = 0, \tag{11}$$

$$M_{n+1} = \sigma^2 L_n + M_n. \tag{12}$$

Subtracting term by term from equation (10) the analogous relation between L_n and L_{n-1}, we get

$$L_{n+1} - L_n = \frac{c^2}{(l + b + c)^2}(L_n - L_{n-1}).$$

The coefficient on the right is less than 1, so that the differences $L_{n+1} - L_n$ decrease sufficiently rapidly that the sequence L_n has a finite limit L. Passing to the limit in equation (10), we get

$$L = \frac{(l + b)^2 c + bc^2}{(l + b + c)^2 - c^2}.$$

At the end of Chapter 6, §12, we showed that the last expression was equal to l. Hence

$$\lim_{n \to \infty} L_n = l. \tag{13}$$

From (11) and (12) we have

$$M_n = \sigma^2(L_0 + L_1 + \cdots + L_{n-1}). \tag{14}$$

It follows from (8) and (13)–(14) that

$$\lim_{n \to \infty} \frac{w^n(\mu, \varphi)}{n} = \begin{cases} -\sigma^2 l & \text{for } \alpha < \infty, \\ -\infty & \text{for } \alpha = \infty. \end{cases} \tag{15}$$

Now (3) follows from (7) and (15).

Finally note that in view of (3) and (7) the asymptotic value of the model is equal to

$$v(x) = -\sigma^2 l. \tag{16}$$

§13. Models with Finitely Many States and Infinite Action Sets

This section is devoted to the proof of the following result: *in a semicontinuous model Z with a finite space X, for any $\varepsilon > 0$ there exists a stationary strategy which is asymptotically optimal in the sense of lower values,* i.e., such that, for all x of X,

$$w(x, \varphi) = \underline{w}(x, \varphi) \geq \underline{v}(x) - \varepsilon. \tag{1}$$

Example 1 of §8 shows that for $\varepsilon = 0$ this is not so. Lower values are quite natural in the problem of maximization of the reward: they estimate the guaranteed average reward per unit time for any sufficiently extended interval of time.

Making use of the definition of a semicontinuous model in §4 of Chapter 2, we may characterize the class of models of interest to us by the following conditions:

(A) The space X is finite.
(B) The fibres $A(x)$ are compact metric spaces.
(C) The transition probability $p(x|a)$ from a into x is a continuous function of a.
(D) The reward $q(a)$ is bounded above and semicontinuous.

In preparation we consider some properties of values of stationary strategies, valid *in any model Z with finite state space X and a reward q which is bounded above.* The analysis of such models, if one fixes the strategy $\varphi = \psi^\infty$, where ψ is a selector of the correspondence $A(x)$, differs in no way from the analysis of finite models presented in §3. In particular, putting

$$P_\psi(x, y) = p(y \,|\, \psi(x)) \quad \text{and} \quad q_\psi(x) = q(\psi(x)),$$

we have

$$w(\varphi) = w(\varphi) = M_\psi q_\psi, \tag{2}$$

where

$$M_\psi = \lim_{n \to \infty} \frac{1}{n}(E + P_\psi + P_\psi^2 + \cdots + P_\psi^{n-1}) \tag{3}$$

Put

$$v^s(x) = \sup_\varphi w(x, \varphi) \qquad (x \in X), \tag{4}$$

where φ runs through the class of all stationary strategies (since q is bounded above, v^s is finite). In (4) we may replace w by $\underline{w}$, so that

$$v^s \leq v \tag{5}$$

(equality does not always hold here—see Example 2 of §8). We shall show that:

(a) *For any $\varepsilon > 0$ there exists a stationary strategy φ such that*

$$w(\varphi) \geq v^s - \varepsilon. \tag{6}$$

(b) *The vector v^s satisfies the condition*

$$v^s = V\Pi v^s. \tag{7}$$

(cf. the canonical equation 2.1).

By the definition of v^s, for any state x there exists a selector ψ_x of the correspondence $A(x)$ such that

$$w(x, \varphi_x) \geq v^s(x) - \varepsilon, \tag{8}$$

where $\varphi_x = \psi_x^\infty$. The collection ψ_x $(x \in X)$ of selectors uses from the whole action space A only some finite subset $\tilde{A}$. Cutting A down to $\tilde{A}$, we obtain from Z a finite model $\tilde{Z}$. From what was proved in §§1–7, the model $\tilde{Z}$ has a stationary asymptotically optimal strategy φ. For that strategy $w(x, \varphi) \geq w(x, \varphi_x)$ for all x. Assertion (a) follows from this and (8).

In order to prove (b), we consider two arbitrary selectors, ψ and χ, of the correspondence $A(x)$. Applying the fundamental equation to the value of the strategy $\chi\psi^\infty$ on the control interval $[0, n]$, we get

$$w^n(\chi\psi^\infty) = q_\chi + P_\chi w^{n-1}(\psi^\infty).$$

After division by n and a passage to the limit as $n \to \infty$, we obtain for the asymptotic values the equality

$$w(\chi\psi^\infty) = P_\chi w(\psi^\infty) \tag{9}$$

(that the limit on the right exists follows from (3), and hence the limit on the left exists as well). As above, the pair χ, ψ of selectors distinguishes a finite submodel $\tilde{Z}$ in Z. This $\tilde{Z}$ has a stationary asymptotically optimal strategy σ, and we have $w(\chi\psi^\infty) \leq w(\sigma) \leq v^s$. Therefore it follows from (9) that

$$P_\chi w(\varphi) \leq v^s \tag{10}$$

where $\varphi = \psi^\infty$ is any stationary strategy. Taking as φ the strategy constructed in (a), we get the inequality $P_\chi v^s \leq v^s + \varepsilon$ from (6) and (10). Hence $P_\chi v^s \leq v^s$ and

$$V\Pi v^s = \sup_\chi P_\chi v^s \leq v^s.$$

To obtain the reverse inequality we put $\chi = \psi$ in (9). Then we get $w(\varphi) = P_\psi w(\varphi) \leq P_\psi v^s \leq V\Pi v^s$ so that

$$v^s(x) = \sup_\varphi w(x,\varphi) \leq V\Pi v^s(x).$$

Assertion (b) is proved.

Let Z be any (nontrivial) model with reward q which is bounded above, and let Z_β be the model obtained from Z by introducing the discount coefficient β. Then

$$\underline{v} \leq v^\mathcal{D}, \tag{11}$$

where

$$v^\mathcal{D}(x) = \varlimsup_{\beta \uparrow 1} (1 - \beta)v_\beta(x) \tag{12}$$

and v_β is the value of Z_β.

This result follows easily from the following Tauberian theorem.*

Theorem. *For any sequence $\{c_t\}$ of real numbers, having an upper bound,*

$$\varlimsup_{n \to \infty} \frac{1}{n} \sum_{t=1}^{n} c_t \leq \varlimsup_{\beta \uparrow 1} (1 - \beta) \sum_{t=1}^{\infty} c_t \beta^{t-1}.$$

(The sum on the right hand side, which is either finite or equal to $-\infty$ has a meaning for all $\beta \in (0,1)$.)

Applying the theorem to the numbers $c_t = P_x^\pi q(a_t)$ we get

$$\underline{w}(x,\pi) \leq \varlimsup_{\beta \uparrow 1} (1 - \beta)w_\beta(x,\pi), \tag{13}$$

where w_β is the value of the strategy in the model Z_β. Inequality (13) only becomes stronger if we replace w_β by v_β and in view of the arbitrariness of π we get (11).

* See Widder [1]

By (6), to prove the basic result of this section we need to show that $v^s = v$. By virtue of (5) and (11) it suffices to verify that *under conditions* (A)—(D)

$$v^{\mathcal{D}} \leq v^s \tag{14}$$

in fact, in view of (5) we have equality both here and in (11).

By the results of §7 of Chapter 6, in a semicontinuous model Z_β with finite state space X and coefficient $\beta < 1$, there exists a stationary optimal strategy $\varphi_\beta = \psi_\beta^\infty$. By (6.5)

$$v_\beta = w_\beta(\varphi_\beta) = N_\beta q_{\psi_\beta} \tag{15}$$

where

$$N_\beta = \sum_{t=0}^{\infty} \beta^t P_{\psi_\beta}^t \tag{16}$$

All the elements of the matrix $(1 - \beta)N_\beta$ lie in the segment $[0,1]$. Therefore, from conditions (A) and (B), there is a sequence $\beta \uparrow 1$ such that the limits

$$\psi = \lim \psi_\beta, \qquad N = \lim(1 - \beta)N_\beta \geq 0 \tag{17}$$

exist (throughout what follows β takes on values only from this sequence). It follows from (C) and (D) that

$$P_\psi = \lim P_{\psi_\beta}, \qquad q_\psi \geq \overline{\lim}\, q_{\psi_\beta} \tag{18}$$

It follows from (12), (15), and (18) that

$$v^{\mathcal{D}} \leq N q_\psi. \tag{19}$$

It is clear from (16) that

$$N_\beta P_{\psi_\beta} = \frac{1}{\beta}(N_\beta - E),$$

E being the identity matrix. Multiplying both sides by $(1 - \beta)$ and taking account of (17) and (18), we find in the limit that $NP_\psi = N$. Hence $NP_\psi^t = N$ for any integer t, so that $NM_\psi = N$ (see (3)). Using this equation and formulas (2) and (19), we get

$$v^{\mathcal{D}} \leq NM_\psi q_\psi = Nw(\psi^\infty) \leq Nv^s. \tag{20}$$

Finally, it follows from (7) that $P_\chi v^s \leq v^s$ for any selector χ. In particular $P_{\psi_\beta} v^s \leq v^s$, so that, by induction $P_{\psi_\beta}^t v^s \leq v^s$, whence

$$N_\beta v^s = \sum_{t=0}^{\infty} \beta^t P_{\psi_\beta}^t v^s \leq \sum_{t=0}^{\infty} \beta^t v^s = \frac{v^s}{1 - \beta}.$$

It follows from this and (17) that $Nv^s \leq v^s$. Along with (20), this yields the desired inequality (14).

Part III
Some Applications

Chapter 8

Models with Incomplete Information

§1. Description of the Model

Up to this point we have been supposing that we observe completely the trajectory of the controlled process:

$$x_m \xrightarrow{a_{m+1}} x_{m+1} \xrightarrow{a_{m+2}} \cdots x_{t-1} \xrightarrow{a_t} x_t \to \cdots . \tag{1}$$

Now we suppose that *the state of the system at the time t is described by a pair $x_t y_t$, the first of these components becoming known to us and the second not.* Thus, the real course of the process is given by the trajectory

$$x_m y_m \xrightarrow{a_{m+1}} x_{m+1} y_{m+1} \xrightarrow{a_{m+2}} \cdots x_{t-1} y_{t-1} \xrightarrow{a_t} x_t y_t \to \cdots , \tag{2}$$

of which we observe, as before, the chain (1). The actions a_t and the observed states x_{t-1} are connected as before by a projection j.

The unobserved states y_t are elements of some sets Y_t. They influence both the transition mechanism into the next state and the resulting reward. The transition function p now gives a probability distribution for the state $x_t y_t$ of the product space $X_t \times Y_t$, depending on y_{t-1} and a_t (inasmuch as $x_{t-1} = j(a_t)$, the introduction of the additional argument x_{t-1} would not give anything new). The running reward q on the tth stage depends on the same pair $y_{t-1} a_t$. The final reward at time n depends on the pair $x_n y_n$*. Here we intend that the reward $q(y_{t-1} a_t)$ at the tth step should be paid out at the end of the control process. If we were to receive this reward immediately, then its numerical value would give us additional information on the unobserved states of the system, and the elements of the model would have to be reconstructed so as to include the number $q(y_{t-1} a_t)$ in the observed state x_t[†].

---

* A more general case, when the reward on the tth step depends on the elements $x_{t-1} y_{t-1} a_t x_t y_t$, reduces to the case we are considering by the introduction of a new payoff equal to the mathematical expectation of $q(x_{t-1} y_{t-1} a_t x_t y_t)$ relative to the distribution $p(\cdot | y_{t-1} a_t)$.

[†] And then we would have a special case of the situation of which we spoke in the previous footnote: $q(x_{t-1} y_{t-1} a_t x_t y_t) = q(x_t)$.

In order to define a measure on the space of trajectories, it is necessary to give the initial distribution μ and the strategy π.

The rôle of the initial distribution μ here is somewhat different from what it was in the models with complete information. In taking μ to be known, we thus suppose known as well the probability distribution for the unobserved initial state y_m (although we are not given the value of y_m). In mathematical statistics one distinguishes the Bayesian approach, in which an "apriori" probability distribution for the unknown parameter y is introduced, and the minimax approach, in which the statistical decisions are evaluated according to the "worst" of the possible values of y. In supposing that μ is known, we choose the Bayesian approach.

The strategy π cannot depend on the unobserved values $y_m, y_{m+1}, \ldots$. However it can take account, besides the observed states $x_m, x_{m+1}, \ldots$ and the actions $a_{m+1}, a_{m+2}, \ldots$ already used, of the initial distribution as well. Inasmuch as the value of x_m becomes known to us, for the choice of the actions it is not the joint distribution μ of the initial pair $x_m y_m$ that is essential for us, but rather the conditional distribution v_m for y_m given the observed x_m. We include the distribution v_m into the observed history on which the next action depends. The pair $x_m v_m$ plays the rôle of the initial state. Here x_m is any point of the space X_m and v_m is any probability measure on Y_m.

> The pairs x_m, μ and x_m, v_m are closely connected, but one can not express one in terms of the other uniquely. It is more convenient to deal with the second pair.

The values $w(\mu, \pi)$ and $v(\mu)$ are defined in the usual way in terms of the measure P_μ^π. The statement of the problem of optimal control does not change.

We note that if each of the spaces Y_t consists of one point, then we obtain the complete information model presented in the preceding chapters.

§2. Reduction to a Model with Complete Information. The Finite Case

For each model with incomplete information we shall construct a model with complete information in such a way that the values of the corresponding strategies coincide. Then we may apply the results of the preceding chapters and obtain theorems on the existence of optimal strategies in models with incomplete information.

Our plan consists in introducing new state spaces, taking as the state at the time t all the information at our disposal necessary for the further action. At the initial time m this information is described by the observed state x_m and the *apriori* distribution v_m for the unobserved state y_m. At any time $t > m$, it is natural to describe it by the pair $x_t v_t$, where v_t is the "*a posteriori*" probability distribution for the state y_t, taking account of the entire observed history.

We begin with the case when all the spaces X_t, Y_t are finite. In this case the probability of a chain $l = x_m y_m a_{m+1} x_{m+1} y_{m+1} \cdots a_n x_n y_n$ is defined by the formula*

$$P_\mu^\pi(l) = \mu(x_m y_m)\pi(a_{m+1} | x_m v_m)p(x_{m+1} y_{m+1} | y_m a_{m+1})$$
$$\cdots \pi(a_n | x_m v_m a_{m+1} x_{m+1} \cdots a_{n-1} x_{n-1})p(x_n y_n | y_{n-1} a_n), \tag{1}$$

where μ is the initial distribution, π is the applied strategy, and the distribution v_m is computed from the formula

$$v_m(y_m) = v_m(y_m | x_m) = \frac{\mu(x_m y_m)}{\sum_{z \in Y_m} \mu(x_m z)} \tag{2}$$

if the denominator is not 0, and v_m is an arbitrary probability measure on Y_m, if the denominator is zero.

The construction of the auxiliary model with complete information begins with the choice of the state spaces $\tilde{X}_t$. Put $\tilde{X}_t = X_t \times N_t$, where N_t is the set of all probability measures on the set Y_t.

The actions in the new model remain as before. Now the same action a_t is possible for different states $\tilde{x}_{t-1} = x_{t-1} v_{t-1}$, differing in the distributions v_{t-1}. If we wish the fibres $\tilde{A}(x)$ not to intersect, then we have to consider a pair $v_{t-1} a_t = \tilde{a}_t$ as an action (see the analogous remark in §2 of Chapter 1).

In order to construct the new transition function $\tilde{p}$, we need to assign to each pair $v_{t-1} a_t$ a probability distribution on the space $X_t \times N_t$. The original transition function defines a distribution on the space $X_t \times Y_t$, as a function of $y_{t-1} a_t$. It is natural to assign to this pair a distribution on the space $X_t \times Y_t$ given by the formula

$$\tilde{p}(x_t y_t | v_{t-1} a_t) = \sum_{y_{t-1} \in Y_{t-1}} p(x_t y_t | y_{t-1} a_t)v_{t-1}(y_{t-1}). \tag{3}$$

This distribution splits into a distribution on X_t and a conditional distribution Y_t:

$$\tilde{p}(x_t y_t | v_{t-1} a_t) = \tilde{p}(x_t | v_{t-1} a_t)v_t(y_t | v_{t-1} a_t x_t); \tag{4}$$

here

$$\tilde{p}(x_t | v_{t-1} a_t) = \sum_{y_t \in Y_t} \tilde{p}(x_t y_t | v_{t-1} a_t) \tag{5}$$

and

$$v_t(y_t | v_{t-1} a_t x_t) = \frac{\tilde{p}(x_t y_t | v_{t-1} a_t)}{\tilde{p}(x_t | v_{t-1} a_t)}; \tag{6}$$

if the denominator in (6) vanishes, we take as v_t the arbitrary fixed measure v_t^0 on Y_t. For fixed $v_{t-1} a_t$ formula (6) defines a mapping $v_t = F(x_t)$ of X_t into N_t. Consider the probability distribution of the point $x_t F(x_t) \in X_t \times N_t$ which corresponds to the distribution of x_t given by (5). We thus have a probability distribution on the

---

* In the definition of the strategy π we have to add the requirement of measurability in the argument of v_m.

space $X_t \times N_t = \tilde{X}_t$, depending on $v_{t-1}a_t = \tilde{a}_t$, i.e. a transition function $\tilde{p}$ of $\tilde{A}_t$ into $\tilde{X}_t$.

According to our plan, the distributions v_t ought to be "*a posteriori*" distributions for y_t, taking account of all the observations made up to the time t. In other words, the formula

$$v_t(y) = P^\pi_\mu\{ y_t \mid x_m a_{m+1} x_{m+1} \cdots a_t x_t\}$$

$$= \frac{P^\pi_\mu(x_m a_{m+1} x_{m+1} \cdots a_t x_t y_t)}{\sum_{z \in Y_t} P^\pi_\mu(x_m a_{m+1} x_{m+1} \cdots a_t x_t z)}$$

has to be satisfied. When $t = m$ this is true in view of formula (2), and if $t > m$ it is verified by induction using (1) and (3)–(6).

The new running reward is given by the formula

$$\tilde{q}(v_{t-1}a_t) = \sum_{y_{t-1} \in Y_{t-1}} q(y_{t-1}a_t)v_{t-1}(y_{t-1}), \tag{7}$$

and the new final reward by the formula

$$\tilde{r}(x_n v_n) = \sum_{y_n \in Y_n} r(x_n y_n)v_n(y_n). \tag{8}$$

* * *

Starting from model Z with incomplete information, we have constructed a new model $\tilde{Z}$ with complete information, in which the state and action spaces are uncountable. Let us show that the model $\tilde{Z}$ is semicontinuous (see Chapter 2, §4).

A probability measure v on a space of s points is described by the choice of s nonnegative numbers adding to 1. This is a bounded closed set in s-dimensional coordinate space and is accordingly compact. Therefore all the spaces $\tilde{X}_t = X_t \times N_t$ and $\tilde{A}_t = N_{t-1} \times A_t$ are compact, hence condition 2.4.A is satisfied.

Now we shall verify the quasi-continuity of the correspondence $\tilde{x} \to \tilde{A}(\tilde{x})$ (condition 2.4.B). Suppose that the sequence of states $\tilde{x}_n = x_n v_n$ converges to a state $\tilde{x} = xv$ and that the actions $\tilde{a}_n$ belong to the fibres $\tilde{A}(\tilde{x}_n)$. Then $x_n \to x$ and $v_n \to v$. As to the $\tilde{a}_n$, we have $\tilde{a}_n = v_n a_n$, where $a_n \in A(x_n)$. Because the whole action space A is finite, infinitely many of the a_n must be equal to the same element $a \in A$, i.e. $a_{n_k} = a$ on some subsequence $\{n_k\}$. Evidently the sequence $\tilde{a}_{n_k} = v_{n_k} a$ converges to $va \in \tilde{A}(\tilde{x})$.

Condition 2.4.C requires that the transition function $\tilde{p}$ carry continuous bounded functions f on $\tilde{X}_t$ into continuous functions g on $\tilde{A}_t$ (see also 2.4.C'). For the transition function $\tilde{p}$ constructed above we have

$$g(v_{t-1}a_t) = \sum_{x_t \in X_t} f(x_t v_t)\tilde{p}(x_t \mid v_{t-1}a_t), \tag{9}$$

where the measures v_t are calculated according to formula (6). Since the sets A_t and X_t are finite, we need only verify that each term of the sum (9) is continuous in v_{t-1}. What we have is the product of two functions, of which the second, $\tilde{p}(x_t | v_{t-1} a_t)$, is continuous everywhere (see (5) and (3)), and the first, $f(x_t v_t)$, is bounded and continuous everywhere where the second is different from 0 (see (6)). Clearly such a product is a continuous function.

The continuity and boundedness of the reward functions (condition 2.4.D) are clear from formulas (7) and (8).

If the model Z is time homogeneous, then the model $\tilde{Z}$ is also homogeneous.

$$* \quad * \quad *$$

The initial distribution μ in the model Z splits into a distribution in X_m and a conditional distribution in Y_m:

$$\mu(x_m y_m) = \mu(x_m) v_m(y_m | x_m), \tag{10}$$

where

$$\mu(x_m) = \sum_{y_m \in Y_m} \mu(x_m y_m), \tag{11}$$

and v_m is found according to formula (2). Formula (2) establishes a mapping $v_m = F(x_m)$ of X_m into N_m. Denote by $\tilde{\mu}$ the probability distribution of the point $x_m F(x_m) \in \tilde{X}_m = X_m \times N_m$ which corresponds to the distribution of x_m given by (11). The measure $\tilde{\mu}$ may be considered as the initial distribution in the model $\tilde{Z}$ corresponding to the initial distribution μ in the model Z.

Given an arbitrary history $h = x_m v_m a_{m+1} x_{m+1} a_{m+2} x_{m+2} \cdots a_t x_t$ in the model Z, we may calculate $v_{m+1}, v_{m+2}, \ldots v_t$ recurrently from (6) and obtain the corresponding history $\tilde{h} = x_m v_m a_{m+1} x_{m+1} v_{m+1} \cdots a_t x_t v_t$ for the model $\tilde{Z}^*$). This allows us to assign to each strategy π in the model $\tilde{Z}$ a strategy $\tilde{\pi}$ in the model Z in the following way: we obtain a probability distribution $\tilde{\pi}(\cdot | h)$ for the next action by substituting into $\pi(\cdot | \tilde{h})$ the value of $\tilde{h}$ constructed above starting from h. It is clear that one can obtain in this way any strategy $\tilde{\pi}$ in the model Z: it suffices to put $\pi(\cdot | x_m v_m a_{m+1} x_{m+1} v_{m+1} \cdots a_t x_t v_t) = \tilde{\pi}(x_m v_m a_{m+1} x_{m+1} \cdots a_t x_t)$, dropping the arguments $v_{m+1} \cdots v_t$ in the right hand side.

In order to reduce the control problem in the model Z with incomplete information to the analogous problem for the model $\tilde{Z}$, we need to show that the value $w(\mu, \tilde{\pi})$ of the strategy $\tilde{\pi}$ in the model Z coincides with the value $\tilde{w}(\tilde{\mu}, \pi)$ of the strategy π in the model $\tilde{Z}$. For this it suffices to verify that

$$P_{\tilde{\mu}}^{\pi} \tilde{q}(v_t a_{t+1}) = P_{\mu}^{\tilde{\pi}} q(y_t a_{t+1}), \qquad P_{\tilde{\mu}}^{\pi} \tilde{r}(x_n v_n) = P_{\mu}^{\tilde{\pi}} r(x_n y_n). \tag{12}$$

---

* Formally we should write v_s two times when $s < t$, as a component of the state $\tilde{x}_s$ and as a component of the action $\tilde{a}_{s+1}$.

Both of these formulas follow from the following general fact: *for any function f, any initial distribution μ, in the model Z and any strategy π in the model $\tilde{Z}$*

$$P^{\tilde{\pi}}_{\mu} f(h_t y_t a_{t+1}) = P^{\pi}_{\tilde{\mu}} \tilde{f}(h_t v_t a_{t+1}), \tag{13}$$

where $h_t = x_m v_m a_{m+1} \cdots a_t x_t$ is the observed history at the time t and

$$\tilde{f}(h_t v_t a_{t+1}) = \sum_{y_t \in Y_t} f(h_t y_t a_{t+1}) v_t(y_t). \tag{14}$$

Let us prove (13). According to formula (1.3.2)

$$P^{\pi}_{\tilde{\mu}}(x_m v_m a_{m+1} x_{m+1} v_{m+1} a_{m+2} \cdots x_t v_t a_{t+1})$$
$$= \tilde{\mu}(x_m v_m) \pi(a_{m+1} | x_m v_m) \tilde{p}(x_{m+1} v_{m+1} | v_m a_{m+1})$$
$$\times \pi(a_{m+2} | x_m v_m a_{m+1} x_{m+1} v_{m+1}) \cdots \tilde{p}(x_t v_t | v_{t-1} a_t)$$
$$\times \pi(a_{t+1} | x_m v_m a_{m+1} x_{m+1} v_{m+1} \cdots a_t x_t v_t)^*. \tag{15}$$

It follows from the definitions of $\tilde{\mu}$ and $\tilde{p}$ that this probability is equal to 0 except in the case when the measure v_m is the function of μ and x_m given by formula (2) and the v_s for $s > m$ are the functions of v_{s-1}, a_s and x_s given by formula (6). By the definition of the strategy $\tilde{\pi}$, for such "admissible" chains one can rewrite (15) in the form

$$P^{\pi}_{\tilde{\mu}}(x_m v_m a_{m+1} x_{m+1} v_{m+1} a_{m+2} \cdots x_t v_t a_{t+1})$$
$$= \mu(x_m) \tilde{\pi}(a_{m+1} | x_m v_m) \tilde{p}(x_m | v_m a_{m+1}) \tilde{\pi}(a_{m+2} | x_m v_m a_{m+1} x_{m+1})$$
$$\cdots \tilde{p}(x_t | v_{t-1} a_t) \tilde{\pi}(a_{t+1} | x_m v_m a_{m+1} x_{m+1} \cdots a_t x_t). \tag{16}$$

Compare this with the formula

$$P^{\tilde{\pi}}_{\mu}(x_m y_m a_{m+1} x_{m+1} y_{m+1} a_{m+2} \cdots x_t y_t a_{t+1})$$
$$= \mu(x_m y_m) \tilde{\pi}(a_{m+1} | x_m v_m) p(x_{m+1} y_{m+1} | y_m a_{m+1})$$
$$\times \tilde{\pi}(a_{m+2} | x_m v_m a_{m+1} x_{m+1}) \cdots p(x_t y_t | y_{t-1} a_t)$$
$$\times \tilde{\pi}(a_{t+1} | x_m v_m a_{m+1} x_{m+1} \cdots a_t x_t), \tag{17}$$

which follows from (1) (here v_m is also the function of μ and x_m given by (2)). Making use of (16) and (17), we shall prove formula (13) by induction on t. For $t = m$ we need to show that

$$P^{\tilde{\pi}}_{\mu} f(x_m v_m y_m a_{m+1}) = P^{\pi}_{\tilde{\mu}} \sum_{y_m \in Y_m} f(x_m v_m y_m a_{m+1}) v_m(y_m). \tag{18}$$

* Formally we ought to have written $\pi(v_s a_{s+1} | x_m \cdots x_s v_s)$; we omit the first component v_s of the action $v_s a_{s+1}$, equal to the second component of the preceding state $x_s v_s$.

In view of (2), for any admissible chain $x_m v_m y_m a_{m+1}$ we have

$$\mu(x_m y_m)\tilde{\pi}(a_{m+1}|x_m v_m) = \mu(x_m)\tilde{\pi}(a_{m+1}|x_m v_m)v_m(y_m).$$

Multiplying both sides of this equation by $f(x_m v_m y_m a_{m+1})$, summing on x_m, y_m and a_{m+1}, and taking account of formulas (16) and (17), we get (18).

Further, in view of (17), the left side of formula (13) is equal to $P_\mu^{\tilde{\pi}} f_1(h_{t-1} y_{t-1} a_t)$, where

$$f_1(h_{t-1} y_{t-1} a_t) = \sum_{x_t y_t a_{t+1}} p(x_t y_t|y_{t-1} a_t)\tilde{\pi}(a_{t+1}|h_t)f(h_t y_t a_{t+1}),$$

and its right side, in view of (16), reduces to $P_{\tilde{\mu}}^{\pi} \tilde{f}_1(h_{t-1} v_{t-1} a_t)$, where

$$\begin{aligned}
\tilde{f}_1(h_{t-1} v_{t-1} a_t) &= \sum_{x_t a_{t+1}} \tilde{p}(x_t|v_{t-1} a_t)\tilde{\pi}(a_{t+1}|h_t)\tilde{f}(h_t v_t a_{t+1}) \\
&= \sum_{x_t y_t a_{t+1}} \tilde{p}(x_t|v_{t-1} a_t)\tilde{\pi}(a_{t+1}|h_t)v_t(y_t)f(h_t y_t a_{t+1}).
\end{aligned}$$

In order to obtain (13) from the induction hypothesis, it remains to verify that

$$\tilde{f}_1(h_{t-1} v_{t-1} a_t) = \sum_{y_{t-1}} f_1(h_{t-1} y_{t-1} a_t)v_{t-1}(y_{t-1}), \tag{19}$$

i.e. that (14) is satisfied when t is replaced by $t - 1$. Since we are dealing only with admissible chains, for which v_t is connected with v_{t-1} by formulas (3)–(4), then

$$\sum_{y_{t-1}} p(x_t y_t|y_{t-1} a_t)v_{t-1}(y_{t-1}) = \tilde{p}(x_t|v_{t-1} a_t)v_t(y_t).$$

Multiplying both sides by $\tilde{\pi}(a_{t+1}|h_t)f(h_t y_t a_{t+1})$ and summing on x_t, y_t, and a_{t+1}, we arrive at (19).

$$* \quad * \quad *$$

Let us review the situation. We have a mapping $\pi \to \tilde{\pi}$ of the strategy set for the model $\tilde{Z}$ onto the strategy set for Z such that

$$w(\mu,\tilde{\pi}) = \tilde{w}(\tilde{\mu},\pi) \tag{20}$$

for any initial distribution μ and corresponding initial distribution $\tilde{\mu}$. Hence it follows that $v(\mu) = \tilde{v}(\tilde{\mu})$, and that the strategy $\tilde{\pi}$ is optimal for the process Z_μ if and only if the strategy π is optimal for the process $\tilde{Z}_\mu$. Accordingly, for the strategy $\tilde{\pi}$ to be uniformly optimal in the model Z it suffices that the strategy π be optimal relative to the model $\tilde{Z}^*$.

---

* The converse is not always true since not every probability distribution on $X_m \times N_m$ can be obtained from some distribution μ on $X_m \times Y_m$. It holds if $\tilde{v}(\tilde{\mu}) = \tilde{\mu}v$.

We have verified that the model $\tilde{Z}$ is semicontinuous. If the interval $[m,n]$ of control is finite, then, according to the results of Chapter 2, the model $\tilde{Z}$ has a simple uniformly optimal strategy $\varphi = \psi_{m+1}\psi_{m+2}\cdots\psi_n$, where the ψ_t are (measurable) mappings of the pairs $x_{t-1}v_{t-1}$ into the $A(x_{t-1})$. In other words, there exist measurable functions

$$a_{t+1} = \psi_{t+1}(x_t v_t) \tag{21}$$

assigning, an action a_{t+1}, to each observed state x_t and to any probability distribution v_t for the nonobserved state y_t (independently of all the other information about the preceding history), and such that the strategy $\varphi = \psi_{m+1}\psi_{m+2}\cdots\psi_n$ is uniformly optimal in the model $\tilde{Z}$. This provides for a model with incomplete information the following method of construction of a strategy which is optimal for all initial distributions μ: we need only *at each stage to choose the action* $a_{t+1} = \psi_{t+1}(x_t v_t)$ *where* x_t *is the observed state and* v_t *the probability distribution for the nonobserved state* y_t, *calculated from* v_{t-1} *by* (6) *in the case* $t > m$ *and from* μ *by* (2) *in the case* $t = m$.

Now consider the control interval $[0,\infty)$. For the existence of a simple optimal strategy $\varphi = \psi_1\psi_2\cdots\psi_t\cdots$ in the model $\tilde{Z}$ it suffices to require in addition, for example, that the series

$$\sum_{t=1}^{\infty} \max_{y_{t-1}a_t} |q(y_{t-1}a_t)| \tag{22}$$

converge (see Chapter 5, §6).

This series converges in particular, *if the model Z is homogeneous and the discount coefficient β is less than 1.* According to Chapter 6, §6, in the case *the model $\tilde{Z}$ has a stationary optimal strategy* (i.e. the selector ψ_t is the same for all the times t).

* * *

In concrete problems it is often necessary to deal with the case when the fibres $A(x)$ intersect for different x, and the transition function and running reward function at the step t depend on x_{t-1} as well as on y_{t-1} and a_t. This case reduces to the one already investigated by the introduction of the new actions $a'_t = x_{t-1}a_t$ (cf. Chapter 1, §2).

Now suppose that the fibre $A(x)$ does not depend on x, the transition function $p(x_t y_t | x_{t-1} y_{t-1} a_t)$ and the running reward $q(x_{t-1} y_{t-1} a_t)$ do not depend on x_{t-1}, i.e. that the observed state does not affect either the possibility of control nor the further evolution of the system and the future reward. In this case the operator T in the model $\tilde{Z}$ carries any function of x, v into a function of v alone. Therefore the value $\tilde{v}(x,v)$ of the model $\tilde{Z}$ does not depend on x. It is easy to see that the selectors ψ_t in the formula (21), yielding the optimal control, may also be chosen independently of x_t.

We make use of these remarks in the next section.

§3. The Two-armed Bandit Problem

One of the simplest examples of control with incomplete data is known in the literature as *the two-armed bandit problem*. This is the name for a slot machine having two levers. After putting his money in the slot, the player pulls one of the levers. The money is either lost, or is returned with a definite gain, not depending on the lever. The levers have different gain probabilities, which we shall denote by p_1 and p_2, supposing $p_1 > p_2$. The problem is that the player does not know which of the levers is the "good" one, i.e. the one with the higher probability p_1. All he learns, at each stage, is whether he has been paid off at that stage or not.

One supposes that at the beginning of the game there is an *a priori* probability distribution for the better lever. After each try, the player calculates the *a posteriori* distribution for that lever. The fundamental result for this problem, remarkable for its simplicity and clarity, is the following: *Independently of the duration of the game, the player should at each stage pull the lever whose probability of being the good one appears at that moment to be the higher.*

In order to obtain this result, we construct a corresponding homogeneous model with incomplete information corresponding to our problem. The unobserved state y_t does not depend on t. We give it the value 1 if the left lever is the good one, and the value 2 if the right lever is the good one. We shall regard the observed state x_t at the tth stage as equal to 1 in the case of a gain and equal to 2 in the case of a loss. The action at each time consists in the choice of the left or right lever. The choice of the left lever will be denoted by $a_t = 1$, and that of the right lever by $a_t = 2$. Thus the observed and unobserved state spaces, and the action space, consist each of two elements: $X = Y = A = \{1,2\}$.

The transition function p defines the probability distribution for $x_t y_t$ depending on $y_{t-1} a_t{}^*$. It is convenient to denote by $p_1(x)$ the probability of the outcome x for the good lever and by $p_2(x)$ the same for the bad lever, so that

$$p_i(1) = p_i, \qquad p_i(2) = 1 - p_i, \qquad i = 1, 2. \tag{1}$$

The transition function is expressed in terms of the $p_i(x)$ by the formula

$$p(xy'|ya) = \begin{cases} p_1(x) & \text{if } y' = y = a, \\ p_2(x) & \text{if } y' = y \neq a, \\ 0 & \text{if } y' \neq y. \end{cases} \tag{2}$$

The gain at each play may take on two values, depending on the construction of the machine. We denote them by d_1 and d_2, supposing that $d_1 > d_2$. In accordance with the footnote on page 201, one may replace the gain at the tth step by

---

* Since the control spaces (fibres) $A(x)$ intersect, in fact coincide, at different states x, then the probability distribution for $x_t y_t$ could have depended not only on $y_{t-1} a_t$, but also on x_{t-1} (see the corresponding remark in §2). In our case the value of x_{t-1} obviously does not affect this distribution.

its mathematical expectation relative to the distribution $p(\cdot \,|\, y_{t-1} a_t)$, and introduce the running reward function

$$q(ya) = \sum p_i(x) d_x, \quad \text{where } i = \begin{cases} 1 & \text{if } y = a \\ 2 & \text{if } y \neq a \end{cases} . \tag{3}$$

The exact values of d_1 and d_2 are inessential for the analysis. One obtains the most compact formulas if one chooses them in such a way that

$$q(ya) = \begin{cases} p_1 - p_2 & \text{if } y = a, \\ p_2 - p_1 & \text{if } y \neq a; \end{cases} \tag{4}$$

it suffices to put $d_1 = 2 - p_1 - p_2$, $d_2 = -p_1 - p_2$. The terminal reward payoff is equal to zero.

In accordance with the general results of §2, we must pass to the model $\tilde{Z}$ with complete information. Here we are dealing with a case when the fibre $A(x)$, the transition function $p(\cdot \,|\, xya)$, and the reward function payoff $q(xya)$, all do not depend on x. The remark at the end of §2 is therefore applicable, and in the construction of the optimal controls in the model $\tilde{Z}$ we may consider the action of the operator T on functions in the space N. In accordance with the formulas of §2 and of Chapter 1, §6, we have

$$Tf(v) = \max[U_1 f(v), U_2 f(v)], \tag{5}$$

where

$$U_a f(v) = \tilde{q}(va) + \sum_{x=1}^{2} \tilde{p}(x \,|\, va) f(v') \quad (a = 1,2), \tag{6}$$

and v' is the distribution for the unobserved parameter $y_t = y_0$, into which the distribution v transforms after applying the action a in the observed state x. By formulas (2.3)–(2.6) and (2)–(4) we have

$$\tilde{p}(x \,|\, v1) = p_1(x) v(1) + p_2(x) v(2),$$
$$\tilde{p}(x \,|\, v2) = p_2(x) v(1) + p_1(x) v(1),$$
$$v'(y \,|\, v1x) = \frac{p_y(x) v(y)}{\tilde{p}(x \,|\, v1)},$$
$$v'(y \,|\, v2x) = \frac{p_y(x) v(y)}{\tilde{p}(x \,|\, v2)}, \tag{7}$$
$$\tilde{q}(va) = (p_1 - p_2)[v(a) - v(\bar{a})],$$

where $\bar{a} = 3 - a$.

The distribution v is defined entirely by the number

$$\delta = v(2) - v(1); \tag{8}$$

in fact,

$$v(1) = \frac{1-\delta}{2}, \qquad v(2) = \frac{1+\delta}{2}. \tag{9}$$

Therefore the space N of distributions on Y may be identified with the segment $\delta \in [-1,1]$.

Taking account of formulas (7)–(9), we may rewrite formula (6), defining the operators U_a, in the form

$$U_1 f(\delta) = -2R\delta + (Q_1 - R\delta)f\left(\frac{-R + Q_1\delta}{Q_1 - R\delta}\right)$$

$$+ (Q_2 + R\delta)f\left(\frac{R + Q_2\delta}{Q_2 + R\delta}\right),$$

$$U_2 f(\delta) = 2R\delta + (Q_1 + R\delta)f\left(\frac{R + Q_1\delta}{Q_1 + R\delta}\right) \tag{10}$$

$$+ (Q_2 - R\delta)f\left(\frac{-R + Q_2\delta}{Q_2 - R\delta}\right),$$

where

$$Q_x = \frac{p_1(x) + p_2(x)}{2},$$

$$R = \frac{p_1(1) - p_2(1)}{2} = \frac{p_2(2) - p_1(2)}{2} > 0. \tag{11}$$

(Here, in the right sides of the expressions for the $U_a f$, the argument v' is replaced by $\delta' = v'(2) - v'(1)$, calculated using (7)).

We wish to prove the optimality of the stationary strategy defined by the selector

$$\psi(\delta) = \begin{cases} 1 & \text{if } \delta < 0, \\ 2 & \text{if } \delta \geq 0. \end{cases} \tag{12}$$

To this end we need to verify that for any n

$$T_\psi^n 0 = T^n 0. \tag{13}$$

Note that in view of (12)

$$T_\psi f(\delta) = \begin{cases} U_1 f(\delta) & \text{if } \delta < 0, \\ U_2 f(\delta) & \text{if } \delta \geq 0. \end{cases} \tag{14}$$

Put

$$f_n = T^n 0. \tag{15}$$

For the proof of (12) it suffices to verify that for any n

$$T_\psi f_n = T f_n. \tag{16}$$

Put

$$g_n = U_2 f_n - U_1 f_n. \tag{17}$$

In view of (14) and (5) formula (16) will be proved if we establish that

$$\delta g_n(\delta) \geq 0, \qquad \delta \in [-1,1]. \tag{18}$$

This last is established quite easily in the special case when $p_2 = 1 - p_1$, i.e. the probability of gain for the good lever is equal to the probability of loss for the bad lever. Indeed, it is clear from (1) and (11) that in this case $Q_1 = Q_2$, and it follows from (10) that $\delta g_n(\delta) = 4R\delta^2$. In the general case it is more convenient to establish by induction on n the following somewhat stronger statement:
(A): $g_n(\delta)$ *is nondecreasing, and*

$$g_n(0) = 0. \tag{19}$$

In order to carry out the induction we will need the following properties of the operators U_a:

a) $U_1 U_2 = U_2 U_1$;
b) The operator U_2 carries a nondecreasing function into a nondecreasing function.

Assertion a) is verified by an elementary calculation using formula (10). It has the following interpretation: if we play twice, pulling first one lever and then the other, the result does not depend on the order of these choices.
Since the function $2R\delta$ is nondecreasing, it suffices to verify assertion b) for the operator

$$Sf(\delta) = U_2 f(\delta) - 2R\delta = \lambda(\delta)f[\alpha(\delta)] + \mu(\delta)f[\beta(\delta)],$$

where

$$\lambda(\delta) = Q_1 + R\delta, \qquad \mu(\delta) = Q_2 - R\delta,$$

$$\alpha(\delta) = \frac{R + Q_1\delta}{\lambda(\delta)}, \qquad \beta(\delta) = \frac{-R + Q_2\delta}{\mu(\delta)}.$$

We verify directly that on the segment $[-1,1]$ the functions λ, α, and β are increasing, that $\alpha \geq \beta$ and $\lambda + \mu = 1$. We have depicted the graphs of α and β in figure 8.1. If $-1 \leq \delta_1 < \delta_2 \leq 1$ we have $\lambda(\delta_2) - \lambda(\delta_1) = \mu(\delta_1) - \mu(\delta_2)$, so that

$$Sf(\delta_2) - Sf(\delta_1) = [\lambda(\delta_2) - \lambda(\delta_1)]\{f[\alpha(\delta_2)] - f[\beta(\delta_2)]\}$$
$$+ \lambda(\delta_1)\{f[\alpha(\delta_2)] - f[\alpha(\delta_1)]\}$$
$$+ \mu(\delta_1)\{f[\beta(\delta_2)] - f[\beta(\delta_1)]\}.$$

If f is a nondecreasing function, then all the terms in square brackets are non-negative, so that the function Sf is also nondecreasing.

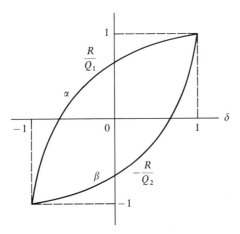

Figure 8.1

Now we turn to the proof of proposition (A). For $n = 0$ we have $f_0 = 0$ and $g_0 = U_2 0 - U_1 0 = 4R\delta$; this function satisfies (A).

Suppose that (A) is valid for some $n \geq 0$, and let us prove that then it is true for $n + 1$ as well. It is clear from (10) that

$$U_a(F_2 - F_1) = (-1)^a 2R\delta + U_a F_2 - U_a F_1.$$

Therefore, using a), we may rewrite the function $g_{n+1} = U_2 f_{n+1} - U_1 f_{n+1} = U_2 T f_n - U_1 T f_n$ in the form

$$g_{n+1} = U_2 \Phi + U_1 \Psi, \tag{20}$$

where

$$\Phi = T f_n - U_1 f_n, \qquad \Psi = U_2 f_n - T f_n. \tag{21}$$

From the induction hypothesis and formulas (5) and (17), it follows that

$$\Phi(\delta) = -\Psi(-\delta) = \begin{cases} 0 & \text{for } \delta \leq 0, \\ g_n(\delta) & \text{for } \delta \geq 0, \end{cases} \tag{22}$$

and that the function Φ is nondecreasing. Formulas (22) and (10) imply that

$$U_1 \Psi(\delta) = -U_2 \Phi(-\delta),$$

so that formula (20) takes on the form

$$g_{n+1}(\delta) = U_2 \Phi(\delta) - U_2 \Phi(-\delta). \tag{23}$$

In view of b) the function $U_2\Phi$ is nondecreasing, and it follows from (23) that the function g_{n+1} has this property as well. If $\delta = 0$ we find from (23) that $g_{n+1}(\delta) = 0$. Thus proposition (A) is valid for $n+1$ as well.

The optimality of the stationary strategy generated by the selector (12) is proved.

§4. Reduction to a Model with Complete Information. The General Case

Up to now we have been supposing that the state and action spaces are finite. This assumption is too restrictive since the most natural applications lead to more general spaces (as in most of the examples considered in the preceding chapters). The basic idea of §2 was that of considering the pair xv as a state, where x was the observed state and v was the probability distribution on the space of unobserved states. This idea is applicable also in the general case, but its realization is technically more complicated, since in place of elementary calculations with conditional probabilities in finite spaces it is necessary to deal with the more complicated theory of conditional distributions presented in Appendix 4.

So, we suppose that X_t, Y_t, and A_t are arbitrary Borel spaces, that the transition function $p_t(dx_t\,dy_t | y_{t-1}a_t)$ and the running reward function $q(y_{t-1}a_t)$ are measurable in $y_{t-1}a_t$, and that the terminal reward function $r(x_n y_n)$ is measurable in $x_n y_n$. We shall also suppose that these spaces and functions have all of the additional properties contained in requirements 2.2.α)–2.2.ε). The strategy π now depends not only on the observed history, but also on the initial distribution v_m on the set Y_m, and must be measurable in the set of all these arguments.

The distribution in the space of paths corresponding to the initial distribution μ and the strategy π, is given by the formula

$$
\begin{aligned}
P_\mu^\pi(dx_m\,dy_m\,da_{m+1}\,dx_{m+1}\,dy_{m+1}&\cdots da_n\,dx_n\,dy_n)\\
&= \mu(dx_m\,dy_m)\pi(da_{m+1}|x_m v_m)p(dx_{m+1}\,dy_{m+1}|y_m a_{m+1})\\
&\quad\cdots \pi(da_n|x_m v_m a_{m+1}x_{m+1}\cdots a_{n-1}x_{n-1})p(dx_n\,dy_n|y_{n-1}a_n). \quad (1)
\end{aligned}
$$

Here v_m is the conditional distribution of y_m for a given x_m. This is a measurable function of x_m, satisfying the equation

$$
\mu(dx_m\,dy_m) = \mu(dx_m)v_m(dy_m|x_m) \tag{2}
$$

(see Appendix 4; formula (2) does not define the function v_m uniquely, but we fix a version of this function).

The space N_t of probability measures on Y_t is also a Borel space (see Appendix 5). As in the discrete case, the new transition function assigns to each value $\tilde{a}_t = v_{t-1}a_t$ a distribution in the space $X_t \times N_t$, concentrated on the pairs $x_t v_t$, where v_t is a single valued function of x_t (whose structure depends on $\tilde{a}_t$). As in §2, we start out from the distribution

$$\tilde{p}(dx_t\, dy_t\,|\,\tilde{a}_t) = \int_{Y_{t-1}} p(dx_t\, dy_t\,|\,y_{t-1}a_t)v_{t-1}(dy_{t-1}) \tag{3}$$

in the space $X_t \times Y_t$ (see formula (2.3)). Formulas (2.4)–(2.5) are replaced by

$$\tilde{p}(dx_t\, dy_t\,|\,\tilde{a}_t) = \tilde{p}(dx_t\,|\,\tilde{a}_t)v_t(dy_t\,|\,\tilde{a}_tx_t) \tag{4}$$

and

$$\tilde{p}(dx_t\,|\,\tilde{a}_t) = \tilde{p}(dx_t \times Y_t\,|\,\tilde{a}_t). \tag{5}$$

In place of the elementary formula (2.6) for the definition of the measures $v_t(\cdot\,|\,\tilde{a}_tx_t)$ we must now make use of the results of Appendix 4. In view of Lemma 2 of §1 of Appendix 5 the measure (3) and hence also the measure (5), depend in a measurable way on $\tilde{a}_t$. Therefore $v_t(\cdot\,|\,\tilde{a}_tx_t)$ may be regarded as measurable relative to $\tilde{a}_t$ and x_t taken together (see the footnote on page 263. Hence one easily deduces that the measure in the space $X_t \times N_t$ given by the pair $\tilde{p}(dx_t\,|\,\tilde{a}_t)$, $v_t(\cdot\,|\,\tilde{a}_tx_t)$ is also measurable relative to $\tilde{a}_t$.

Let $\mu_a(dx)$ be a measure on X, depending measurably on a, and $\varphi_a(x)$ be a measurable mapping of the product $A \times X$ into the space E. Then the image $\tilde{\mu}_a$ of the measure μ under the mapping φ_a depends measurably on a. To see this suppose that f is any measurable function on the space E. Then $f[\varphi_a(x)]$ is measurable in xa, and according to Lemma 2 of §1 of Appendix 5 the function

$$F(a,a') = \int_X f[\varphi_a(x)]\mu_{a'}(dx) \tag{6}$$

is measurable in aa'. Accordingly, $F(a,a)$ is measurable in a. But if $f = \chi_\Gamma$ we have $F(a,a) = \tilde{\mu}_a(\Gamma)$.

The running and terminal reward in the new model are given by the formulas

$$\begin{aligned}
\tilde{q}(v_{t-1}a_t) &= \int_{Y_{t-1}} q(y_{t-1}a_t)v_{t-1}(dy_{t-1}), \\
\tilde{r}(x_n v_n) &= \int_{Y_n} r(x_n y_n)v_n(dy_n).
\end{aligned} \tag{7}$$

It is easy to verify that the functions $\tilde{p}$, $\tilde{q}$, and $\tilde{r}$ satisfy conditions 2.2.α)–2.2.ε). Denote the Borel model with complete information defined by them by $\tilde{Z}$. If Z is homogeneous, then $\tilde{Z}$ is homogeneous as well. We leave it to the reader to verify that if Z is nontrivial then $\tilde{Z}$ is also nontrivial.

* * *

We assign to each initial distribution μ in the model Z an initial distribution $\tilde{\mu}$ in the model $\tilde{Z}$, as follows. Suppose that $v_m(\cdot\,|\,x_m)$ is the measure in the space

Y_m introduced at formula (2). The measurable mapping

$$x_m \to v_m(\cdot \mid x_m)$$

of the space X_m into the space N_m induces the measurable mapping

$$x_m y_m \to x_m v_m(\cdot \mid x_m) \tag{8}$$

of the product $X_m \times Y_m$ into the product $X_m \times N_m$. The image of the measure μ under the mapping (8) is the desired initial distribution $\tilde{\mu}$ in $\tilde{Z}$.

The mapping $\pi \to \tilde{\pi}$ of strategies in the model $\tilde{Z}$ into strategies in the model Z is given as in §2, except that now v_t is found recurrently from the decomposition (4). In order to obtain the basic equation

$$w(\mu, \tilde{\pi}) = \tilde{w}(\tilde{\mu}, \pi), \tag{9}$$

we need to show that for any bounded measurable function f and any μ, π

$$P_\mu^{\tilde{\pi}} f(h_t y_t a_{t+1}) = P_{\tilde{\mu}}^{\pi} \tilde{f}(h_t v_t a_{t+1}), \tag{10}$$

where $h_t = x_m v_m a_{m+1} x_{m+1} \cdots a_t x_t$ and

$$\tilde{f}(h_t v_t a_{t+1}) = \int_{Y_t} f(h_t y_t a_{t+1}) v_t(dy_t). \tag{11}$$

This is proved in the same way as the analogous assertion in §2, with sums replaced by integrals.

$$* \quad * \quad *$$

As in §2, it follows from equation (9) that *if the strategy π is optimal in the auxiliary model $\tilde{Z}$ with complete information, then the corresponding strategy $\tilde{\pi}$ is optimal in the model Z*. The same is true for ε-optimal strategies. For the optimality (ε-optimality) of a strategy $\tilde{\pi}$ under the initial distribution μ, it is necessary and sufficient that π should be optimal (ε-optimal) under the corresponding initial distribution $\tilde{\mu}$. If π is stationary, then $\tilde{\pi}$ is stationary as well.

We have proved relation (9) under the assumption that the control interval is finite and that the reward functions q and r are bounded above. It is easy to see that (9) remains valid for nonnegative unbounded reward functions and for an infinite control interval. If the reward function takes values of different signs, then it is useful to consider its positive and negative parts. Note that for any function q

$$\tilde{q}^+ \leq \widetilde{q^+}, \qquad \tilde{q}^- \leq \widetilde{q^-}, \tag{12}$$

the wavy line denoting the operation defined by formula (7). Therefore, if the model Z is μ-summable above (below), the model $\tilde{Z}$ is $\tilde{\mu}$-summable above (below).

An initial state in the model $\tilde{Z}$ is a pair xv, where x is the observed initial state in the model Z and v is the apriori distribution for the unobserved initial state. Therefore, for the model $\tilde{Z}$ to be summable above (below), it is sufficient that for any x and v the model Z should be xv-summable above (below). Taking account of inequalities (12) and the formula (2.12), we see that the boundedness above (below) of the model Z implies the analogous property for the model $\tilde{Z}$. Here, by boundedness above of a model with incomplete information, we mean the existence of positive functions $c_t(xv)$ such that, for sufficiently large t, for any strategy $\tilde{\pi}$

$$P_{xv}^{\tilde{\pi}} q^+ (y_{t-1} a_t) \le c_t(xv) \qquad (x \in X_m, v \in N_m),$$

and the series Σc_t converges at each point xv. (Boundedness below of Z is defined analogously).

Combining these results with the results of Chapters 3–6, we may obtain various conditions for existence of optimal strategies in Z. For example, it follows from Result II′a of Chapter 5, §1 that *if the model Z is μ-summable above and bounded above, then for any $\varepsilon > 0$ there exists a simple strategy φ (in the model $\tilde{Z}$) such that $w(\mu,\tilde{\varphi}) \ge v(\mu) - \varepsilon$.* Or it follows from Chapter 6, §8 that *if in the homogeneous model Z the reward q is bounded and the discount coefficient β is less than 1, then for any $\varepsilon > 0$ and any initial distribution μ there exists a stationary strategy $\tilde{\varphi}$ such that $w(\mu,\tilde{\varphi}) \ge v(\mu) - \varepsilon$.*

§5. The Stabilization Problem

Now we turn to the stabilization problem, treated previously in Chapter 1, §2, Chapter 2, §11, Chapter 6, §12, and Chapter 7, §12. We will suppose that at each time t the state of the system is observed with some error ξ_t. As everywhere else in this chapter, we denote the observed state by x. It is connected with the true state y_t by the formula

$$x_t = y_t + \xi_t, \qquad t = 0, 1, 2, \ldots. \tag{1}$$

The actions and the random perturbations of the system are denoted as before respectively by a_t and s_t. Thus, the recurrence equation describing the actual (unobserved) evolution of the system now has the form

$$y_t = y_{t-1} - a_t + s_t, \tag{2}$$

and the running reward is equal to

$$q(y_{t-1} a_t) = -b(y_{t-1} - a_t)^2 - ca_t^2 \qquad (t = 1, 2, \ldots) \tag{3}$$

(in the case of complete information we have $y_t = x_t$). It remains to assume how does the process start. We will suppose that the control begins at an instant when the system is taken out of equilibrium by a random perturbation s_0, so that

$$y_0 = s_0. \tag{4}$$

Complete results have been obtained only under the hypothesis that *all the random variables* $s_0, \xi_0, s_1, \xi_1, \ldots$ *are normally distributed*. Suppose moreover that *they are mutually independent*, and that

$$Es_t = E\xi_t = 0, \qquad \text{Var } s_t = \sigma^2, \qquad \text{Var } \xi_t = \tau^2. \tag{5}$$

Without loss of generality we may suppose that $\tau = 1$ (changing the units of measurement if necessary).

The linear operations (1) and (2) do not lead out of the class of normal distributions. If (η_1, η_2) is a normally distributed random vector with the parameters

$$E\eta_i = c_i, \qquad E(\eta_i - c_i)(\eta_j - c_j) = b_{ij} \qquad (i, j = 1,2),$$

then the conditional distribution of η_2 for a known value of η_1 is also normal with the parameters*.

$$E(\eta_2|\eta_1) = c_2 + \frac{b_{12}}{b_{11}}(\eta_2 - c_1),$$

$$\text{Var}(\eta_2|\eta_1) = b_{22}\left(1 - \frac{b_{12}^2}{b_{11}b_{22}}\right) \tag{6}$$

Therefore we have to deal only with the normal distributions v_t for the unobserved states y_t. A normal distribution is defined by two parameters—the mathematical expectation m and the variance D—so that the space N_t may be identified with the half-plane $N = \{(m,D) : D \geq 0\}$.

Let us describe the remaining elements of the auxiliary model $\tilde{Z}$ introduced in §2 and §4. The initial distribution $\tilde{\mu}$ in the space $X \times N$ is constructed starting from the joint distribution μ of the pairs $(x_0, y_0) = (s_0 + \xi_0, s_0)$. This last is normal and in view of (5) has the parameters

$$Ex_0 = Ey_0 = 0, \qquad \text{Var } x_0 = \sigma^2 + 1, \qquad \text{Var } y_0 = \sigma^2, \qquad Ex_0 y_0 = \sigma^2. \tag{7}$$

Therefore, according to formulas (6), the conditional distribution $v_0(\cdot|x_0)$ has the parameters

$$m_0 = \frac{\sigma^2}{\sigma^2 + 1}x_0, \qquad D_0 = \frac{\sigma^2}{\sigma^2 + 1}. \tag{8}$$

The transition function $\tilde{p}$ assigns to each pair $v_{t-1}a_t = (m_{t-1}, D_{t-1})a_t$ a probability distribution for x_t and a conditional distribution $v_t(\cdot|x_t) = (m_t(x_t), D_t(x_t))$. In view of formulas (1)–(2) and (5) the normal distribution $\tilde{p}(\cdot|v_{t-1}a_t)$ of the pair

---

* See H. Cramér [1], Chapter 21, Section 12.

$x_t y_t$ has the parameters

$$
\begin{aligned}
Ex_t = Ey_t &= m_{t-1} - a_t, \\
\operatorname{Var} y_t &= D_{t-1} + \sigma^2, \\
\operatorname{Var} x_t &= D_{t-1} + \sigma^2 + 1, \\
E(x_t - Ex_t)(y_t - Ey_t) &= D_{t-1} + \sigma^2.
\end{aligned}
\tag{9}
$$

The parameters of the normal distribution $\tilde{p}(dx_t | v_{t-1} a_t)$ are contained in formulas (9), and the parameters of the normal distribution $v_t(dy_t | v_{t-1} a_t x_t)$ are, by formulas (6) and (9), equal to

$$
m_t = m_{t-1} - a_t + \frac{D_{t-1} + \sigma^2}{D_{t-1} + \sigma^2 + 1}(x_t - m_{t-1} + a_t),
\tag{10}
$$

$$
D_t = \frac{D_{t-1} + \sigma^2}{D_{t-1} + \sigma^2 + 1}.
\tag{11}
$$

Formulas (10)–(11) are valid for $t = 0$ as well, if we put

$$
m_{-1} = a_0 = D_{-1} = 0.
\tag{12}
$$

For the running reward $\tilde{q}$, according to formulas (4.7) and (3) we have the expression

$$
\tilde{q}(v_{t-1} a_t) = Eq(y_{t-1} a_t) = -bD_{t-1} - b(m_{t-1} - a_t)^2 - ca_t^2.
\tag{13}
$$

According to the general theory, for control at the instant t it is essential to know only a_s, x_s, m_s, and D_s for $s < t$. The variances D_t are computed according to formulas (8) and (11), independently of the observations. On the other hand, by formulas (8) and (10) we may express $x_0, x_1, \ldots, x_t$ in terms of $m_0, m_1, \ldots, m_t$ and $a_1, \ldots, a_t$. Therefore it suffices to watch the evolution of m_t only. It follows from formulas (1), (2), (8), (10) and (11) that

$$
m_t = m_{t-1} - a_t + \tilde{s}_t \qquad (t = 0, 1, 2, \ldots),
\tag{14}
$$
where
$$
\tilde{s}_t = D_t(y_{t-1} - m_{t-1} + s_t + \xi_t) \qquad (t = 0, 1, 2, \ldots).
\tag{15}
$$

(Here we are supposing that $y_{-1} = 0$.) Since the constant terms in the running reward do not affect the difference $w(x, \pi) - w(x, \rho)$ of the values of any two strategies, then in the search for an optimal strategy these terms may be dropped, and the running reward (13) replaced by

$$
\tilde{\tilde{q}}(m_{t-1} a_t) = -b(m_{t-1} - a_t)^2 - ca_t^2.
\tag{16}
$$

Formulas (14)–(16) define a model with complete information, in which the states are the numbers m_t; this is the stabilization problem with complete information which we studied earlier, except that now the perturbations $\tilde{s}_t$ are different.

In the preceding chapters we supposed that the perturbations were independent, identically distributed, and had zero mathematical expectations. We shall show that all of these properties are satisfied for the $\tilde{s}_t$, except the property of being identically distributed.

The difference

$$z_t = y_t - m_t$$

is normally distributed with parameters $(0,D_t)$. Indeed, since v_t is the conditional distribution for y_t under the observed history h, then

$$m_t = E(y_t|h), \qquad D_t = E[(y_t - m_t)^2|h].$$

Therefore

$$Em_t = EE(y_t|h) = Ey_t \tag{17}$$

and

$$D_t = EE[(y_t - m_t)^2|h] = E(y_t - m_t)^2 = Ez_t^2. \tag{18}$$

It is easy to deduce formulas (17)–(18) also by induction from the recurrence relation for z_t which follows from (1)–(2) and (10)–(11):

$$z_t = (1 - D_t)z_{t-1} + \zeta_t \qquad (t = 0,1,2,\ldots), \tag{19}$$

where

$$z_{-1} = 0 \tag{20}$$

and

$$\zeta_t = (1 - D_t)s_t - D_t\xi_t. \tag{21}$$

Using formulas (17)–(21), we show that the random variables $\tilde{s}_t$ are non-correlated and accordingly independent. Put

$$Q_t^s = \begin{cases} \prod_{k=s}^{t} (1 - D_k) & \text{for } s \le t, \\ 1 & \text{for } s > t. \end{cases} \tag{22}$$

One easily deduces from (19) that for $T > t$

$$z_{T-1} = Q_{T-1}^t z_{t-1} + Q_{T-1}^{t+1}\zeta_t + Q_{T-1}^{t+2}\zeta_{t+1} + \cdots + Q_{T-1}^T \zeta_{T-1}. \tag{23}$$

It follows from (15), (17) and (5) that

$$E\tilde{s}_t = 0. \tag{24}$$

From (15), (21) and (23), using the orthogonality of $z_{t-1}, s_t, \zeta_t, s_t, \ldots, s_{T-1}$ and ζ_{T-1}, and formulas (5), we find for $0 \le t < T$ that

$$\frac{1}{D_t D_T} E\tilde{s}_t\tilde{s}_T = E(z_{t-1} + s_t + \xi_t)(z_{T-1} + s_T + \zeta_T)$$

$$= Q_{T-1}^t Ez_{t-1}^2 + Q_{T-1}^{t+1}E(s_t + \xi_t)\zeta_t$$

$$= Q_{T-1}^{t+1}[(1 - D_t)D_{t-1} + (1 - D_t)\sigma^2 - D_t].$$

In view of (11) the term in square brackets is equal to 0. From (15), (18) and (5) we have

$$\text{Var } \tilde{s}_t = D_t^2(D_{t-1} + \sigma^2 + 1) = \frac{D_t^2}{1 - D_t}; \tag{25}$$

cf. (11).

The original problem has been reduced to the problem of control of a system given by the recurrence relation (14) *with the independent random perturbations* $\tilde{s}_t$. For the case of constant variance of the perturbations this last problem was solved in Chapter 2, §11 for a finite interval of control, in Chapter 6, §12 for an infinite interval of control and a discounted reward, and in Chapter 7, §12 for an averaged reward per unit time.

It is easy to see that in the general case when the random perturbations have nonequal distributions, the optimal strategies remain the same, and the value of the model is changed by a constant. For example, in the problem of maximization of the average reward the asymptotic value v of the model is given by the formula

$$v = -bD - \sigma^2 l, \tag{26}$$

where l is the positive root of the equation

$$l^2 + bl - bc = 0, \tag{27}$$

and $D = \lim_{t \to \infty} D_t$ is the positive root of the equation

$$D^2 + \sigma^2 D - \sigma^2 = 0. \tag{28}$$

(See formula (7.12.16), we leave the verification of this to the reader.)

* * *

The conditional mathematical expectation m_t of the random variable y_t is a natural estimate of y_t in terms of the observed history h (it is a function of h for which the quantity $E[y_t - f(h)]$ is minimal). In this section, we have, on a simple example, obtained the *separation theorem*, which asserts that under rather general conditions the optimal control of a linear Gaussian system with a quadratic loss function splits into: 1) the computation of the best estimates of the unobserved parameters in terms of the observed ones, and 2) the optimal control of the system resulting from the original one by the replacement of the unobserved parameters by their estimates.

Chapter 9

Concave Models.
Models of Economic Development

§1. The Gale Model

We turn to the Gale model described in the Introduction. In this model we are dealing with nonnegative m-dimensional coordinate space, the ith coordinate denoting the quantity of the ith product. The production process ζ is characterized by a pair (ξ, η), ξ being the input vector and η being the output vector. The initial stock of products is denoted by η_0. At each time $t = 1, 2, \ldots, n$ one is given a set $\mathfrak{T}_t$ of production processes which are technically realizable in the period t (the *technology set*). The *plan* is a sequence of production processes $\zeta_t = (\zeta_t, \eta_t)$, satisfying the conditions

$$\zeta_t \in \mathfrak{T}_t, \qquad \xi_t \leq \eta_{t-1}, \qquad t = 1, \ldots, n. \tag{1}$$

The objective of the control is the choice of a plan having the maximum possible utility

$$q_1(\zeta_1) + q_2(\zeta_2) + \cdots + q_n(\zeta_n). \tag{2}$$

We suppose that for each t:

 a) The technology set $\mathfrak{T}_t$ is convex*, closed, and contains the element $(0,0)$.
 b) The set $\mathfrak{T}_t$ is compact or is a cone not containing elements $(0,\eta)$ with $\eta \neq 0$.
 c) The utility function q_t is concave and upper semicontinuous on $\mathfrak{T}_t$[†].

The concavity and semicontinuity properties are preserved if one extends the function q_t to the set R_+^{2m} of all nonnegative $2m$-dimensional vectors, putting $q_t = -\infty$ outside $\mathfrak{T}_t$.

---

* A set $\mathfrak{T}$ in a linear space is said to be *convex* if along with any two vectors ζ' and ζ'' it contains all of its linear combinations $\alpha\zeta' + \beta\zeta''$ with $\alpha, \beta \geq 0$ and $\alpha + \beta = 1$. It is said to be a *cone* if moreover, along with any ζ, it contains all the vectors $\alpha\zeta$ with $\alpha > 0$. A function q given on $\mathfrak{T}$ is said to be *concave* if, with α and β as above, one has $q(\alpha\zeta' + \beta\zeta'') \geq \alpha q(\zeta') + \beta q(\zeta'')$.

[†] Suppose that $\mathfrak{T}$ is a convex set in a Euclidean space. Denote by X the minimal Euclidean space which contains $\mathfrak{T}$. We say that x is an interior point of $\mathfrak{T}$ if, for some $\varepsilon > 0$, all points of X of a distance smaller than ε from x belong to $\mathfrak{T}$. If f is a concave function on $\mathfrak{T}$, then f is continuous at all interior points of $\mathfrak{T}$, but may have discontinuities on the boundary of $\mathfrak{T}$. If $\mathfrak{T}$ is a segment, then q is necessarily lower semicontinuous; however already in the two-dimensional case this is not so. See e.g. Rockafellar [1], §10.

The condition (1) defining the plan may be rewritten in the form

$$\zeta_t \in \mathfrak{T}_t(\eta_{t-1}), \tag{3}$$

where $\mathfrak{T}_t(c)$ denotes the set of production processes $(\xi, \eta) \in \mathfrak{T}_t$ with input ξ not exceeding c.

Now we suppose that the technical possibilities and our evaluations of the utility are subjected to the influence of random factors. We will describe their action during the period t by the parameter s_t, taking on values from some Borel space S_t. The set $\mathfrak{T}_t$ and the function q_t depend on the "history" $s^t = s_1 s_2 \cdots s_t$. *We shall suppose that this dependence is measurable*, that conditions* a)–c) *are satisfied, and that the functions* $q_t(s^t, \zeta_t)$ *are bounded above.* The joint distribution of $s_1, s_2, \ldots, s_n$ is regarded as known.

A collection of measurable functions $\zeta_t = \psi_t(s^t)$ is said to be a *plan*, if conditions (1) are satisfied for all values of the parameters $s_1, s_2, \ldots, s_n$. The plan is said to be *optimal* if the mathematical expectation of the sum (2) takes on a maximum on it. The existence of an optimal plan will be deduced in the next section from a more general result. From the same result there follows the existence of an optimal plan in a *non-closed Gale model*. In such a model, besides the sets $\mathfrak{T}_t$ and functions q_t, there are given measurable functions $\Delta_t(s^t)$, $t = 1, 2, \ldots, n$, whose values are nonnegative m-dimensional vectors. These vectors are interpreted as resources received from outside. In the definition of the plan the conditions

$$\xi_t(s^t) \leq \eta_{t-1}(s^{t-1})$$

are replaced by

$$\xi_t(s^t) \leq \eta_{t-1}(s^{t-1}) + \Delta_t(s^t), \qquad t = 1, 2, \ldots n; \tag{4}$$

(for $t = 1$ the term $\eta_0(s^0)$ means the initial deterministic stock η_0).

§2. Concave Models

Consider the control model, described by the following schema. Sets $C_0, C_1, \ldots, C_n$ are given. To each ζ of C_{t-1} there corresponds a nonempty subset $Z_t(\zeta)$ of the set C_t. The initial state ζ_0 of C_0 is fixed, and a function q_t on the set C_t is given for each $t = 1, 2, \ldots, n$. Among the collections $\zeta_1, \zeta_2, \ldots, \zeta_n$ satisfying the conditions

$$q_t(\zeta_1) + q_2(\zeta_2) + \cdots + q_n(\zeta_n)$$

we are required to find one for which the sum

$$\zeta_t \in Z_t(\zeta_{t-1}), \qquad t = 1, 2, \ldots, n,$$

takes on its largest value.

---

* In accordance with §5 of Chapter 2 we will say that the set $\mathfrak{T}(s)$ depends measurably on s if the distance of $\mathfrak{T}(s)$ from any fixed point is a measurable function of s.

We will say that this model is *concave*, if:

a) $C_0, C_1, \ldots, C_n$ are convex closed sets in finite-dimensional vector spaces;
b) The functions q_t are concave and upper semicontinuous;
c) The sets of pairs (ζ_{t-1}, ζ_t) satisfying the conditions

$$\zeta_t \in Z_t(\zeta_{t-1})$$

are convex;
d) The correspondences $\zeta \to Z_t(\zeta)$, $\zeta \in C_{t-1}$, are quasi-continuous (cf. Chapter 2, §4).

The Gale model is a special case of a concave model. The rôle of ζ_0 is played by the vector η_0 of initial resources and the rôle of ζ_t for $t > 0$ is played by the production process (ξ_t, η_t). Here $C_t = R_+^{2m}$ and

$$Z_t(\zeta) = \mathfrak{T}_t(\eta) \quad \text{for } \zeta = (\xi, \eta) \in C_{t-1}.$$

Conditions a), b), and c) are obviously satisfied. Condition d) requires that if

$$(\xi_n, \eta_n) \in \mathfrak{T}_t, \qquad (\xi_n', \eta_n') \in \mathfrak{T}_{t-1}, \qquad \xi_n \le \eta_n', \qquad (\xi_n', \eta_n') \to (\xi', \eta'), \tag{1}$$

there should exist a subsequence $\{(\xi_{n_k}, \eta_{n_k})\}$ converging to the point

$$(\xi, \eta) \in \mathfrak{T}_t, \qquad \xi \le \eta'.$$

The inequality $\xi \le \eta'$ is obtained by passing to the limit from the inequalities $\xi_n \le \eta_n'$. The fact that the point (ξ, η) belongs to the set $\mathfrak{T}_t$ follows from the fact that $\mathfrak{T}_t$ is closed. The existence of the convergent subsequence $\{(\xi_{n_k}, \eta_{n_k})\}$ is trivial if $\mathfrak{T}_t$ is compact (see condition b) of §1). Now we suppose that $\mathfrak{T}_t$ is a cone not containing elements of the form $(0, \eta)$ with $\eta \ne 0$. It suffices to verify that the sequence $\{(\xi_n, \eta_n)\}$ is bounded. The boundedness of the sequence $\{\xi_n\}$ follows from (1). Consider the vectors

$$(\xi_n^0, \eta_n^0) = \frac{1}{|\eta_n|}(\xi_n, \eta_n),$$

where $|\eta|$ denotes the length of the vector η. If the sequence $\{\eta_n\}$ were not bounded, then it would contain a subsequence $\{\eta_{n_l}\}$ for which $|\eta_{n_l}| \to \infty$, and $\eta_{n_l}^0$ has a limit η^*. Obviously $\xi_n^0 \to 0$. Since the cone $\mathfrak{T}_t$ contains all the elements (ξ_n^0, η_n'), then it contains also the limit vector $(0, \eta^*)$. Inasmuch as $|\eta^*| = 1$, this contradicts condition b) of §1.

Now suppose that a probability distribution on the product $S_1 \times S_2 \times \cdots \times S_n$ of Borel spaces is given, and that the functions q_t and mappings Z_t depend (in a measurable way) on the history $s^t = s_1 s_2 \cdots s_t$. Suppose that for each value of s^t conditions a)–d) are satisfied, and that the functions $q_t(s^t, \zeta_t)$ are bounded above. A *plan* is a sequence of measurable functions $\zeta_t(s^t)$ satisfying for all values of the random parameters the condition

$$\zeta_t(s^t) \in Z_t(s^t, \zeta_{t-1}(s^{t-1})) \qquad (t = 1, 2, \ldots, n). \tag{2}$$

Stochastic variants of the Gale model (Closed and nonclosed) may be obtained as special cases of concave models: formulas (1.4) follow from (2) if one puts

$$Z_t(s^t,\zeta) = \mathfrak{Z}_t(s^t,\eta + \varDelta_t(s^t)). \tag{3}$$

The model is said to be a *Markov model*, if $s_1, s_2, \ldots, s_n$ is a Markov chain and if q_t and Z_t depend only on s_t i.e. do not depend on $s_1, \ldots, s_{t-1}$. The general case easily reduces to the Markovian one: it suffices to take the history $s^t = s_1 s_2 \cdots s_t$ as the characterization of the situation at the time t, thus replacing the spaces S_t by $S^t = S_1 \times S_2 \times \cdots \times S_t$. In view of the results of Appendix 4, there exist a distribution $\mu(ds_1)$ and conditional distibutions $p_t(ds^{t+1}|s^t)$. These are the initial distribution and the transition function of the Markov chain $s^1, s^2, \ldots, s^n$. *In this and the following sections we deal only with Markov models.*

In order to include the control process described above into the general schema investigated in the preceding chapters, we will consider, in the chain $\zeta_0 s_1 \zeta_1 s_2 \cdots$ $\zeta_{n-1} s_n \zeta_n$, the pairs $\zeta_t s_{t+1}$ as the states x_t, and the pairs $s_t \zeta_t$ as the actions a_t. More precisely, the class of actions $A(x_{t-1})$ possible in the state $x_{t-1} = \zeta_{t-1} s_t$ consists of the pairs $a_t = s_t \zeta_t$, where $\zeta_t \in Z_t(\zeta_{t-1} s_t)$, and the action $a_t = s_t \zeta_t$ takes the system into the state $x_t = \zeta_t s_{t+1}$, where s_{t+1} has the distribution $p(ds_{t+1}|s_t)$. The operators U_t and V_t (cf. Chapter 2, §4) are given in our case by the formulas

$$U_t f(s_t \zeta_t) = q_t(s_t \zeta_t) + \int_{S_{t+1}} f(\zeta_t s_{t+1}) p_t(ds_{t+1}|s_t), \tag{4}$$

$$V_t g(\zeta_{t-1} s_t) = \sup_{\zeta_t \in Z_t(\zeta_{t-1} s_t)} g(s_t \zeta_t) \tag{5}$$

(the terminal reward being equal to 0). To each plan there corresponds some strategy in the sense of the preceding chapters. This is a strategy of a special form, assigning to each history $h = \zeta_0 s_1 \zeta_1 s_2 \cdots \zeta_{t-1} s_t$ not a probability distribution in the action space, but rather a uniquely defined action $a_t = s_t \zeta_t$. On the other hand, to each strategy of the indicated special form there corresponds a plan gotten by successively eliminating $\zeta_1, \ldots, \zeta_{t-1}$ from the expression of ζ_t in terms of the history h. We shall construct for our model a simple uniformly optimal strategy $\varphi = \psi_1 \psi_2 \cdots \psi_n$. Here ψ_t is a measurable selector of the correspondence $x_{t-1} \rightarrow A(x_{t-1})$. Each such selector has the form

$$\psi_t(\zeta_{t-1} s_t) = s_t F_t(\zeta_{t-1} s_t), \tag{6}$$

where F_t is a measurable selector of the correspondence $Z_t(\zeta_{t-1} s_t)$. The formulas

$$\zeta_t = F_t(\zeta_{t-1} s_t) \tag{7}$$

give us an optimal plan. We shall call a plan *Markovian* if it is defined by formulas (7) for some set of measurable functions F_t. Thus, from the existence of a simple strategy it follows that there exists a Markovian optimal plan.

The proof of the existence of a simple optimal strategy is based on the same idea which we used in Chapter 2. We introduce on each of the spaces $A_t = S_t \times C_t$

and $X_t = C_t \times S_{t+1}$ a class of functions $\mathscr{L}(A_t)$, $\mathscr{L}(X_t)$, with the following properties:

A) $\quad U_t \mathscr{L}(X_t) \subset \mathscr{L}(A_t), \qquad V_t \mathscr{L}(A_t) \subset \mathscr{L}(X_{t-1}) \qquad (t = 1, 2, \ldots, n).$

B) If $g_t \in \mathscr{L}(A_t)$, then there exists a measurable selector ψ_t of the correspondence $A(x)$ such that

$$V_t g_t = g_t(\psi_t).$$

Making use of these properties, we may construct a simple optimal strategy as follows (cf. Chapter 2, §5): We calculate the functions u_t and v_t according to the recurrence formulas

$$v_n = 0, \qquad u_t = U_t v_t, \qquad v_{t-1} = V_t u_t \qquad (t = 1, 2, \ldots, n). \tag{8}$$

Then we find the selectors ψ_t from the equations

$$u_t(\psi_t) = v_t$$

(the existence of a measurable solution is guaranteed by property B). Integrating the value $v_0(\zeta_0 s_1)$ with respect to the initial distribution μ for s_1, we obtain the expression

$$v_0(\zeta_0) = \int_{S_t} v_0(\zeta_0 s_1) \mu(ds_1) \tag{9}$$

for the maximal utility possible under the initial state ζ_0; (cf. the formula $v(\mu) = \mu v$ of Chapter 2, §5).

§3. The Spaces $\mathscr{L}$

The proof of assertions A) and B) of the preceding section is based on a series of lemmas, which it is convenient to formulate in neutral terms.

Suppose that M is a closed convex set in k-dimensional Euclidean space, E an arbitrary measurable space. We denote by $\mathscr{L} = \mathscr{L}(E \times M)$ the set of real-valued functions $f(y,z)$, with $y \in E$ and $z \in M$, bounded above, measurable in y, concave and upper semicontinuous in z.

Lemma 1. *Let M' be an everywhere dense subset of M and let $Q \subset M$ be open in M. If f is a concave function on M, then the supremum of f on Q coincides with the supremum of f on $Q \cap M'$.*

Indeed, let a be the supremum of f on $Q \cap M'$. It suffices to prove that $f \leq a$ everywhere on Q. This inequality holds at interior points of M^* since f is continuous at these points (see footnote on p. 222). Every point z of M is an end of an open interval I consisting of interior points of M. If $z \in Q$ and if I is small enough, then $I \subset Q$. Hence $f \leq a$ on I. Being concave, f is lower semi-continuous at z. Hence $f(z) \leq a$.

---

* Without any loss of generality, we may assume that dim $M = k$ and therefore interior points of M do exist.

Lemma 2. *Each function f of the class $\mathscr{L}$ is the limit of a nonincreasing sequence of functions continuous in z and measurable in y.*

Choose in M an everywhere dense sequence $\{z_m\}$, and put

$$f_{mn}(y,z) = f(y,z_m) - n|z - z_m|,$$
$$f_n = \sup_m f_{mn},$$

It is clear that $f_{n+1} \leq f_n$ and that f_n is measurable in y. Inasmuch as $|z - z_m| - |z' - z_m| \leq |z - z'|$, then

$$f_{mn}(y,z') \leq f_{mn}(y,z) + n|z - z'|$$

so that

$$f_n(y,z') \leq f_n(y,z) + n|z - z'|.$$

This inequality, and the inequality obtained from it by exchanging z and z', show that the function f_n is continuous in z. It is easy to see that the distance $|z' - z|$ of z' from z is, for each fixed z, convex in z'^*. Therefore the function $f(y,z') - n|z - z'|$ is concave in z'. Then, in view of Lemma 1,

$$f_n(y,z) = \sup_{z'} \left[f(y,z') - n|z - z'| \right] \geq f(y,z).$$

Fix z and y, and choose $z_{m(n)}$ for each n so that

$$\frac{1}{n} + f(y,z_{m(n)}) - n|z - z_{m(n)}| \geq f_n(y,z) \geq f(y,z). \tag{1}$$

Since f is bounded above, then $|z - z_{m(n)}| \to 0$ as $n \to \infty$. Using the upper semi-continuity of f and inequality (1), we get

$$f(y,z) \geq \varlimsup_{n \to \infty} f(y,z_{m(n)}) \geq \varlimsup_{n \to \infty} f_n(y,z) \geq f(y,z).$$

Hence f_n converges to f.

Lemma 3. *All the functions of the class $\mathscr{L}$ are measurable in (y,z).*

This follows from Lemma 2 and the fact that a function of two variables which is measurable in one of them and continuous in the second is measurable in the pair of the two variables[†].

---

* Indeed, putting $z^* = \alpha z_1' + \beta z_2'$, where $\alpha \geq 0$, $\beta \geq 0$, $\alpha + \beta = 1$, we get $|z^* - z| = |\alpha(z_1' - z) + \beta(z_2' - z)| \leq \alpha|z_1' - z| + \beta|z_2' - z|$.

[†] Indeed, for every $\delta > 0$, M can be represented as a finite union of disjoint subjects $M_{\delta k}$ with diameters smaller than δ. Let $z_{\delta k}$ be a point of $M_{\delta k}$. If a function $f(y,z)$ is measurable in y and continuous in z, then $f = \lim_{\delta \to 0} f_\delta$ where

$$f_\delta(y,z) = f(y,z_{\delta k}) \quad \text{for } z \in M_{\delta k}.$$

Lemma 4. *Suppose that Q is a measurable compact- and convex-valued correspondence of E into M. If $f \in \mathscr{L}$, then:*

a) *The function*

$$\bar{f}(y) = \sup_{z \in Q(y)} f(y,z)$$

is measurable;

b) *The formula*

$$\hat{Q}(y) = \{z : z \in Q(y), f(y,z) = \bar{f}(y)\}$$

defines a measurable compact-valued correspondence of E into M;

c) *The correspondence $\hat{Q}$ admits a measurable selection.*

For the proof we choose in M an everywhere dense sequence $\{z_m\}$, and consider the open sets $Q^n(y) \downarrow Q(y)$ constructed in the measurability criterion (cf. Chapter 2, §6). Put

$$f_m^n(y) = \begin{cases} f(y,z_m) & \text{for } z_m \in Q^n(y), \\ -\infty & \text{for } z_m \notin Q^n(y). \end{cases}$$

and

$$f^n(y) = \sup_m f_m^n(y).$$

It follows from 2.6.a) that the functions f_m^n are measurable, so that for the proof of a) it suffices to verify that

$$\bar{f}(y) = \lim_{n \to \infty} f^n(y). \tag{2}$$

By Lemma 1, $f^n(y)$ is equal to the supremum of $f(y,z)$ over all $z \in Q^n(y) \cap M$. Because of 2.6.b and semi-continuity of f, this implies (2).

That $\hat{Q}(y)$ is nonempty and compact follows from the compactness of $Q(y)$ and the semicontinuity of f relative to z (see the beginning of §4 in Chapter 2). In order to prove the measurability of $\hat{Q}$ we consider the functions $f_n \downarrow f$ constructed in Lemma 2, which were continuous in z, and put

$$\hat{Q}^n(y) = \left\{ z : |z - Q(y)| < \frac{1}{n}, f_n(y,z) > \bar{f}(y) - \frac{1}{n} \right\}.$$

It is easy to verify that these sets satisfy all the requirements of the measurability criterion of Chapter 2, §6. Condition 2.6.a) follows from the measurability of the functions f_n and $\bar{f}$ and the correspondence Q; condition 2.6.b) is deduced from the compactness of $\hat{Q}(y)$, the continuity of the f_n in z, and the relation $f_n \downarrow f$. Point b) is proved.

Point c) follows from b) and Theorem 2.6.B.

* * *

Now we proceed to the realization of the program outlined in the section 2. Put $\mathscr{L}(A_t) = \mathscr{L}(E \times M)$ with $E = S_t$, $M = C_t$, and $\mathscr{L}(X_t) = \mathscr{L}(E \times M)$ with $E = S_{t+1}$, $M = C_t$ (in accordance with the notations of the preceding sections, we will write the argument $\zeta_t \in C_t$ as the first entry in the functions $f \in \mathscr{L}(X_t)$, and the argument $s_{t+1} \in S_{t+1}$ as the second).

Now we shall prove assertions 2.A and 2.B. Suppose that $f \in \mathscr{L}(X_t)$. Then $f(\zeta_t, s_{t+1})$ is bounded above and is measurable in s_{t+1}, so that the integral in formula (2.4) has a meaning. Since q_t and p_t are measurable in s_t, the function $U_t f$ has the same property. That $U_t f$ is concave in the argument ζ_t and that it is bounded follow immediately from the analogous properties of q_t and f. Further, suppose that $\zeta_t^n \to \zeta_t$. By Fatou's lemma*, for any sequence of measurable functions F_n bounded above, and any probability measure v on the measurable space S,

$$\varlimsup_{n \to \infty} \int_S F_n(s) v(ds) \leq \int_S \varlimsup_{n \to \infty} F_n(s) v(ds).$$

Applying this lemma to the functions $F_n(s_{t+1}) = f(\zeta_t^n s_{t+1})$ and the measure $p_t(\cdot \mid s_t)$, and using the semicontinuity of the functions f and q_t relative to ζ_t, we have

$$\varlimsup_{n \to \infty} U_t f(s_t, \zeta_t^n) \leq U_t f(s_t, \zeta_t).$$

Therefore the function $U_t f$ is upper semicontinuous relative to ζ_t and belongs to $\mathscr{L}(A_t)$.

Further, suppose that $g \in \mathscr{L}(A_t)$. We may consider g as a function of the three variables $\zeta_{t-1} s_t \zeta_t$, independent of ζ_{t-1}. Obviously $g \in \mathscr{L}(E \times M)$ for $E = C_{t-1} \times S_t$, $M = C_t$. The sets $Q(y) = Z_t(\zeta_t, s_t)$, where $y = \zeta_{t-1} s_t$, are convex and compact in view of conditions 2a) and 2c). Consider the function

$$f(y) = f(\zeta_{t-1} s_t) = -\rho(Z_t(\zeta_{t-1} s_t), \zeta_t) = -\rho(Q(y), \zeta_t),$$

where ζ_t is any fixed point of the set C_t. In view of the measurability of Z_t in s_t, f is measurable in s_t as well. One easily deduces from 2d) that f is upper semicontinuous in ζ_{t-1}, and from 2c) that f is concave relative to ζ_{t-1}. By virtue of Lemma 3, f is measurable in the pair $y = \zeta_{t-1} s_t$, i.e. the correspondence Q is measurable. Applying Lemma 4 to this correspondence and to the function g, we find that

$$V_t g(\zeta_{t-1} s_t) = g(s_t, F_t(\zeta_{t-1}, s_t)), \tag{3}$$

where F_t is some measurable mapping of $C_{t-1} \times S_t$ into C_t, satisfying the condition

$$F_t(\zeta_{t-1}, s_t) \in Z_t(\zeta_{t-1}, s_t). \tag{4}$$

Defining ψ_t using formula (2.6), we get the result 2.B). It remains to verify that $V_t g \in \mathscr{L}(X_{t-1})$. That this function is bounded above is obvious. That it is measurable in s_t is clear from Lemma 4a), or else from formula (3) and Lemma 3. One easily deduces, from (3)–(4), condition 2d), and the semicontinuity of g relative to

---

* See e.g. Neveu [1], Chapter II, Section 3.

ζ_t, that $V_t g$ is upper semicontinuous relative to ζ_{t-1}, and from (3)–(4), condition 2c) and the concavity of g relative to ζ_t, that $V_t g$ is concave in ζ_{t-1}. Assertion 2.A) is completely proved.

§4. Stimulating Prices

From general concave models, we now turn to the stochastic model of economic development introduced in §1, the Gale model. This model makes it possible to understand the significance of a goal-seeking system of prices as a powerful tool for the control of an economy. Using prices it is possible to replace the global criterion: Maximize the mathematical expectation of the cumulative utility"

$$E \sum_{t=1}^{n} q_t(\zeta_t),$$

by a simpler and clearer local criterion: "Act at each time t, and in each random situation s^t, guiding yourself by the immediate gain." Here by "immediate gain" we mean the so-called reduced utility of the production process, equal to the sum of its utility and the expected profits.

Let $\pi = (\pi^1, \ldots, \pi^m)$ be a nonnegative m-dimensional vector. By the *cost of the goods bundle* $\xi = (\xi^1, \ldots, \xi^m)$ *with the prices* π we mean the scalar product $\pi\xi = \pi^1\xi^1 + \cdots + \pi^m\xi^m$. The profit of a production process $\zeta = (\xi,\eta)$ is equal to the difference $\pi\eta - \pi\xi$ between the costs of the output and the input. Two improvements should be introduced into this formula. First, the prices change in time. If a production process (ξ,η) is carried out in the period t, then, the input is due at the beginning and the output appears at the end of this period, so the profit should rather be written in the form $\pi_{t+1}\eta - \pi_t\xi$. Further, the prices π_t should depend on the random situation s^t. But at the beginning of the period t the value of s_{t+1} is unknown, and so is the profit $\pi_{t+1}(s^{t+1})\eta - \pi_t(s^t)\xi$. Replacing the price π_{t+1} by its forecast

$$\bar{\pi}_{t+1}(s^t) = E(\pi_{t+1}|s^t) = \int_{S^{t+1}} \pi_{t+1}(s^{t+1})p_t(ds^{t+1}|s^t), \qquad (2)$$

depending on s^t only, we introduce the expected profit

$$\bar{\pi}_{t+1}(s^t)\eta - \pi_t(s^t)\xi.$$

Here $p_t(\cdot|s^t)$ is a transition function of s^t into s^{t+1} which exists according to Appendix 4). In order that formula (2) should have a meaning, we have to suppose that the function $\pi_t(s^t)$ is measurable. The *reduced utility* of the production process $\zeta = (\xi,\eta)$ at the time period t is the sum

$$\Pi_t(\zeta) = \Pi_t(s^t,\zeta) = q_t(s^t,\zeta) + \bar{\pi}_{t+1}(s^t)\eta - \pi_t(s^t)\xi \qquad (3)$$

of the utility and the expected profit (for conciseness of notation we will frequently drop the argument s^t).

We shall say that *the prices $\pi_t(s^t)$ stimulate the plan $\zeta_t^*(s^t)$*, if, with probability 1:

A. For all t and all $\zeta \in \mathfrak{T}_t(s^t)$

$$\Pi_t(s^t, \zeta^*(s^t)) \geq \Pi_t(s^t, \zeta).$$

B. For all t

$$\pi_t(s^t)[\eta_{t-1}^*(s^{t-1}) - \xi_t^*(s^t)] = 0.$$

Condition A means that it is not possible to increase the "immediate gain" by deviating from the plan ζ^* (even if we are not bound by the restriction $\xi_t \leq \eta_{t-1}$ and may acquire any goods bundle at the prices π_t). Condition B requires that zero prices be assigned to oversupplied goods (i.e. to goods which are not entirely used under the plan ζ_t^*).

Our aim is to construct prices stimulating an optimal plan. Under these prices one achieves the consistency referred to above between the global criterion (1) and the immediate benefit at each time t, for almost all random situations s^t. If all the functions q_t are strictly concave, then we may assert something more: by maximizing the reduced utility at each random situation, we inevitably arrive at a plan which is optimal from the point of view of the global criterion. Thus, the stimulating prices not only localize in time the problem of optimal planning, but also make it possible in planning the next step in the situation s^t to take account only of the actual prices π_t and the forecast $\bar{\pi}_{t+1}$ of the prices for the next period. It is not necessary to analyze the situation s^t in detail, nor even to know the probability mechanism for the phenomena. In this sense π_t and $\bar{\pi}_{t+1}$ may be considered as sufficient statistics of the problem.

The existence of prices stimulating an optimal plan will be proved in the following section. In preparation we prove a simple lemma, from which it follows in particular that *if a price system stimulates any plan, that plan must be optimal.*

By a *price system* we understand a collection of measurable functions $\pi_1(s^1), \ldots, \pi_n(s^n)$ with values in R_+^m. (We note that $\bar{\pi}_{n+1}$ still appears in formula (3) for $t = n$. We will assume, by definition, that π_{n+1} and therefore $\bar{\pi}_{n+1}$ as well are equal to zero.*)

We note first of all that if $\zeta_t(s^t) = (\xi_t(s^t), \eta_t(s^t)) \in \mathfrak{T}_t(s^t)$, $t = 1, 2, \ldots, n$, is any collection of production processes, depending measurably on the random situation, and

$$\Pi_t = \Pi_t(\zeta_t),$$

then

$$E \sum_{t=1}^n \Pi_t = E \sum_{t=1}^n [q_t(\zeta_t) + \pi_t(\eta_{t-1} - \xi_t)] - E\pi_1\eta_0. \tag{4}$$

Indeed, it follows from (3) that

$$\sum_{t=1}^n \Pi_t = \sum_{t=1}^n [q_t(\zeta_t) + \bar{\pi}_t\eta_{t-1} - \pi_t\xi_t] - \bar{\pi}_1\eta_0. \tag{5}$$

---

* This condition must be dropped if a certain stock must remain at the end of the planning period (formally, one considers only plans with a fixed value of η_n). In this case a free choice of $\pi_{n+1}(s^{n+1})$ is necessary, and the price system consists of $\pi_1, \pi_2, \ldots \pi_n, \pi_{n+1}$.

Since

$$E\pi_{t+1}\eta_t = EE(\pi_{t+1}\eta_t | s^t) = E[E(\pi_{t+1} | s^t)\eta_t(s^t)] = E\bar{\pi}_t\eta_t, {}^*$$

the mathematical expectation of the sum (5) coincides with the right side of (4).

Now we shall show that *if the price system π_t stimulates a plan, then it stimulates all the optimal plans and only optimal plans* (Hence one may consider stimulating prices without relating them to any particular plan.)

Indeed, suppose that the prices π_t stimulate the plan ζ_t^*. Put

$$\Pi_t^* = \Pi_t(\zeta_t^*). \tag{6}$$

Applying formula (4) to the plan ζ_t^* and using B, we have

$$E \sum_{t=1}^{n} \Pi_t^* = E \sum_{t=1}^{n} q_t(\zeta_t^*) - E\pi_1\eta_0.$$

It follows from (4) and (6) that for any collection of production processes $\zeta_t(s^t) \in \mathfrak{T}_t(s^t)$

$$E \sum_{t=1}^{n} [q_t(\zeta_t^*) - q_t(\zeta_t)] = E \sum_{t=1}^{n} (\Pi_t^* - \Pi_t) + E \sum_{t=1}^{n} \pi_t(\eta_{t-1} - \xi_t). \tag{7}$$

If the collection ζ_t is a plan, then this last sum is nonnegative, and we find from (7) and A that

$$E \sum_{t=1}^{n} [q(\zeta_t^*) - q(\zeta_t)] \geq 0. \tag{8}$$

This means that the plan ζ_t^* is optimal. If ζ_t is another optimal plan, then (8) is satisfied with equality. From this equation and from (7), taking into account A and that $\pi_t(\eta_{t-1} - \xi_t)$ is nonnegative, we find that, with probability 1, $\Pi_t^* = \Pi_t$ and $\pi_t(\eta_{t-1} - \xi_t) = 0$, which means that the plan ζ_t is optimal.

§5. The Existence of Stimulating Prices

In order to construct stimulating prices, we need two conditions. Their economic meaning is the following:

a) It is possible to impose a penalty for interruption in deliveries, proportional to the cost of the undelivered goods, and compensating in all cases the damage caused to the production.

b) The damage estimate takes into account the possible decrease in utility of the production process.

---

* Here and in §5 we are using the properties of the mathematical expectation presented in §2 of Appendix 4.

In order to translate these conditions into mathematical language, we suppose that instead of the goods bundle η, a bundle η' is supplied. Consider the vector $(\eta - \eta')_+$ of nondelivered goods (the subscript $+$ means that all negative co-ordinates are replaced by zero) and calculate their cost

$$\delta(\eta,\eta') = c(\eta - \eta')_+ \tag{1}$$

with some fixed set of prices c. Suppose that the expenses of the passage from the productive process ζ to the productive process ζ' are equal to $d(\zeta,\zeta')$. In this esti-mate, along with the costs of reequipping and so forth, there enter the losses $\delta(\eta,\eta')$ from the change in the output. Therefore it is natural to suppose that

$$d(\zeta,\zeta') \geq \delta(\eta,\eta') \quad \text{for } \zeta = (\xi,\eta),\ \zeta' = (\xi',\eta'). \tag{2}$$

It is also natural to assume that
$$d(\zeta,\zeta) = 0. \tag{3}$$

Now we will give a precise mathematical formulation of conditions a) and b). *Prices c and a nonnegative function $d(\zeta,\zeta')$ $(\zeta,\zeta' \in R_+^{2m})$ may be introduced in such a way that conditions (1)–(3) are satisfied and*:

A. *For any t, s^t and $\tilde{\xi} \geq 0$, for each $\zeta = (\xi,\eta) \in \mathfrak{T}_t(s^t)$ there exists a $\zeta' = (\xi',\eta') \in \mathfrak{T}_t(s^t,\tilde{\xi})$ such that $d(\zeta,\zeta') \leq K\delta(\xi,\tilde{\xi})$, where K is some constant inde-pendent of t, s^t and $\tilde{\xi}$.*
B. *For any $\zeta,\ \zeta' \in \mathfrak{T}_t(s^t)$ and any s^t*

$$q_t(s^t,\zeta) - q_t(s^t,\zeta') \leq d(\zeta,\zeta').$$

It follows obviously from A that

$$\sup_{\zeta \in \mathfrak{T}_t(s^t,\xi)} \inf_{\zeta' \in \mathfrak{T}_t(s^t,\xi')} d(\zeta,\zeta') \leq K\delta(\xi,\xi'). \tag{4}$$

Besides conditions A and B we need one further technical requirement:

C. *There exist functions $j_t(s^t)$ such that $Ej_t(s^t) < \infty$ and $|\zeta| < j_t(s^t)$ for all $\zeta \in \mathfrak{T}_t(s^t)$.*

The aim of this section is to prove that *if conditions A, B, and C are fulfilled, then there exist bounded stimulating prices.*

The proof is based on the study of the change in the maximal utility when the resources are varied.

Suppose that the change of resources in the period t in the situation s^t is de-scribed by the vector $\Delta_t(s_t)$. Denote by Q the set of all collections $\Delta = \{\Delta_1,\ldots,\Delta_n\}$ such that the functions $\Delta_t(s_t)$ are measurable and

$$\|\Delta\| = E\sum_{t=1}^{n} |\Delta_t| < \infty. \tag{5}$$

A collection of measurable functions $\zeta_t(s^t) = (\xi_t(s^t),\eta_t(s^t))$ will be called a Δ-*plan* if, for all values of s^t,

$$\zeta_t(s^t) \in \mathfrak{T}_t(s^t), \qquad \xi_t(s^t) \le \eta_{t-1}(s^{t-1}) + \Delta_t(s^t). \tag{6}$$

Put $\Delta \in \tilde{Q}$ if the set of Δ-plans is nonempty. Denote by $\lambda(\Delta)(\Delta \in \tilde{Q})$ the supremum of the utilities

$$E \sum_{t=1}^{n} q_t(\zeta_t)$$

over all Δ-plans $\{\zeta_t\}$. Using assumptions a) and c) of §1, we easily verify that $\tilde{Q}$ is a convex set and λ a concave function*.

We shall prove the theorem in three steps:

1. We prove the estimate

$$\lambda(\Delta) - \lambda(0) \le b\|\Delta\| \qquad (\Delta \in \tilde{Q}), \tag{7}$$

where b is some constant.

2. We deduce from (5) the existence in the space Q of a linear functional $l^\dagger$ such that

$$\lambda(\Delta) - \lambda(0) \le l(\Delta) \qquad (\Delta \in \tilde{Q}), \tag{8}$$

$$l(\Delta) \le b\|\Delta\| \qquad (\Delta \in Q). \tag{9}$$

and we establish that l may be rewritten in the form

$$l(\Delta) = E \sum_{t=1}^{n} \pi_t \Delta_t, \tag{10}$$

where $\pi_t(s^t)$ are bounded measurable functions with values from R_+^m.

3. Finally, we show that the π_t are stimulating prices.

STEP 1. Note that every Δ-plan is a Δ_+-plan, so that $\lambda(\Delta) \le \lambda(\Delta_+)(\Delta \in \tilde{Q})$. On the other hand, $\|\Delta_+\| \le \|\Delta\|$, so that it suffices to prove the estimate (7) for nonnegative collections Δ. For $\Delta \ge 0$ we get a non-closed Gale model of the type considered in §§1,2, and for the calculation of $\lambda(\Delta)$ we may use formula (2.9), where μ is the distribution of the parameter s_1 and where the initial state ζ_0 is described by the initial stock η_0. The function v_0 entering into (2.9) may be expressed by formula (2.8) in terms of the operators U_t and V_t given by formulas (2.4)–(2.5). Taking account of the fact that these formulas are written out for the Markovian case and that we are here dealing with the general case, we need to pass, in all the formulas, from the parameters s_t to the parameters s^t. Using (2.3),

---

* The set Q is a linear space relative to the natural operations of addition and multiplication by real numbers.

$\dagger$ A scalar-valued function l in a linear space Q is said to be a *linear functional*, if $l(c_1 f_1 + c_2 f_2) = c_1 l(f_1) + c_2 l(f_2)$ for any scalars c_1, c_2 and any f_1, f_2 of Q.

we arrive at the recurrence relations

$$v_n^A = 0, \tag{11}$$

$$u_t^A(s^t, \zeta_t) = q_t(s^t, \zeta_t) + \int_{S^{t+1}} v_t^A(\eta_t, s^{t+1}) p_t(ds^{t+1} \,|\, s^t), \tag{12}$$

$$v_{t-1}^A(\eta_{t-1}, s^t) = \sup_{\zeta \in \mathfrak{T}_t(s^t, \eta_{t-1} + A_t(a^t))} u_t^A(s^t, \zeta). \tag{13}$$

Here, reflecting the dependence of u_t and v_t on Δ, we have written Δ as a superscript. Since the set Z_t depends only on η_{t-1} and not on ξ_{t-1}, the functions v_t^A have the same property. In view of (2.9), for the proof of inequality (7) it suffices to verify that

$$v_0^A(\eta, s^1) - v_0(\eta, s^1) \le \text{const} \sum_{t=1}^{n} E(|A_t(s^t)| \,|\, s^1).$$

We shall prove by induction from t to $t - 1$ the more general inequality

$$v_t^A(\zeta', s^{t+1}) - v_t(\zeta, s^{t+1}) \le \text{const}[d(\zeta', \zeta) + \beta_{t+1}^{t+1}], \tag{14}$$

where

$$\beta_t^r(s^t) = \sum_{i=r}^{n} E(|A_i(s^i)| \,|\, s^t). \tag{15}$$

For $t = n$ (14) is valid, since $v_n = v_n^A = 0$. Further, by virtue of (13),

$$v_{t-1}^A(\zeta', s^t) - v_{t-1}(\zeta, s^t) = \sup_{\zeta_t' \in \mathfrak{T}_t(s^t, \eta' + A_t(s^t))} \inf_{\zeta_t \in \mathfrak{T}_t(s^t, \eta)} [u_t^A(s^t, \zeta_t') - u_t(s^t, \zeta_t)], \tag{16}$$

where $\zeta = (\xi, \eta)$, $\zeta' = (\xi', \eta')$. In view of (12)

$$u_t^A(s^t, \zeta_t') - u_t(s^t, \zeta_t) = E[v_t^A(\zeta_t', s^{t+1}) - v_t(\zeta_t, s^{t+1}) \,|\, s^t] + q_t(s^t, \zeta_t') - q_t(s^t, \zeta_t). \tag{17}$$

From (17), the induction hypothesis (14) and B, and taking account of the equation $E(\beta_{t+1}^{t+1} \,|\, s^t) = \beta_t^{t+1}(s^t)$, we have

$$u_t^A(s^t, \zeta_t') - u_t(s^t, \zeta_t) \le \text{const}[d(\zeta', \zeta) + \beta_t^{t+1}(s^t)]. \tag{18}$$

It follows from (16), (18) and (4) that

$$v_{t-1}^A(\zeta', s^t) - v_{t-1}(\zeta, s^t) \le \text{const} |\delta(\eta' + A_t(s^t), \eta) + \beta_t^{t+1}(s^t)]. \tag{19}$$

Using (1), (2) and the nonnegativity of Δ, we get

$$\delta(\eta' + A_t(s^t), \eta) \le \text{const}[|A_t(s^t)| + \delta(\eta', \eta)]$$
$$\le \text{const}[|A_t(s^t)| + d(\zeta', \zeta)]. \tag{20}$$

It follows from (19) and (20) that

$$v^{\Delta}_{t-1}(\zeta',s^t) - v_{t-1}(\zeta,s^t) \leq \text{const}\left[d(\zeta',\zeta) + \beta^{t+1}_t(s^t) + |\Delta_t(s^t)|\right].$$

In order to get the estimate (14) for the value $t - 1$ it remains now only to note that $\beta^{t+1}_t + \Delta_t = \beta^t_t$.

STEP 2. Now we make use of the following general theorem from functional analysis: if B and C are nonintersecting convex sets in the Banach space* $\mathscr{L}$, and B is open, then there exists a nonzero linear functional l in the space $\mathscr{L}$ and a number α such that $l(f) \geq \alpha$ on B and $l(f) \leq \alpha$ on C. The set Q with the norm (5) may be considered as a Banach space, if one identifies collections Δ, Δ' such that $\|\Delta - \Delta'\| = 0$. We shall denote by Q_1 the Cartesian product of Q and the real line R, and consider in Q_1 the convex sets

$$B = \{(\Delta,r):b\|\Delta\| < r\},$$

$$C = \{(\Delta,r):\Delta \in \tilde{Q}, \lambda(\Delta) - \lambda(0) \geq r\}$$

(we recall that $\tilde{Q}$ is convex and that λ is concave on $\tilde{Q}$). In view of (7), the sets B and C do not intersect. The set B is open. Therefore there exists a number α and a nonzero linear functional l_1 in Q_1 such that $l_1 \geq \alpha$ on B and $l_1 \leq \alpha$ on C. The functional l_1 has the form

$$l_1(\Delta,r) = l_0(\Delta) + ar,$$

where l_0 is a linear functional in Q and a a real number. Since $(0,1) \in B$, then $l_1(0,1) = a \geq \alpha$. For any $\Delta \in \tilde{Q}$ we have

$$(\Delta,\lambda(\Delta) - \lambda(0)) \in C$$

so that

$$l_0(\Delta) + a[\lambda(\Delta) - \lambda(0)] \leq \alpha \qquad (\Delta \in \tilde{Q}). \tag{21}$$

Analogously, from the fact that $(\Delta,b\|\Delta\| + \varepsilon) \in B$ for any $\varepsilon > 0$, it follows that

$$l_0(\Delta) + ab\|\Delta\| \geq \alpha \qquad (\Delta \in Q). \tag{22}$$

The set $\tilde{Q}$ contains all the nonnegative elements of Q. Putting $\Delta = 0$ in (21) and (22), we verify that $\alpha = 0$, so that $a \geq 0$. If $a = 0$, then it follows from (21) and (22) that $l_0 = 0$, contradicting the fact that $l_1 \neq 0$. This means that $a > 0$, and we may put $l = -l_0/a$. (8)–(9) then follows from (21)–(22).

---

* A linear space $\mathscr{L}$ is said to be a *Banach space*, if to each f of $\mathscr{L}$ one may assign a nonnegative number $\|f\|$, such that: a) for each scalar c, $\|cf\| = |c|\,\|f\|$; b) The formula $\rho(f,g) = \|f - g\|$ defines a metric in $\mathscr{L}$ relative to which $\mathscr{L}$ is a complete metric space.

Details on Banach spaces may be found in any textbook of functional analysis, for instance Kolmogorov and Fomin [1] or Dunford and Schwartz [1]. The proof of the theorem cited may be found for instance in Bourbaki [1], Chapter 2, §3, Proposition 1.

Every linear functional in the space Q satisfying inequality (8) may be written in the form (10), where $\pi_t = \pi_t(s^t)$ are measurable vector-valued functions with $\|\pi_t\| \leq b^*$.

STEP 3. Suppose that $\{\zeta_t^*\}$ is an optimal plan and that $\{\zeta_t\}$ is any Δ-plan. From the definition of the function $\lambda(\Delta)$ and formulas (8) and (10), it follows that

$$E \sum_{t=1}^{n} q_t(\zeta_t) \leq \lambda(\Delta) \leq \lambda(0) + l(\Delta) = E \sum_{t=1}^{n} \pi_t \Delta_t + \lambda(0). \tag{23}$$

The zero vectors ζ_t form a Δ-plan for any $\Delta \geq 0$ of Q. Therefore, from the boundedness of the functions q_t and (23), it follows that $E\sum \pi_t \Delta_t$ is bounded below on the set of nonnegative elements of the space Q. Therefore it follows that $\pi_t(s^t) \geq 0$ with probability 1. Since the values of the functions π_t may be altered in any way on a set of measure zero, we may suppose that these functions are nonnegative.

Now suppose that $\{\zeta_t\} = \{(\xi_t, \eta_t)\}$ is any collection of production processes. Put

$$\Delta_t = \xi_t - \eta_{t-1} \qquad (t = 1, \dots, n).$$

In view of condition C, the collection $\Delta = \{\Delta_t\}$ belongs to Q. Obviously, the collection $\{\zeta_t\}$ is a Δ-plan, and therefore that inequality (23) is satisfied for it. We rewrite this last in the form

$$E \sum_{t=1}^{n} q_t(\zeta_t) + E \sum_{t=1}^{n} \pi_t(\eta_{t-1} - \xi_t) \leq E \sum_{t=1}^{n} q_t(\zeta_t^*). \tag{24}$$

Putting $\zeta_t = \zeta_t^*$, we arrive at the inequality

$$E \sum_{t=1}^{n} \pi_t(\eta_{t-1}^* - \xi_t^*) \leq 0.$$

This implies the condition 4.B, since $\eta_{t-1}^* \geq \xi_t^*$ and $\pi_t \geq 0$.

Now we prove 4.A. It follows from (4.4) and (24) that

$$E \sum_{t=1}^{n} [\Pi_t(\zeta_t^*) - \Pi_t(\zeta_t)|] \geq 0. \tag{25}$$

Written out in detail, this means that

$$\Pi_t(\zeta) = \Pi_t(s^t, \zeta) = q_t(\zeta, s^t) + \bar{\pi}_{t+1}(s^t)\eta - \pi_t(s^t)\xi.$$

This function belongs to the class $\mathscr{L}(E \times M)$ for $E = S^t$, $M = R_+^{2m}$ (see §3), and the mapping $Q(y) = \mathfrak{T}_t(s^t)$ satisfies the conditions of Lemma 3.4. By this lemma,

---

* See Dunford and Schwartz [1], Chapter 4, §8, Theorem 5.

one may choose measurable functions $\zeta_t = \zeta_t(s^t) \in \mathfrak{T}_t(s^t)$ such that

$$\Pi(s^t, \zeta_t(s^t)) = \sup_{\zeta \in \mathfrak{T}_t(s^t)} \Pi_t(s^t, \zeta). \qquad (26)$$

Inequality (25) is satisfied for $\{\zeta_t\}$ as well. It follows from (25) and (26) that with probability 1

$$\Pi(s^t, \zeta_t^*) = \sup_{\zeta \in \mathfrak{T}_t(s^t)} \Pi_t(s^t, \zeta) \qquad (t = 1, \ldots, n),$$

This is condition 4.A.

Appendix 1

Borel Spaces

§1. Introduction

A measurable space B is called *Borelian*, or a *Borel space*, if it is isomorphic to a measurable subset of a Polish (i.e. complete separable metric) space E^*. (We recall that the σ-algebra $\mathscr{B}(E)$ of measurable sets in E is the minimal σ-algebra containing all open sets.) The following are examples of Borel spaces:

    1. A finite or countable space S, with the σ-algebra of all subsets.
    2. The unit interval I with the σ-algebra of all Borel sets.

These spaces are Polish relative to the metric $\rho(x,y) = 1$ for $x \neq y$, in the case of S, and the metric $\rho(x,y) = |x - y|$ in the case of I.

    Our objective is to prove that *every Borel space is isomorphic either to S or to I*.

    For finite and countable spaces B this assertion is trivial: in a metric space E the one-point sets are closed and therefore measurable, so that every subset of B is measurable. Thus, it suffices to prove that *all uncountable Borel spaces are isomorphic to one another*. In the proof of this assertion special rôles are played by two spaces: the product H of a countable number of segments I, the so-called Hilbert cube, and the product M of a countable number of two-point sets $\{0,1\}$. We shall show that:

    a) *Any Borel space admits an isomorphism into H*, i.e. it is isomorphic to a measurable subset of the space H.
    b) *The space M may be mapped isomorphically into any uncountable Borel space*.
    c) *There exists an isomorphism of H into M*.

Assertions a)–c) will be proved in §§2–4. It follows from them that any two uncountable Borel spaces may be isomorphically imbedded into one another. This suffices to assert their isomorphism. Indeed, the following general proposition is true.

    Let E and E' be arbitrary measurable spaces. If there exists an isomorphism f of E into E' and an isomorphism g of E' into E, then E and E' are isomorphic.

    For the proof we consider the sets $X = g(E')$ and $Y = gf(E)$. Obviously $Y \subseteq X \subseteq E$ and the mapping $\varphi = gf$ is an isomorphism of E onto Y. Inasmuch as E' is isomorphic to X, it suffices to verify that X is isomorphic to E.

---

  * A one-to-one mapping f of a measurable space E onto a measurable space E' is called an *isomorphism* if both f and f^{-1} are measurable.

Consider two sequences of measurable sets

$$E_0 = E, \qquad E_{n+1} = \varphi(E_n),$$
$$X_0 = X, \qquad X_{n+1} = \varphi(X_n),$$

$n = 0, 1, \ldots,$ and put

$$E_\infty = \bigcap E_n, \qquad X_\infty = \bigcap X_n.$$

Clearly

$$E_0 \supseteq X_0 \supseteq E_1 \supseteq X_1 \supseteq \cdots \supseteq E_n \supseteq X_n \supseteq E_{n+1} \supseteq \cdots,$$

so that $E_\infty = X_\infty$. The isomorphism ψ of the space E onto the space X that we need is defined by the formula

$$\psi(x) = \begin{cases} \varphi(x) & \text{for } x \in \bigcup_n (E_n \backslash X_n), \\ x & \text{for } x \in \bigcup_n (X_n \backslash E_{n+1}) \cup E_\infty. \end{cases}$$

§2. Imbedding of a Borel Space into the Hilbert Cube

A Borel space B is by definition isomorphic to a measurable subset of a Polish space E, which means that it suffices to imbed E into H isomorphically.

Suppose that $\{z_n\}$ is a sequence which is everywhere dense in E. The functions

$$f_n(x) = \frac{\rho(x, z_n)}{1 + \rho(x, z_n)}, \qquad x \in E, \tag{1}$$

are continuous, and consequently measurable. Therefore the formula

$$f(x) = \{f_1(x), f_2(x), \ldots, f_n(x), \ldots\}$$

yields a measurable mapping of E into H. If $f(x) = f(y)$, then $\rho(x, z_n) = \rho(y, z_n)$ for all n, so that $x = y$. Thus f is an injection (i.e. one to one mapping) of E into H.

It remains to verify that f maps measurable sets of the space E into measurable sets of the space H. We shall show that the situation reduces to the verification of the measurability of the set $f(E)$. We shall put $A \in \mathscr{A}$ if $A \subseteq E$ and $f(A)$ is measurable. If $E \in \mathscr{A}$, then $\mathscr{A}$ is a σ-algebra. For any $\varepsilon \in (0,1)$ the image of the set

$$U_\varepsilon(z_n) = \{x : \rho(x, z_n) < \varepsilon\} = \left\{ x : f_n(x) < \frac{\varepsilon}{1 - \varepsilon} \right\}$$

is equal to the intersection of $f(E)$ with the set of points $h = h_1 h_2 \cdots$ of the space H such that $h_n < \varepsilon/(1 - \varepsilon)$. Therefore if $E \in \mathscr{A}$ all the sets $U_\varepsilon(z_n)$ also belong to $\mathscr{A}$. But these sets generate $\mathscr{B}(E)$, so that that $\mathscr{B}(E) \subseteq \mathscr{A}$.

Now we turn to the proof of the measurability of $f(E)$. We introduce a metric into H by the formula

$$\rho(h, h') = \sum_{n=1}^{\infty} \frac{|h_n - h'_n|}{2^n}. \tag{2}$$

This metric is consistent with the measurable structure of H as a product of segments*.

The mapping f^{-1} is continuous on $f(E)$ in the metric (2). Indeed, if $f(x_m) \to f(x)$, then $\rho(x_m, z_n) \to \rho(z_n, x)$ for each n. Choosing z_n in a small neighborhood of x, we easily deduce from the inequality $\rho(x_m, x) \le \rho(x_m, z_n) + \rho(z_n, x)$ that $\rho(x_m, x) \to 0$.

For each point $f(x)$ from $f(E)$ and any integer m there exists an open sphere S in the space H with center at $f(x)$ and such that the diameters of S and $f^{-1}(S)$ are smaller than $1/m$. Construct for each $f(x) \in f(E)$ such a sphere, and denote by G_m their union. The set G_m is open and contains $f(E)$. We show that the intersection of the sets G_m over all m coincides with $f(E)$.

Indeed, if the point h belongs to such an intersection, then for each m there exists a sphere U_m covering h with center $h_m = f(x_m)$ and such that the diameters of U_m and $V_m = f^{-1}(U_m)$ are smaller than $1/m$. It is clear that $h_m \to h$, so that h belongs to the closure of the set $f(E)$. Therefore for any k and m there is in the neighborhood $U_k \cap U_m$ of the point h a point h' lying in $f(E)$. Then the point $x' = f^{-1}(h')$ belongs to $V_k \cap V_m$ and

$$\rho(x_k, x_m) \le \rho(x_k, x') + \rho(x', x_m) < \frac{1}{k} + \frac{1}{m}.$$

Hence it follows that the sequence $\{x_m\}$ is a Cauchy sequence, so that it converges to some limit x in the complete space E. It follows from (1) and (2) that the mapping f is continuous, so that $f(x) = \lim f(x_m) = \lim h_m = h$. Accordingly, $h \in f(E)$.

So, the set $f(E)$ coincides with the intersection of the open sets G_m and is therefore measurable.

§3. Imbedding of the Space of Dyadic Sequences into an Uncountable Borel Space

As in the case of the space H, it is convenient to treat measurable sets of the space M as Borel sets relative to the metric (2.2). In this metric the space M is compact. We shall show that:

a) *The space M may be mapped continuously and injectively into any uncountable Polish space E.*

b) *Any Borel space B is the image under a continuous injection of some Polish space E[†].*

It follows from a) and b) that M may be continuously injected into any uncountable Borel space B. Under such a mapping f the images and preimages of

---

* Indeed, the space H is separable in the metric ρ, and the function ρ is measurable in each argument. Thus all open sets, as countable unions of spheres, are measurable. On the other hand, all sets $(a_1, b_1) \times (a_2, b_2) \times \cdots \times (a_n, b_n) \times I \times I \times \cdots$ are open in the metric ρ and generate $\mathscr{B}(H)$. We note that the space H is complete in the metric (2) and is therefore a Polish space.

† By definition B is isomorphically imbedded in some Polish space E', and in B one considers the metric induced by the metric E'. It is easy to see that the σ-algebra of measurable sets in B is generated by the closed (or open) subsets of the space B.

compact sets are compact. Since M is compact, the systems of closed sets in M and $f(M)$ coincide with the systems of compact sets. Hence the mappings f and f^{-1} carry closed sets into measurable sets, and f is an isomorphism of M into B.

Now we turn to the proof of point a). We choose in E an everywhere dense sequence $\{z_k\}$ and put $U_{kn} = \{x : \rho(x,z_k) < 1/n\}$. Consider those of the spheres U_{kn} containing no more than a countable number of points, and denote by Y their union. The set Y is countable, so that the set $X = E \backslash Y$ is measurable and uncountable.

Any neighborhood U of any point x of X contains an uncountable number of points of X. Indeed, in the contrary case U would be a countable set, and then the point x would be covered by one of the spheres U_{kn} appearing in the complement Y of the set X.

Choose in X two arbitrary distinct points x_0 and x_1, and describe about them nonintersecting spheres U_0 and U_1 of radii less than 1. Choose two distinct points x_{00} and x_{01} in the uncountable set $U_0 \cap X$, and describe about them nonintersecting spheres U_{00} and U_{11} of radii less than $\frac{1}{2}$ and lying in U_0. Analogously we choose distinct points x_{10} and x_{11} in $U_1 \cap X$ and corresponding spheres U_{10} and U_{11} in U_1. Continuing in this way, we obtain, for each integer $n = 1, 2, \ldots$, a mapping, defined on sequences $A_n = m_1 m_2 \cdots m_n$ of length n of 0's and 1's, and sending each such sequence into a point $x_{A_n} \in X$. The image of the collection $\{A_n\}$, with n fixed, thus consists of a collection $\{x_{A_n}\}$ of 2^n points, each the center of a sphere U_{A_n} of radius less than $1/n$, and the U_{A_n} are pairwise disjoint. Finally, if the sequence A_k is an initial segment of the sequence A_n, then $U_{A_n} \subseteq U_{A_k}$.

Suppose now that $m = m_1 m_2 \cdots m_s \cdots$ is any point of the space M, and suppose that $A_n = m_1 m_2 \cdots m_n$. Then if (k,n) is any pair of integers with $k \leq n$, we have $x_{A_n} \in U_{A_n} \subseteq U_{A_k}$, so that $\rho(x_{A_k}, x_{A_n}) \leq 1/k$. The sequence $\{x_{A_n}\}$ is accordingly a Cauchy sequence, and has a limit point x in the complete space E. We write $f(m) = x$. Obviously $f(m') \neq f(m)$ if $m' \neq m$ and $\rho(f(m), f(m')) < 2/n$ if $\rho(m,m') < 1/2^n$. Accordingly the mapping f of the space M into the space E is a continuous injection. Assertion a) is proved.

Now consider any Polish space E'. Denote by $\mathscr{F}$ the class of all of its subsets which are images under continuous injections of Polish spaces. Obviously proposition b) will be proved if we verify that the σ-algebra $\mathscr{B}(E')$ is contained in the system $\mathscr{F}$.

First of all, $\mathscr{F}$ contains all open sets $B \subset E'$. Indeed, in this case we may take the set B itself as the space E, with the new metric

$$\tilde{\rho}(x,y) = \rho(x,y) + \frac{g(x,y)}{1 + g(x,y)},$$

where

$$g(x,y) = \left| [\rho(x,E'\backslash B)]^{-1} - [\rho(y,E'\backslash B)]^{-1} \right|.$$

The triangle inequality for $\tilde{\rho}$ follows from the obvious triangle inequality for g and the fact that the function $F(u) = u/(1 + u)$ is increasing, zero at the origin, and concave for $u \geq 0$. It is easy to see that if x and x_n belong to B and $\rho(x_n,x) \to 0$, then also $\tilde{\rho}(x_n,x) \to 0$. Therefore the intersection of B with any sequence $\{z_m\}$

which is everywhere dense in E' is a sequence everywhere dense in E, and so the space E is separable. Finally, if $\{x_n\}$ is a Cauchy sequence in E, then in view of the inequality $\rho \leq \tilde{\rho}$ it is a Cauchy sequence in E' as well, so that it has a limit x in E'. We have to prove that $x \in B$. Now if $x \in E' \backslash B$, then we would have $\rho(x_n, E' \backslash B) \to 0$, so that $g(x_k, x_n) \to \infty$ as $n \to \infty$ for any fixed k. That means that $\tilde{\rho}(x_k, x_n) \to 1 + \rho(x_k, x)$ as $n \to \infty$, contradicting the hypothesis that the sequence $\{x_n\}$ is a Cauchy sequence in the metric $\tilde{\rho}$. Thus the space E is Polish. The identity mapping $f(x) = x$ of the space E onto the space B is continuous, since $\rho(x, y) \leq \tilde{\rho}(x, y)$.

Further, consider a sequence of sets B_n of the class $\mathscr{F}$, and suppose for each n that f_n is a continuous 1-1 mapping of the Polish space E_n onto B_n. We shall show that the intersection of the B_n, and also the union of the B_n (in the case when the B_n are pairwise disjoint) also belong to $\mathscr{F}$.

It is easy to see that the direct product $E_\infty = E_1 \times E_2 \times \cdots$ is also a Polish space in the metric

$$\rho(x_1 x_2 \cdots x_n \cdots, y_1 y_2 \cdots y_n \cdots) = \sum_{n=1}^{\infty} 2^{-n} \frac{\rho_n(x_n, y_n)}{1 + \rho(x_n, y_n)}.$$

In view of the continuity of all the mappings f_n, the subspace of the space E_∞ on which $f_1(x_1) = f_2(x_2) = \cdots = f_n(x) = \cdots$ is closed, so that it is also a Polish space. The formula $f(x_1 x_2 \cdots x_n \cdots) = f_1(x_1)$ yields a continuous surjection of the space E_∞ onto the intersection of the B_n.

Now suppose that the sets B_n are pairwise disjoint. Without loss of generality we may suppose that the diameters of the spaces E_n do not exceed 1 (a situation we may achieve by introducing the new equivalent distance function $\rho' = \rho/(1 + \rho)$). Denote by E the union of the spaces E_n, taking $\rho(x, y) = 2$ when x and y belong to distinct E_n. The formula

$$f(x) = f_n(x_n) \quad \text{for } x \in E_n, n = 1, 2, \ldots$$

defines a continuous surjection of the space E onto the union of the B_n.

The inclusion $\mathscr{B}(E') \subseteq \mathscr{F}$ follows now from the following lemma.

Lemma. *If the class $\mathscr{F}$ of sets of the metric space X contains all open sets and is invariant relative to countable intersections and countable unions of pairwise disjoint sets, then $\mathscr{F}$ contains all the measurable sets in X.*

Indeed, any closed set is the intersection of a countable number of its ε-neighborhoods, and so belongs to $\mathscr{F}$. Accordingly the class $\mathscr{F}_1$ of sets Γ such that both Γ itself and its complement $\bar{\Gamma} = X \backslash \Gamma$ belong to $\mathscr{F}$ contains all open sets. If we show that the class $\mathscr{F}_1$ is invariant relative to countable unions and intersections, then we will find that $\mathscr{B}(X) \subseteq \mathscr{F}_1 \subseteq \mathscr{F}$, and the lemma will be proved.

If $\Gamma_1, \Gamma_2, \ldots, \in \mathscr{F}_1$, then $\Gamma = \bigcap_n \Gamma_n \in \mathscr{F}$ and

$$\bar{\Gamma} = \bigcup_n (\Gamma_1 \cap \Gamma_2 \cap \cdots \cap \Gamma_{n-1} \cap \bar{\Gamma}_n) \in \mathscr{F},$$

so that $\Gamma \in \mathcal{F}_1$. Analogously

$$\varDelta = \bigcup_n \Gamma_n = \bigcup_n (\bar{\Gamma}_1 \cap \bar{\Gamma}_2 \cap \cdots \cap \bar{\Gamma}_{n-1} \cap \Gamma_n) \in \mathcal{F}$$

and $\bar{\varDelta} = \bigcap_n \bar{\Gamma}_n \in \mathcal{F}$, so that $\varDelta \in \mathcal{F}_1$ as well.

§4. Imbedding of the Hilbert Cube into the Space of Dyadic Sequences

It is easy to see that if φ is an isomorphism of X into Y, then the formula

$$f(x_1 x_2 \cdots) = \varphi(x_1)\varphi(x_2) \cdots$$

defines an isomorphism of the product $X^\infty = X \times X \times \cdots$ into the product $Y^\infty = Y \times Y \times \cdots$. Inasmuch as $H = I^\infty$, it suffices to:

  a) Construct an isomorphism φ of the segment I into the space M;
  b) Prove that the spaces M and M^∞ are isomorphic.

The formula

$$\psi(m) = \sum_{k=1}^\infty 2^{-k} m_k$$

defines a measurable mapping of the space M onto I, each term of the series appearing there being obviously measurable. This mapping is not 1–1. In fact to each dyadic rational on $(0,1)$ there correspond two sequences $m = m_1 m_2 \cdots$, one ending in 0's, and the other in 1's. We write $m \in M'$ if the sequence $m = m_1 m_2 \cdots$ contains zeros, but only finitely many of them. The set M' is countable, so that the set $N = M - M'$ is measurable. It is easy to see that ψ is a measurable injection of N onto I. We shall show that the inverse mapping $\varphi = \psi^{-1}$ of the segment I onto N is also measurable. To this end it suffices to verify that the preimages $\varphi^{-1}(\Gamma_k) = \psi(\Gamma_k)$ of the sets $\Gamma_k = \{m : m_k = 0, m \in N)$ are measurable. But such a preimage is equal to the union of the intervals $[n/2^k, (n+1)/2^k]$, where n runs through all even values from 0 to $2^k - 2$. Thus, φ is an isomorphism of I into the space M.

Now we will construct an isomorphic mapping F of the space M^∞ onto the space M. Each point of the space M^∞ is a sequence $m^1 m^2 \ldots$, where m^n, in turn, is a sequence $m_1^n m_2^n \cdots$ of 0's and 1's. Writing out the sequences in the form of a rectangular display

$$m^1 = m_1^1 \quad m_2^1 \quad m_3^1 \quad \cdots$$
$$m^2 = m_1^2 \quad m_2^2 \quad m_3^2 \quad \cdots$$
$$m^3 = m_1^3 \quad m_2^3 \quad m_3^3 \quad \cdots$$
$$\vdots$$

we form the element

$$m = m_1^1 m_2^1 m_1^2 m_3^1 m_2^2 m_1^3 \cdots$$

of the space M. It is obvious that we obtain in this way a $1-1$ mapping F of the space M^∞ onto M. Denote by B_k^n the subset of the space M^∞ defined by the condition $m_k^n = 0$. Obviously the sets B_k^n generate the σ-algebra $\mathscr{B}(M^\infty)$, and the sets $F(B_k^n)$ the σ-algebra $\mathscr{B}(M)$. Therefore the mappings F and F^{-1} are measurable, so that F is an isomorphism.

From the results obtained above it is easy to deduce that *the direct product $E = E_0 \times E_1 \times \cdots$ of a countable number of Borel spaces is also a Borel space*, a fact that we needed in Chapter 5, §4. Indeed, each of the spaces E_n may be interpreted as a Borel set in the space M. Then E will be a measurable set in the product M^∞, isomorphic to M, and therefore itself a Borel space.

Appendix 2

Analytic Sets

§1. Introduction

Suppose that B and B' are Borel spaces and f a measurable mapping of B into B'. The main purpose of this appendix is to prove the result used in Chapter 3, that *the image of a measurable set in B is universally measurable in B'*. In order to prove this assertion, it suffices to introduce into each Borel space B a class $\mathscr{A}(B)$ of sets with the following properties:

1) If $\Gamma \in \mathscr{A}(B)$, then $f(\Gamma) \in \mathscr{A}(B')$;
2) $\mathscr{A}(B)$ contains all measurable sets of the space B;
3) All the sets of the class $\mathscr{A}(B)$ are universally measurable.

It is not possible to take as $\mathscr{A}$ the class of all measurable sets; that class does not satisfy property 1) (see §5). It is also not possible to take as $\mathscr{A}$ the class of all universally measurable sets; P. S. Novikov constructed a universally measurable set on the plane, for which it is impossible to prove the universal measurability of its projection onto the line in the framework of the current axiomatic theory of sets*. The class $\mathscr{A}(B)$ that we need is formed by the so-called *analytic sets*. They may be defined as *the measurable images of Borel spaces in other Borel spaces*. It is clear that conditions 1) and 2) will then be satisfied, and the problem reduces to the verification of property 3).

We first show that *the analytic sets of a Borel space B may be defined in the same way, as images of Polish spaces under continuous mappings of these spaces into B*. Suppose that A is an analytic set of the space B, so that A is the image of a Borel space B' under a measurable mapping f. We imbed B and B' in corresponding Polish spaces E and E', with metrics ρ and ρ'. The direct product $E' \times E$ of these measurable spaces becomes a Polish space in the metric

$$d(x_1 y_1, x_2 y_2) = \rho'(x_1, x_2) + \rho(y_1, y_2) \qquad (x_1, x_2 \in E'; y_1, y_2 \in E).$$

---

* See P. S. Novikov [1] (this result was announced by Goedel as far back as 1938). Thus the proposition that the class of all universally measurable sets has property 1) cannot be proved. On the other hand, in view of the results of Solovay [1], this proposition cannot be refuted. It must be said that the result of Solovay was obtained by adjoining an additional postulate to the usual axiomatics of the theory of sets, that of the existence of an unattainable cardinal.

Consider in $E' \times E$ the graph Γ of the mapping f:

$$\Gamma = \{xy : x \in E', y = f(x)\}.$$

It is easy to see that

$$\Gamma = \bigcap_{k=1}^{\infty} \bigcup_{n=1}^{\infty} [U_{nk} \times f^{-1}(U_{nk})],$$

where $\{U_{1k}, U_{2k}, \ldots\}$ is a decomposition of the set B into measurable sets of diameter less than $1/k$. Accordingly, Γ is measurable, and, being a Borel space, is the continuous image of some Polish space X (see Appendix 1, §3, Assertion b)). But A is the image of the graph Γ under the projection of $E' \times E$ onto E. A projection is a continuous mapping. Therefore A is a continuous image of X^*.

The universal measurability of an analytic set will be proved in §3, after we have in preparation established that every analytic set may be obtained by means of the so-called 𝒜-operation. The object of §4 is to show that *a 1–1 measurable mapping of a Borel space onto a Borel space is an isomorphism.* This result will be deduced from the possibility of separating two nonintersecting analytic sets by Borel sets. In §5 we present an example of a nonmeasurable analytic set.

§2. The 𝒜-Operation

Suppose that in a set F there have been chosen a countable number of subsets $F_1, F_2, \ldots, F_{n_1}, \ldots$, and in each set F_{n_1} a countable number of subsets $F_{n_1 1}$, $F_{n_1 2}, \ldots, F_{n_1 n_2}, \ldots$ and so forth out to infinity, so that the sets $F_{n_1 n_2 \ldots n_k}$ are defined for any finite collections of integers $n_1 n_2 \cdots n_k$, and $F_{m_1 m_2 \cdots m_k m_{k+1}} \subset F_{n_1 n_2 \ldots n_k}$ if $m_1 = n_1, m_2 = n_2, \ldots, m_k = n_k$. Then one says that the sets $F_{n_1 n_2 \ldots n_k}$ form an *array* $\mathscr{F}$.

To any sequence $n = n_1 n_2 \cdots$ of natural numbers there corresponds a sequence of nested sets $F_{n_1} \supset F_{n_1 n_2} \supset \cdots$ of the array $\mathscr{F}$. Their intersection is denoted by F_n. The union of the sets F_n over all sequences n of natural numbers is said to be *the result of the 𝒜-operation, applied to the array* $\mathscr{F}$; we will denote that set by $\mathscr{A}\mathscr{F}$.

We shall show that *any analytic set A of a Borel space B is the result of the 𝒜-operation, applied to some array $\mathscr{F}$ made up of measurable sets of the space B*[†].

Let f be a continuous mapping of a Polish space E onto A. We set up an array $\mathscr{E}$ in E by dividing E into a countable number of sets E_{n_1} of diameter less than 1, each E_{n_1} into a countable number of sets $E_{n_1 n_2}$ of diameter less than $1/2$, each $E_{n_1 n_2}$ into a countable number of sets $E_{n_1 n_2 n_3}$ of diameter less than $1/3$, and so forth out to infinity (That such a procedure is possible follows from the separability of the space E.). Then, to each point x of the space E there will correspond a unique

---

[*] In §3 of Appendix 1 we proved that the intersection of continuous images of Polish spaces is also a continuous image of a Polish space; the proof there remains valid also without the assumption there that the mappings were 1-1. Therefore the intersection of analytic sets is again an analytic set, a remark that we shall find of use in §5.

[†] The converse assertion is also valid, but we will not need it; see Kuratowski [1], §38, subsection IX.

sequence $n = n(x)$ such that $x = E_n$. Conversely, each E_n is either a one-point set or is the empty set (empty sets may also appear in the array).

Further, we define in B an array $\mathscr{F}$, putting $F_{n_1 n_2 \cdots n_k}$ equal to the closure of the image $f(E_{n_1 n_2 \cdots n_k})$. The sets of the array $\mathscr{F}$ are measurable, and we shall show that $\mathscr{A}\mathscr{F} = A$.

Each point y of A has at least one preimage x in the space E, and there corresponds to the point x a sequence n such that $x = E_n$. Then $y = f(x) = f(E_n) \subset F_n \subset \mathscr{A}\mathscr{F}$. Hence $A \subseteq \mathscr{A}\mathscr{F}$.

Conversely, if y belongs to $\mathscr{A}\mathscr{F}$, then there exists a sequence $n = n_1 n_2 \cdots$ such that $y \in F_{n_1 n_2 \cdots n_k}$ for all $k = 1, 2, \ldots$. Since $F_{n_1 n_2 \cdots n_k}$ is the closure of the set $f(E_{n_1 n_2 \cdots n_k})$, it is possible to choose in $f(E_{n_1 n_2 \cdots n_k})$ a point y_k a distance from y of less than $1/k$. Suppose that x_k is one of the preimages of y_k. Then the x_k belong to the nested sets $E_{n_1 n_2 \cdots n_k}$ with diameters tending to zero, and so form a Cauchy sequence. This sequence has a limit point x in the complete space E. By the continuity of f we have $f(x) = \lim f(x_k) = \lim y_k = y$. Hence $y \in f(E) = A$, so that $\mathscr{A}\mathscr{F} \subseteq A$.

§3. Universal Measurability of Analytic Sets

For the proof of the universal measurability of an analytic set it suffices to show that *if $\mathscr{F}$ is an array made up of measurable sets of the measurable space B, then the set $\mathscr{A}\mathscr{F}$ is μ-measurable for any probability distribution μ in the space B.*

For each set A of the space B we introduce outer and inner measures

$$v(A) = \inf_{\Gamma \supseteq A} \mu(\Gamma), \qquad \lambda(A) = \sup_{\Gamma \subseteq A} \mu(\Gamma), \qquad (1)$$

Γ running through the measurable sets. We note that the infimum in (1) is necessarily attained: if $\Gamma_n \supset A$, $\mu(\Gamma_n) < v(A) + 1/n$, then for the intersection Γ of the sets Γ_n we have $\Gamma \supseteq A$, $\mu(\Gamma) \leq v(A)$, so that $\mu(\Gamma) = v(A)$. Analogously, the supremum in the definition of λ is attained as well. Therefore for the μ-measurability of the set A it suffices that

$$\lambda(A) = v(A). \qquad (2)$$

(Obviously, condition (2) is necessary as well.)

We note the following properties of the outer measure:

  a) If $A_1 \subset A_2$, then $v(A_1) \leq v(A_2)$.
  b) If $A_1 \subset A_2 \subset \cdots \subset A_n \subset \cdots$, and $A = \bigcup_n A_n$, then

$$v(A) = \lim_{n \to \infty} v(A_n). \qquad (3)$$

The first of these properties is obvious. In order to prove the second we consider measurable sets Γ_n such that $A_n \subseteq \Gamma_n$ and $\mu(\Gamma_n) = v(A_n)$, and put

$$\Gamma = \bigcup_{n=1}^{\infty} \bigcap_{m=n}^{\infty} \Gamma_m.$$

Since $A_n \subseteq \Gamma_m$ for $m \geq n$, then $A_n \subset \Gamma$, so that $A \subseteq \Gamma$. Therefore

$$v(A) \leq \mu(\Gamma) = \lim_{n \to \infty} \mu\left(\bigcap_{m=n}^{\infty} \Gamma_m\right) \leq \varlimsup_{n \to \infty} \mu(\Gamma_n) = \lim_{n \to \infty} v(A_n).$$

But in view of a)

$$\lim_{n \to \infty} v(A_n) \leq v(A)$$

and (3) is proved.

We now turn to the proof of relation (2) for the set $A = \mathscr{A}\mathscr{F}$. Obviously, it suffices, for any $\varepsilon > 0$, to construct a measurable subset Γ of the set A such that

$$\mu(\Gamma) > v(A) - \varepsilon. \tag{4}$$

We denote by $F^{m_1 m_2 \cdots m_k}$ the union of the sets $F_{n_1 n_2 \cdots n_k}$ over all values $n_1 \leq m_1$, $n_2 \leq m_2, \ldots, n_k \leq m_k$, and put

$$A_{n_1 n_2 \cdots n_k} = A \cap F_{n_1 n_2 \cdots n_k}, \qquad A^{m_1 m_2 \cdots m_k} = A \cap F^{m_1 m_2 \cdots m_k}.$$

Clearly

$$A = \bigcup_{n_1 = 1}^{\infty} A_{n_1};$$

and by property b), for any $\varepsilon > 0$ there exists an index m_1 such that

$$v(A^{m_1}) > v(A) - \frac{\varepsilon}{2}.$$

Further

$$A^{m_1} = \bigcup_{n_2 = 1}^{\infty} \left[\bigcup_{n_1 = 1}^{m_1} A_{n_1 n_2}\right]$$

and by property b) there exists an index m_2 such that

$$v(A^{m_1 m_2}) > v(A^{m_1}) - \frac{\varepsilon}{4}.$$

Continuing this construction, we obtain an infinite sequence $m_1 m_2 \cdots$ of natural numbers such that

$$v(A^{m_1 m_2 \cdots m_{k-1} m_k}) > v(A^{m_1 m_2 \cdots m_{k-1}}) - \frac{\varepsilon}{2^k}. \tag{5}$$

Write

$$A(k) = A^{m_1 m_2 \cdots m_k}, \qquad F(k) = F^{m_1 m_2 \cdots m_k}.$$

It follows from (5) that $v(A(k)) > v(A) - \varepsilon$ for any k. The sets $F(k)$ are measurable, $A(k) \subseteq F(k)$, and therefore

$$\mu(F(k)) = v(F(k)) \geq v(A(k)) > v(A) - \varepsilon.$$

Obviously $F(1) \supseteq F(2) \supseteq \cdots$, so that their intersection Γ satisfies inequality (4). It remains to be proved that Γ is contained in A.

If $x \in \Gamma$, then, for any integer k, $x \in F(k)$, so that $x \in F_{n_1 n_2 \cdots n_k}$ for some collection $n_1 n_2 \cdots n_k$ satisfying the condition $n_1 \leq m_1$, $n_2 \leq m_2, \ldots, n_k \leq m_k$. We shall call such collections $n_1 n_2 \cdots n_k$ x-collections. By the definition of the array $\mathscr{F}$, any initial segment $n_1 n_2 \cdots n_l$ of the x-collection $n_1 n_2 \cdots n_l \cdots n_k$, $1 \leq l < k$, is also an x-collection. According to what was said above, there exist x-collections of arbitrary length k. We shall say that a collection is *good* if it coincides with an initial segment of x-collections of arbitrarily high length. There exists at least one good collection n_1 of length 1, since otherwise the lengths of all x-collections would be bounded. Analogously, for the good collection n_1 there exists at least one good continuation $n_1 n_2$ of length 2, and so forth out to infinity. The infinite sequence $n_1 n_2 \cdots n_k \cdots = n$ resulting from this procedure has the property that all of its initial segments are x-collections. But then $x \in F_n = \bigcap_k F_{n_1 n_2 \cdots n_k}$ and therefore $x \in A$. Accordingly, $\Gamma \subseteq A$.

Thus, all the sets gotten by $\mathscr{A}$-operations from measurable sets are μ-measurable.

Actually, we have proved here *the general theorem on the extension of capacities.* Suppose that $\mathbb{C}$ is some class of sets closed relative to the union of a finite number of sets and the intersection of a countable number of sets, and suppose that the function v is defined for all sets and satisfies conditions a), b), and the condition

c) $v(F(k)) \to v(\Gamma)$, if $F(1) \supseteq F(2) \supseteq \cdots$ belong to the system $\mathbb{C}$ and $\Gamma = \bigcap_k F(k)$.

If $\mathscr{F}$ is an array of the sets belonging to $\mathbb{C}$ and $A = \mathscr{A}\mathscr{F}$, then $v(A)$ is equal to the supremum $v(C)$ over all sets $C \in \mathbb{C}$ contained in A. In our case $\mathbb{C}$ is the system of all measurable sets, and property c) follows from the fact that the function v coincides with the measure μ on $\mathbb{C}$.

§4. Separability of Analytic Sets

We show in this section that *two nonintersecting analytic sets A_1 and A_2 can be separated by some pair B_1, B_2 of measurable sets* (i.e. this means that $A_1 \subseteq B_1$, $A_2 \subseteq B_2$, and $B_1 \cap B_2 = 0$).

First of all we note that *if every set of the sequence A_m can be separated from every set of the sequence C_n, then the union $A = \bigcup_m A_m$ can be separated from the union $C = \bigcup_n C_n$.* Indeed, if Γ_{mn} and Δ_{mn} are the measurable sets separating A_m from C_n, then the sets

$$\Gamma = \bigcup_m \bigcap_n \Gamma_{mn}, \qquad \Delta = \bigcup_n \bigcap_m \Delta_{mn}$$

separate A from C.

Now suppose that the nonintersecting analytic sets A_1 and A_2 are not separable. This supposition will lead us to a contradiction. According to §1, $A_1 = f(E_1)$, $A_2 = g(E_2)$, where E_1 and E_2 are Polish spaces and f, g continuous mappings. In the spaces E_1 and E_2 we choose arrays $\mathscr{F}$ and $\mathscr{G}$ made up of closed sets, and such that for any k

$$\bigcup_{m_1 m_2 \cdots m_k} F_{m_1 m_2 \cdots m_k} = E_1, \qquad \bigcup_{n_1 n_2 \cdots n_k} G_{n_1 n_2 \cdots n_k} = E_2$$

and the diameters of the sets $F_{m_1 m_2 \cdots m_k}$ and $G_{n_1 n_2 \cdots n_k}$ are less than $1/k$. Since

$$A_1 = \bigcup_{m_1} f(F_{m_1}), \qquad A_2 = \bigcup_{n_1} g(G_{n_1}),$$

it follows from the nonseparability of A_1 and A_2 that some pair $f(F_{m_1})$, $g(G_{n_1})$ is not separable. Since

$$f(F_{m_1}) = \bigcup_{m_2} f(F_{m_1 m_2}), \qquad g(G_{n_1}) = \bigcup_{n_2} g(G_{n_1 n_2}),$$

it follows from the fact that $f(F_{m_1})$ and $g(G_{n_1})$ are not separable that there exists a nonseparable pair $f(F_{m_1 m_2})$, $g(G_{n_1 n_2})$. By induction we obtain two sequences $m_1 m_2 \cdots = m$ and $n_1 n_2 \cdots = n$ such that the sets $f(F_{m_1 m_2 \cdots m_k})$ and $g(G_{n_1 n_2 \cdots n_k})$ are nonseparable for any k.

The closed nested sets $F_{m_1 m_2 \cdots m_k}$ and $G_{n_1 n_2 \cdots n_k}$ with diameters tending to 0 close down in the complete spaces E_1 and E_2 onto limit points x and y. Since $f(x) \in A_1, g(y) \in A_2$, then $f(x) \neq g(y)$ and the points $f(x)$ and $g(y)$ may be enclosed in nonintersecting open spheres U_1 and U_2. From the continuity of the mappings f and g it follows that for sufficiently large k the images $f(F_{m_1 m_2 \cdots m_k})$ and $g(G_{n_1 n_2 \cdots n_k})$ will be contained in U_1 and U_2 respectively, i.e. that they will be separated, which contradicts our supposition.

From the separability just proved it follows that *if the analytic set A has an analytic complement B\A, then A is measurable.* Indeed, A and $B \backslash A$ must be separable, and the measurable sets separating them can only be the sets A and $B \backslash A$ themselves.

Now it is easy to prove that *if the surjection f of the Borel space B_1 onto the Borel space B_2 is measurable, then the inverse mapping f^{-1} is also measurable (f thus becomes an isomorphism).* Indeed, if Γ is a measurable set in B_1, then $f(\Gamma)$ and $f(B_1 \backslash \Gamma)$ are analytic sets in B_2 and complementary to one another, so that $f(\Gamma)$ is measurable.

§5. Example of a Nonmeasurable Analytic Set

In this example we make use of an *analytic set A in the Oxy plane such that all Borel subsets of the line are contained among its x-sections.* (We shall show at the end of the section how one goes about constructing this set, using information from textbooks on the theory of functions of a real variable.) The intersection D of the set A with the diagonal $x = y$ will be an analytic set, so that its projection H on the Oy-axis will be analytic as well (see fig. A.2.1). We shall show that the set H is nonmeasurable.

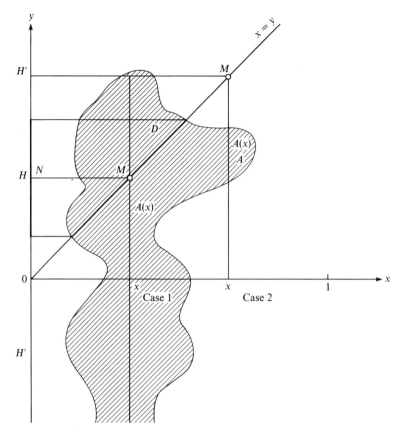

Figure A.2.1

It suffices to verify that the complement H' of the set H relative to the y axis is nonmeasurable. In view of the universality property of the set A, it suffices for this to verify that none of the x-sections $A(x)$ of the set A projects onto H'. For any x we consider a point M on the diagonal with coordinates (x,x) and its projection N on the Oy axis. Two cases are possible: 1) M belongs to $A(x)$; 2) M does not belong to $A(x)$. In the first case M belongs to the set D and accordingly projects into the set H. Therefore the projection of the set $A(x)$ onto the Oy axis is distinct from H'. In the second case M lies on the diagonal $x = y$ outside the set D, and therefore projects into H'. Hence once again the projection of $A(x)$ onto the Oy does not coincide with H'.

In order to construct the set A we make use of the Baire classification of functions and the theorem of Lebesgue on the universal function. One can familiarize himself with the Baire classification by looking, for example, in the textbook [1] of I. P. Natanson, Chapter 15. There one will also find a proof of the Lebesgue theorem, according to which in particular there exists a measurable function $y = F(x,z)$, with $0 \le x$, $z \le 1$, such that every function $y = f(z)$ of Baire class ≤ 2 can be obtained from F by fixing on some x; see Natanson loc. cit. Chapter 15, §3, Theorem 4. The projection of the graph of the function F onto the oxy plane is the set A which we need.

In order to show this we need to verify that any Borel set B on the line may be represented as the set of values of some function $y = f(z)(0 \le z \le 1)$ of Baire class ≤ 2. First we prove that B is a continuous image of the set N of all irrational numbers on the segment $[0, 1]$. In view of §3, b) in Appendix 1, to this end it suffices to show that every Polish space E is a continuous image of N. Every number $z \in N$ may be uniquely represented in the form of an infinite continued fraction

$$z = \cfrac{1}{n_1 + \cfrac{1}{n_2 + \cfrac{1}{n_3 + \cfrac{1}{\ddots}}}},$$

where $n_1, n_2, \ldots$ is a sequence of integers*. The set of all such sequences maps onto the space E with the aid of the $\mathscr{A}$-operation, applied to successive coverings of E by closed sets with diameters tending to 0.

The function f we desire has now been constructed at the irrational points. It remains to extend it to the rational points of the segment $[0,1]$ without increasing its range and without going outside the limits of the second Baire class. Denote by $\tilde{f}(z)(0 \le z \le 1)$ the upper limit of $f(u)$ when u tends to z through the irrationals. The function $\tilde{f}$ coincides with the function f on the set N. It can have discontinuities only at the rational points, and therefore it belongs to a Baire class ≤ 1[†]. It does not leave the Baire class 1 on changing its values at a finite number of points, since there will be once again not more than a countable number of discontinuities. Successively changing the values of $\tilde{f}$ at all the rational points, we obtain a sequence of functions of class ≤ 1, which converges to a function f mapping $[0,1]$ onto B. By the definition, the class of this function does not exceed 2.

* * *

In Chapter 3 we made use of the existence of a plane Borel set B with a non-Borelian projection onto the line. The existence of such a set follows from the existence of a nonmeasurable analytic set H and the fact that any analytic set on the line is the projection of some plane Borel set B.

We prove this assertion as follows. From the definition of an analytic set and the isomorphism of Borel spaces, it follows that any analytic set is a measurable image of the segment $[0,1]$. If B is an analytic set on the line $0y$ and $y = f(x)$ is a measurable mapping of the segment $[0,1]$ onto B, then B is the projection onto the $0y$ axis of the graph of the function f, and this graph is a Borel set in the $0xy$ plane. This was shown in §1 in a more general form.

---

* A. Ya. Khinchin [1], Chapter II, §5, Theorem 14.
† See Natanson ibid., in Chapter 15, §3, Example II.

Appendix 3

Theorems on Measurable Selection

§1. The Lemma of Yankov

Suppose that $Y \xrightarrow{f} X$ is a measurable mapping of the measurable space Y onto the measurable space X. Assigning to each x of X its preimage or fibre $Y(x) = f^{-1}(x)$, we obtain a correspondence f^{-1} of the space X into the space Y. The mapping $X \xrightarrow{\varphi} Y$ is said to be a *selector* of the correspondence f^{-1} if $f(\varphi(x)) = x$, i.e. $\varphi(x) \in Y(x)$ for each y. A measurable selector φ defines a *uniformization* of the correspondence f^{-1}, or a *measurable selection*.

In the case of an arbitrary measurable mapping f of the Borel space Y onto the Borel space X, it is possible that the correspondence f^{-1} does not admit a measurable selection (see §3). However we have the following result, Yankov's Lemma: *For any probability measure μ on X there exists a measurable mapping $X \xrightarrow{\varphi} Y$ such that $f(\varphi(x)) = x$ (a.e. μ).* We shall prove this assertion.

Let us show that it is always possible to metrize X and Y in such a way that:

    a) The measurable sets in X and Y coincide with the Borel sets;
    b) Y becomes a Polish space;
    c) The mapping f is continuous.

By the definition of the Borel spaces X and Y, they may be metrized in such a way that condition a) is satisfied. The product $X \times Y$ is also a metric space, and the graph Γ of the mapping $Y \xrightarrow{f} X$ is a measurable set in $X \times Y$ (see Appendix 2, §1). Hence Γ is a Borel space. In view of point b) of §3 of Appendix 1, the space Γ is a 1–1 image of some Polish space E under a continuous mapping g. Consider the diagram

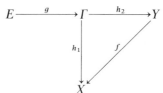

where h_1 and h_2 are projections of the graph Γ onto the spaces X and Y. Obviously $f h_2 = h_1$. The mappings g, h_1, and h_2 are continuous, and, accordingly, measurable. The measurable mappings g and h_2 are 1–1 and are therefore isomorphisms (see Appendix 2, §4). Using the product of these isomorphisms, we identify the spaces E and Y. Then the mapping $Y \overset{f}{\to} X$ goes over into a continuous mapping $E \overset{hg}{\longrightarrow} X$ of the Polish space E onto X. The identification of E and Y is equivalent to the introduction in Y of a new metric, which, along with the old metric in X, satisfies conditions a), b), and c).

It follows from conditions b) and c) that the fibres $Y(x)$ are all closed. Put $\varepsilon_n = 1/n$. We construct measurable sets $Y_1 \subseteq Y$ and $X_1 \subseteq X$ such that $X_1 = f(Y_1)$, $\mu(X_1) = 1$, and all the fibres $Y_1(x) = Y_1 \cap Y(x)$ for $x \in X_1$ are closed and have diameters less than ε_1. Repeating this construction, we obtain sequences $Y \supset Y_1 \supseteq Y_2 \supseteq \cdots$ and $X \supseteq X_1 \supseteq X_2 \supseteq \cdots$ such that $X_n = f(Y_n)$, $\mu(X_n) = 1$, and all the fibres $Y_n(x) = Y_n \cap Y(x)$ are closed and have diameters less than ε_n. Denote by Y_∞ the intersection of the sets Y_n and by X_∞ the intersection of the X_n. It is clear that Y_∞ and X_∞ are measurable, $\mu(X_\infty) = 1$, and $f(Y_\infty) \subseteq X_\infty$. For any $x \in X_\infty$ the fibre $Y_\infty(x) = Y_\infty \cap Y(x)$ is equal to the intersection of the nested subsets $Y_n(x)$ of the complete space Y with diameters tending to 0, and therefore consists of a single point. Therefore $f(Y_\infty) = X_\infty$ and the measurable mapping $Y_\infty \overset{f}{\to} X_\infty$ is 1–1. By §4 of Appendix 2, the inverse mapping $Y_\infty \overset{\varphi}{\to} X_\infty$ is also measurable. Completing the definition of the mapping φ to the set $X \backslash X_\infty$ by the formula $\varphi(x) = y_0$ (y_0 being a fixed point of the space Y), we obtain a measurable mapping $X \overset{\varphi}{\to} Y$ satisfying the required condition $f(\varphi(x)) = x$ for $x \in X_\infty$, $\mu(X_\infty) = 1$.

It remains to describe the construction of Y_1 and X_1. We cover the space Y with a countable number of closed spheres F_n of diameter smaller than ε_1. The images $A_n = f(F_n)$ of these spheres are analytic, and therefore μ-measurable sets in the space X (see Appendix 2). Therefore in each A_n one may select a measurable subset C_n with $\mu(C_n) = \mu(A_n)^*$. Put $D_1 = C_1$ and denote by D_n the collection of points of C_n not falling into $C_1 \cup C_2 \cup \cdots \cup C_{n-1}$. The sets D_n are measurable, nonintersecting, and up to a set of measure 0, cover the entire space X. Putting

$$X_1 = \bigcup_n D_n, \qquad Y_1 = \bigcup_n [f^{-1}(D_n) \cap F_n],$$

so that for $x \in D_n$ the fibre $Y_1(x)$ is equal to $F_n \cap Y(x)$, we obtain sets X_1 and Y_1 possessing all the needed properties.

§2. The Theorem of Blackwell and Ryll-Nardzewski

Suppose that $Y \overset{f}{\to} X$ is a measurable mapping of the Borel space Y onto the Borel space X, and that for each $x \in X$ a finite measure $v(\cdot | x)$ is defined on Y, for which:

1) *The function $v(\Gamma | x)$ is measurable in x for any measurable Γ of Y;*
2) *For any x, the measure $v(\cdot | x)$ is concentrated on the fibre $Y(x) = f^{-1}(x)$;*
3) *$v(Y | x) > 0$ for all x.*

Then the correspondence f^{-1} admits a measurable selection.

This result is proved along the same lines as the Jankov lemma in §1, with however the difference that now $X = X_1 = X_2 = \cdots = X_\infty$.

---

* The measure μ may be regarded as extended to all μ-measurable sets.

The set Y_1 is constructed as follows. We consider a countable covering $\{F_n\}$ of the space Y by closed spheres of diameter less than ε_1, and put

$$C_n = \{x : v(F_n | x) > 0\}. \tag{1}$$

In view of 1) the sets C_n are measurable, and in view of 3) they cover the entire space X. The sets D_n and X_1 are constructed starting from the C_n just as in §1, while now $\bigcup_n D_n = X$. In view of assumption 2) and formula (1)

$$v(Y_1 | x) = v(Y_1 \cap Y(x) | x) = v(F_n \cap Y(x) | x) = v(F_n | x) > 0 \tag{2}$$

for $x \in D_n$, so that all the fibres $Y_1(x) = Y_1 \cap Y(x)$ are nonempty.

Obviously, the measure $v(\cdot | x)$ retains properties 1) and 2) on the replacement of the space Y by its measurable subset Y_1, and according to (2) it also retains property 3). Therefore one may apply the same construction to Y_1 but with the number ε_2, and then proceed as in §1.

§3. Example of a Correspondence Not Admitting a Measurable Selection

In this section we construct *a Borel set D in the space 0xyz, projecting onto the entire plane 0xy and not containing the graph of any measurable function $z = \varphi(x,y)$, $x,y \in (-\infty,\infty)$ whatever*. This construction makes use of delicate results from the descriptive theory of functions, and we shall confine ourselves here to a short presentation of the fundamental idea.

As is well-known, any measurable (i.e. Borel) function $\varphi(x,y)$ belongs to one of the Baire classes. There is no highest class among these classes. No x-section of a function φ of class α can have its class higher than α. Therefore if among the x-sections of D there are graphs of measurable functions $z = f(y)$ of arbitrarily high classes, D contains no graph of a measurable function of two variables.

Now if F is the universal function considered in §5 of Appendix 2, the Borel set

$$\Gamma = \{(x,y,z) : 0 \le x \le 1, \ -\infty < y < +\infty, \ 0 \le z \le 1, \ y = F(x,z)\}$$

has just this last property. In fact, it is possible to construct a function $z = f(y)$ of arbitrarily high class, mapping the real line onto the segment $[0,1]$ in a 1–1 way, and such that the inverse function $y = g(z)$ will have class not exceeding 2*.

---

* Indeed, fix an arbitrary α. Construct a subdivision of the real line into two uncountable Borel sets B and B' of class α, and consider a continuous 1–1 mapping f of the set of irrational numbers of the segment $[0,\frac{1}{2}]$ onto $B \backslash S$ and of the set of irrational numbers of the segment $[\frac{1}{2},1]$ onto $B' \backslash S'$, where S and S' are countable subsets of B and B'. (Here we use the fact, for which see Kuratowski [1], §36, IV, Theorem 2, that any noncountable Borel set is the union of a countable set and a continuous 1–1 image of the space of irrational numbers.) Then f is extended to the rational numbers in such a way that they are mapped in a 1–1 manner onto S and S'. Then same considerations as those in §5 of Appendix 2 show that the resulting function $f(z)$ ($0 \le z \le 1$) is of class not exceeding 2.

Therefore, there exists an x such that $F(x,z) = g(z)$. Then the x-section of the set Γ is equal to the set

$$\{(y,z): y = F(x,z)\} = \{(y,z): y = g(z)\} = \{(y,z): z = f(y)\}.$$

However the set Γ is bad in the sense that it projects into a non-Borel set of the $0xy$ plane (see §5 of Appendix 2). Using the so-called theorem on the character of the set of uniqueness points, we may replace Γ by a Borel set D which projects onto the whole plane and which has identical x-sections with Γ for all x, and such that the equation $y = F(x,z)$ defines a 1–1 correspondence between y and z^*.

An isomorphic mapping of the $0xy$ plane onto the segment $I = [0 \le u \le 1]$ carries D into a Borel set H of the square $I \times [0 \le z \le 1]$, projecting onto I and not containing the graph of any measurable function $z = z(u)$. In addition the correspondence $H(u) = H \cap (u \times [0,1])$ is not uniformizable, since, if ψ is a measurable selector of this correspondence, and k is the projection onto z, then the graph of the measurable function $z = k\psi(u)$ belongs to H.

* See N. N. Luzin [1], pages 216–221.

Appendix 4

Conditional Distributions

§1. Introduction

In this appendix we will show that *if P is a probability measure on the product of Borel spaces $X \times Y$ and μ is the measure induced by it on X, then there exists a transition function $v(dy|x)$ from X into Y such that*

$$P(dx\,dy) = \mu(dx)v(dy|x). \tag{1}$$

More precisely, we shall construct a function $v(\Gamma|x)(\Gamma \in \mathscr{B}(Y),\ x \in X)$ with the following properties:

a) *v is for each x a probability measure on Y*;
b) *v is a measurable function of x for each Γ*;
c) *For any measurable bounded function f on the space $X \times Y$*

$$\int_{X \times Y} f(x,y)P(dx\,dy) = \int_X \mu(dx) \int_Y f(x,y)v(dy|x) \tag{2}$$

(Here properties a) and b) constitute the definition of a transition function, and property c) is an expanded version of (1).) A function v with the indicated properties is called a *conditional probability distribution on Y relative to X*.

In the case of a finite or countable space X the conditions a)–c) are satisfied by

$$v(\Gamma|x) = \begin{cases} \dfrac{P(x \times \Gamma)}{\mu(x)} & \text{if } \mu(x) > 0. \\ \beta(\Gamma) & \text{if } \mu(x) = 0, \end{cases}$$

where β is any fixed probability measure on Y. (We leave the verification to the reader.) The construction of conditional distributions for noncountable spaces X will be carried out in §4. This construction makes use of the concept of conditional mathematical expectation and some of its properties, and also of the existence in a Borel space of the so-called support system of functions. These last questions will be dealt with in §§2,3.

From the theorem on conditional distributions it is easy to obtain the decompositions of probability measures on product spaces used in Chapters 3 and 5 (Theorem F of Chapter 3, §4 and Chapter 5, §4). Suppose that P is a probability measure on a finite or infinite product of Borel spaces:

$$E = E_0 \times E_1 \times E_2 \times \cdots. \tag{3}$$

Denoting the measure induced on the product of a smaller number of factors also by P, and applying our result to $X = E_0 \times E_1 \times \cdots \times E_{t-1}$ and $Y = E_t$, we obtain the formula

$$P(dx_0\, dx_1 \cdots dx_t) = P(dx_0\, dx_1 \cdots dx_{t-1})v(dx_t | x_0 x_1 \cdots x_{t-1}), \tag{4}$$

where $v(dx_t | x_0 x_1 \cdots x_{t-1})$ is a transition function from $E_0 \times E_1 \times \cdots \times E_{t-1}$ into E_t. By an obvious induction on t one deduces from (4) that

$$P(dx_0\, dx_1 \cdots dx_t) = \mu(dx_0)v(dx_1 | x_0) \cdots v(dx_t | x_0 x_1 \cdots x_{t-1}),$$

where $\mu(dx_0) = P(dx_0)$ (here t is any positive integer smaller than the number of factors in (3)).

§2. Conditional Mathematical Expectations

In Appendix 4 we shall, without saying this explicitly each time, consider only measurable bounded functions f.

Suppose that P is a probability measure on a measurable space E. To any decomposition

$$E = E_1 \cup E_2 \cup \cdots \cup E_n \tag{1}$$

of the space E into pairwise disjoint measurable sets there corresponds a probability measure v^x, depending on the point $x \in E$, defined by the formula

$$v^x(B) = \begin{cases} \dfrac{P(E_k \cap B)}{P(E_k)} & \text{if } P(E_k) > 0, \\ \gamma(B) & \text{if } P(E_k) = 0, \end{cases} \tag{2}$$

where the number $k = k(x)$ indicates the set E_k to which x belongs, and γ is any fixed probability measure on E. The integral $v^x f$ of a function f relative to this measure is the *conditional mathematical expectation of f relative to the decomposition* (1). The function $\tilde{f}(x) = v^x f$ is constant on each set E_k, and if $A = E_k$ then

$$\int_A f(x)P(dx) = \int_A \tilde{f}(x)P(dx). \tag{3}$$

This formula remains valid also for a set A equal to the union of several elements of the decomposition. Such unions form a σ-algebra $\mathscr{A}$, all elements of which are measurable sets of the space E.

Now suppose that $\mathscr{A}$ is any σ-algebra made up of measurable sets. The *conditional mathematical expectation of the function f relative to $\mathscr{A}$ is any function $\tilde{f}$, measurable relative to $\mathscr{A}$ and satisfying the relation* (3) *for each set A of $\mathscr{A}$.*

The above definition is satisfied, along with the function $\tilde{f}$, by any function measurable relative to $\mathscr{A}$ and differing from $\tilde{f}$ only on a set of measure zero. We shall denote any such function by $E(f \mid \mathscr{A})$.

One easily deduces the following properties of conditional mathematical expectations from the above definition:

1) For any constant c
$$E(c \mid \mathscr{A}) = c \quad \text{(a.s.)};$$

2) $E(f + g \mid \mathscr{A}) = E(f \mid \mathscr{A}) + E(g \mid \mathscr{A})$ (a.s.);
3) If f is measurable relative to $\mathscr{A}$, then

$$E(fg \mid \mathscr{A}) = f E(g \mid \mathscr{A}) \quad \text{(a.s.)};$$

4) If $\tilde{\mathscr{A}} \subseteq \mathscr{A}$ are two σ-algebras, then

$$E\big[E(f \mid \tilde{\mathscr{A}}) \mid \mathscr{A}\big] = E\big[E(f \mid \mathscr{A}) \mid \tilde{\mathscr{A}}\big] = E(f \mid \tilde{\mathscr{A}}) \quad \text{(a.s.)}.$$

Suppose that g is any function and that $\mathscr{A}_g$ is the minimal σ-algebra containing all sets of the type $\{x : g(x) < c\}$, where c is any constant. The function F is measurable relative to the σ-algebra $\mathscr{A}_g$ if and only if it is representable in the form $F(x) = \varphi(g(x))$, where φ is a function measurable relative to the Borel measurable structure on the real line; one easily deduces this fact from Lemma 1 of Appendix 5. If g is measurable, then $\mathscr{A}_g$ consists of measurable sets, and we define $E(f \mid g)$ to be the conditional mathematical expectation $E(f \mid \mathscr{A}_g)$.

Further, suppose $\mathscr{A}_1 \subseteq \mathscr{A}_2 \subseteq \cdots$ is a sequence of σ-algebras consisting of measurable sets, and let $\mathscr{A}_\infty$ be the minimal σ-algebra containing all the $\mathscr{A}_n$. Suppose that the function f_n is the conditional mathematical expectation of the function f relative to $\mathscr{A}_n$, $n = 1, 2, \ldots$. Denote by C the set of those points x for which the limit

$$\lim_{n \to \infty} f_n(x) \tag{4}$$

exists, and denote by f_∞ a function equal to this limit on C and equal to zero outside C. In the theory of martingales one proves that $P(C) = 1$ and that the function f_∞ is one of the versions of the conditional mathematical expectation of f relative to $\mathscr{A}_\infty$.*

---

* See J. L. Doob [1], Chapter 7, §4, Theorem 4.3.

§3. Support Systems

In each Borel space E there exists a finite or countable system W of bounded measurable functions, having the following properties:

   1) *If for the sequence $\{v_n\}$ of probability measures on E the limit of the integral $v_n f$ exists for all functions of W, then there exists a probability measure v on E such that $v_n f \to vf$ for all $f \in W$;*
   2) *Any system $\mathscr{H}$ of functions containing W and closed relative to addition, multiplication by constants, and bounded passage to the limit*, contains all bounded measurable functions.*

We shall call a system W having the above properties a *support system*.

If the space E is finite, then the indicator functions of one-point sets form a support system. In the case of a countable space E a support system is formed by the indicator functions of one-point sets, except for a certain distinguished point, and the function everywhere equal to 1; we leave the verification to the reader. If the space E is uncountable, then, in view of the isomorphism of Borel spaces, one may regard E as the unit interval. We shall show that in this case one may take as a support system the sequence of functions $1, x, x^2, \ldots, x^m, \ldots$.

Suppose that v_n, $n = 1, 2, \ldots$ are probability measures on $[0,1]$. If the integrals of the function x^m relative to these measures converge for all m, then the integrals of any polynomial converge as well. Using the Weierstrass theorem on the uniform approximation of a continuous function by polynomials and the estimate

$$|v_n f - v_n g| \leq \sup_x |f(x) - g(x)|,$$

one easily verifies that the limit

$$l(f) = \lim_{n \to \infty} v_n f$$

exists for all continuous functions f. Obviously, $l(f_1 + f_2) = l(f_1) + l(f_2)$, $l(cf) = cl(f)$ for constant c, and $l(f) \geq 0$ for $f \geq 0$. According to the Riesz representation theorem[†], there exists a measure v such that $l(f) = vf$ for all continuous functions f, and, in particular, for all powers x^m. Since $v(1) = l(1) = 1$, then v is a probability measure, so that property 1) is satisfied.

Property 2) follows from the lemma on multiplicative systems (see Appendix 5, §1), if one puts $\mathscr{C} = W$ and takes account of the fact that a σ-algebra on the segment $[0,1]$, relative to which the function $f(x) = x$ is measurable, contains all the Borel sets of that segment.

---

   * We shall say that the sequence of functions f_n converges boundedly to a function f if $f_n(x) \to f(x)$ for each x, and if all the functions f_n are bounded by a common constant.
   [†] See Halmos [1], §56, Theorem 4.

§4. Existence of Conditional Distributions

In view of the isomorphism theorem it suffices to consider the case when X is equal to the half-open interval $(0,1]$. We denote by I_n^k the half-open interval $((k-1)/2^n, k/2^n]$, and consider a sequence of decompositions

$$X \times Y = \bigcup_{k=1}^{2^n} (I_n^k \times Y) \tag{1}$$

of the space $X \times Y$. By formula (2.2), to each decomposition there corresponds a measure on $X \times Y$, depending on the point of the space. We shall denote by v_n^{xy} the measure corresponding to the decomposition (1). It is easy to see that it does not in fact depend on y, and we will therefore write v_n^x.

Suppose that $\mathscr{A}_n$ is the σ-algebra in the space $X \times Y$ generated by the decomposition (1). By §2, we may take the function $f_n(x,y) = f_n(x) = v_n^x f$ as the conditional mathematical expectation of the function $f(x,y)$ relative to the σ-algebra $\mathscr{A}_n$. It is easy to see that $\mathscr{A}_1 \subset \mathscr{A}_2 \subset \cdots \subset \mathscr{A}_n \subset \cdots$, and that the σ-algebra $\mathscr{A}_\infty$ generated by the union of all the $\mathscr{A}_n$ coincides with the collection of sets $B \times Y$ for $B \in \mathscr{B}(X)$.

We shall write $(x,y) \in C_f$ if the limit

$$\lim_{n \to \infty} f_n(x) \tag{2}$$

exists. We denote the limit on C_f by $f_\infty(x)$, and extend this value to be zero outside C_f. According to §2, $P(C_f) = 1$, and f_∞ is the conditional mathematical expectation of f relative to $\mathscr{A}_\infty$.

Now fix on a support system W in $X \times Y$, and consider the intersection C of the sets C_f over all $f \in W$. It is clear that C belongs to the σ-algebra $\mathscr{A}_\infty$ and that $P(C) = 1$. Accordingly, $C = X' \times Y$, where $X' \in \mathscr{B}(X)$ and

$$\mu(X') = 1. \tag{3}$$

By property 1) of the support system, for $x \in X'$ there exist probability measures v^x on the space $X \times Y$ such that

$$f_\infty(x) = v^x f \qquad (f \in W, x \in X'). \tag{4}$$

The desired conditional distribution is

$$v(\Gamma \,|\, x) = \begin{cases} v^x(X \times \Gamma) & \text{for } x \in X', \\ \beta(\Gamma) & \text{for } x \in X \backslash X', \end{cases} \tag{5}$$

where $\Gamma \in \mathscr{B}(Y)$ and β is some fixed probability measure on Y. Indeed, property a) of §1 is obviously satisfied.

By (5), for any measurable set Γ of the space Y

$$v(\Gamma \,|\, x) = \begin{cases} v^x f & \text{for } x \in X, \\ \beta(\Gamma) & \text{for } x \in X \backslash X', \end{cases} \tag{6}$$

where f is the indicator of the set $X \times \Gamma$. Property b) will be proved if we show that for any measurable bounded function f the function $v^x f$ is measurable relative to x on the set X'. By property 2) of the support system it suffices to verify this for the functions f of W. But for $f \in W$

$$v^x f = \lim_{n \to \infty} f_n(x) = \lim_{n \to \infty} v_n^x f$$

for all $x \in X'$ (see (4) and the definitions of f_∞ and f_{n}.). The measure v_n^x corresponds, by (2.2), to the decomposition (1) of the space $E = X \times Y$. It is clear from (2.2) that $v_n^x f$ is measurable relative to x^*.

In order to deduce property c), we note that the relation (2.3) defining the conditional mathematical expectation, in the case of the σ-algebra $\mathscr{A}_\infty$, takes on the form[†]

$$\int_{B \times Y} f(x,y) P(dx\,dy) = \int_B f_\infty(x) \mu(dx) \qquad (B \in \mathscr{B}(X)). \tag{7}$$

Write $f \in \mathscr{H}$ if

$$\int_{B \times Y} f(x,y) P(dx\,dy) = \int_B (v^x f) \mu(dx) \qquad (B \in \mathscr{B}(X)). \tag{8}$$

By (3), (4) and (7), the support system W belongs to $\mathscr{H}$. It follows from the second property of a support system that $\mathscr{H}$ contains all bounded measurable functions.

Further, we shall show that for almost all x the measure v^x in the space $X \times Y$ is concentrated on the fibre $x \times Y$. To this end we apply (7) to the indicator of the

---

* Suppose that to each z of a measurable space Z there corresponds a probability measure $P(\cdot\,|\,z)$ on $X \times Y$, and that, for any measurable set A in $X \times Y$, the quantity $P(A\,|\,z)$ is measurable in z. In the decomposition (1.1) the measure μ is defined by the formula $\mu(B\,|\,z) = P(B \times Y\,|\,z)$, and therefore it depends in a measurable way on z. The conditional distribution $v(\cdot\,|\,x)$ constructed above also depends on z. We shall show that for any measurable set Γ in Y the conditional probability $v(\Gamma\,|\,x,z)$ is measurable in the pairs (x,z). Instead of the sets C_f and C in the space $X \times Y$, we consider the analogous sets D_f and D in the product $X \times Y \times Z$. The set D has the form $Q \times Y$, where Q is a measurable subset of $X \times Z$. Denote by Q_z the z-sections of Q. Then $P(Q_z\,|\,z) = 1$ for all z. Formulas (4), (5), and (6) remain valid when x is replaced by the pair (x,z) and X' by Q, and the considerations presented in the text show that $v(\Gamma\,|\,x,z)$ is measurable in the pairs (x,z) for $\Gamma \in \mathscr{B}(Y)$.

† Since

$$\int_{B \times Y} \tilde{f}_\infty(x) P(dx\,dy) = \int_B \tilde{f}_\infty(x) \mu(dx).$$

For the indicators of measurable sets, this relation reduces to the definition of μ as the measure induced on X by the measure P. Any bounded measurable function can be obtained from the indicators of measurable sets by linear operations and uniform passage to the limit.

set $(X\backslash I_n^k) \times Y$ and to the set $B = I_n^k$. We find that

$$0 = \int_{I_n^k} v^x((X\backslash I_n^k) \times Y)\mu(dx).$$

Since the integrand is nonnegative, the set

$$X_n^k = \{x : x \in I_n^k, v^x((X\backslash I_n^k) \times Y) > 0\}$$

has measure 0. Since the set of those x for which

$$v^x((X\backslash x) \times Y) > 0$$

is covered by the union of the X_n^k over all n and k, its measure is also zero.

Now for arbitrary measurable sets $B \subseteq X$ and $\Gamma \subseteq Y$ we have

$$P(B \times \Gamma) = \int_B v^x(B \times \Gamma)\mu(dx) = \int_B v^x(X \times \Gamma)\mu(dx) = \int_B v(\Gamma \mid x)\mu(dx). \quad (9)$$

The first equation follows from (8), the second from the fact that the measure v^x is almost surely concentrated on the fibre $x \times Y$, and the third from (5) and (3). If f is the indicator of the rectangle $B \times \Gamma$, then (1.2) coincides with (8). Applying the lemma 1 of Appendix 5, §1 to the system $\mathscr{C}$ consisting of the indicators of all measurable rectangles $B \times \Gamma$, we obtain (1.2) for all bounded measurable functions f.

Appendix 5

Some Lemmas on Measurability

§1. The Lemma on Multiplicative Systems

Often one has to prove that it is possible, starting with a stock of functions on hand, to obtain all measurable functions by using linear operations and passage to the limit. Various propositions about this problem are in the books of Doob [1], Dynkin [1], and Meyer [1]. We shall present the formulation of Meyer*.

Lemma 1. *Suppose that $\mathscr{H}$ is any system of bounded functions on the space E, closed with respect to addition, multiplication by constants, and bounded passage to the limit. Suppose that $\mathscr{C}$ is a system of functions on E, closed with respect to multiplication, and that $\sigma(\mathscr{C})$ is the minimal σ-algebra in E with respect to which all the functions of $\mathscr{C}$ are measurable. If the system $\mathscr{H}$ contains $\mathscr{C}$ and the constants, then it contains all bounded functions measurable relative to $\sigma(\mathscr{C})$.*

We note one of the consequences of this lemma.

Lemma 2. *Suppose that E_1, E_2, E_3 are measurable spaces, $F(x_1,x_3)$ a bounded measurable function on $E_1 \times E_3$, and $\mu(dx_3,x_2)$ a measure on E_3 depending measurably on x_2. Then the formula*

$$\Phi(x_1,x_2) = \int_{E_3} F(x_1,x_3)\mu(dx_3,x_2) \tag{1}$$

defines a measurable function on $E_1 \times E_2$.

For the proof it suffices to apply Lemma 1 to the system $\mathscr{C}$ of all functions of the type

$$F(x_1,x_2) = \chi_A(x_1)\chi_B(x_3) \qquad (A \in \mathscr{B}(E_1),\, B \in \mathscr{B}(E_3)),$$

and to the system $\mathscr{H}$ of bounded measurable functions $F(x_1,x_3)$ for which the integral (1) is measurable in (x_1,x_2). It is easy to see that $\sigma(\mathscr{C}) = \mathscr{B}(E_1) \times \mathscr{B}(E_3) = \mathscr{B}(E_1 \times E_3)$.

---

* See P. Meyer, [1], Chapter I, Theorem 20.

§2. Measurable Structure in the Space of Probability Measures

Suppose that $\mathcal{M}$ is the set of all probability measures on the measurable space E. We will consider $\mathcal{M}$ as a measurable space, by introducing a measurable structure using the functions*

$$F(\mu) = \mu f, \qquad \mu \in \mathcal{M}. \tag{1}$$

Here the f are arbitrary bounded measurable functions on E. We shall show that *if E is Borelian, then $\mathcal{M}$ is also Borelian.*

Consider in E a support system $W = \{f_1, f_2, \ldots, f_n, \ldots\}$ where all the f_n are bounded by unity; see §3 of Appendix 4. The functions

$$F_n(\mu) = \mu f_n, \qquad \mu \in \mathcal{M}, \tag{2}$$

generate the σ-algebra $\mathcal{F}$. Indeed, suppose that $\mathcal{F}'$ is the σ-algebra generated by the functions (2). Obviously $\mathcal{F}' \subseteq \mathcal{F}$. Denote by $\mathcal{H}$ the system of all functions f for which (1) is measurable relative to $\mathcal{F}'$. It follows from property 2) of a support system that $\mathcal{H}$ contains all bounded measurable functions on the space E. Hence $\mathcal{F}' = \mathcal{F}$.

Assigning to each probability measure μ the sequence $h_1 = \mu f_1, h_2 = \mu f_2, \ldots,$ $h_n = \mu f_n, \ldots$, we define a mapping ψ of the set $\mathcal{M}$ into the product H of a countable number of unit intervals. By §2 of Appendix 1, H may be regarded as a Polish space. It follows from property 1) of a support system (see §3 of Appendix 4) that the image $H' = \psi\mathcal{M}$ of $\mathcal{M}$ is closed in H. It follows easily from property 2) that the mapping ψ is 1–1. Under the mapping ψ the sets

$$\{\mu : \mu f_n < c\}$$

go into the sets

$$\{h : h \in H', h_n < c\}.$$

The former generate the σ-algebra $\mathcal{F}$, and the latter the σ-algebra $\mathcal{B}(H')$. Accordingly, ψ is an isomorphism of $\mathcal{M}$ onto H'. Hence $\mathcal{M}$ is a Borel space.

---

* Suppose that $\mathcal{H}$ is any system of functions on any set X, and suppose that $\mathcal{F}$ is the minimal σ-algebra in X with respect to which all the functions of the system $\mathcal{H}$ are measurable. We will then say that $\mathcal{H}$ *generates the σ-algebra* $\mathcal{F}$. To introduce a measurable structure in X using the system of functions $\mathcal{H}$ means to adopt as the measurable sets the elements of the σ-algebra $\mathcal{F}$.

Historical-Bibliographical Notes

These notes do not have as their objective that of giving a full bibliography or an exhaustive estimate of the roles of individual authors. As a rule, we have indicated only those papers which played a basic role in the development of the theory, and those which we have used as source material.

Maximization of the Total Reward (General Theory of Chapters 1–6)

In the Preface we spoke of the basic role of the ideas of Wald [1,2] in the creation of methods of making sequential decisions in a stochastic situation, and also of the significance of the books [1,3] of Bellman, in which he formulated the principles of dynamic programming and demonstrated their usefulness in the solution of numerous applied problems. (Bellman's investigations became widely known and were influential for several years before the publication of the first monograph [1].)

Controlled Markov processes with a finite number of states first appeared in the literature, under the name "Markov decision processes", in the works [1,2] of Bellman. Still earlier Bellman and Blackwell [1], and Shapley [1], studied so-called stochastic games, which were Markov processes controlled by two players with opposed interests. On the other hand, Arrow, Harris, and Marschak [1], and Dvoretzky, Kiefer and Wolfowitz [1], developed the theory of inventory control, in which there arise Markov control processes with nondiscrete state and action spaces. The first attempt to give a description of general models with arbitrary state and action spaces was carried out by Karlin [1].

Interest in Markov control processes as an independent object of investigation was stimulated by the book [1] of Howard, devoted to finite homogeneous models. Howard in particular showed that there exists a best stationary strategy among all stationary strategies in both the case of a discounted total reward and the case of an average reward per unit time, and indicated algorithms for finding these best strategies.

The theory reached its present form in the papers of Blackwell and Strauch. Finite homogeneous models with discounting were studied by Blackwell [2]. Here one finds the first proof of the existence of a stationary optimal strategy (cf.

Chapter 6, §3). The difficulties connected with the passage to general models were investigated by Blackwell [3,4]. He showed that it is possible to avoid them by making use of theorems on measurable selection, and by passing from optimal strategies to strategies which are a.s. ε-optimal. Here conditions were imposed on the running reward function which guaranteed the absolute and uniform convergence of the total reward. Models with arbitrary positive reward function were studied by Blackwell in [5], and those with arbitrary negative reward by Strauch [1]. Our exposition of the general theory in §13 of Chapter 1, Chapter 3, §8 of Chapter 4, and Chapter 5 follows basically the above-mentioned paper of Strauch, and that in Chapter 6, §3 and §8 of the papers [2,4] of Blackwell.

Independently from Blackwell, and using different methods, Krylov [1] showed the existence of a stationary ε-optimal strategy for models with countable state spaces, under conditions on the running reward function which are somewhat weaker than those of Blackwell. Using the same methods, Krylov showed in [2] the existence of a stationary optimal strategy for finite models.

Countable models were also dealt with in the papers of Derman [2] and Maitra [1]. A different approach to countable models is developed by Hordijk [1].

We shall go into some detail in the question of the existence of stationary a.s. ε-optimal strategies. This question was answered positively in Chapter 6, §8 under the hypothesis that the reward q is bounded and the discount coefficient $\beta < 1$. This result was first proved by Blackwell in [4]. If $\beta = 1$ the situation is more complicated. As Blackwell showed in [5], even in the case when $q \geq 0$ and the value v is everywhere finite, stationary a.s. ε-optimal strategies may fail to exist. In [5] Blackwell conjectured that for the existence of such a strategy it should be sufficient that $v(x)$ is bounded and q nonnegative. That this conjecture is in fact true follows from the following result of Frid [1]: if $q \geq 0$ and v is everywhere finite, then for any $\lambda < 1$ there exists a stationary strategy φ such that $w(x,\varphi) \geq \lambda v(x)$ (a.s.).

We note further the paper [6] of Blackwell, where he proves that if the state space is countable, $q \geq 0$ and $v < \infty$, then the existence of any optimal strategy at all implies the existence of a stationary optimal strategy. Orkin [1] generalized this result to the case of an arbitrary Borel state space (but his stationary strategy is only a.s. optimal). The analogous result proved in Chapter 6, §§3,7 for Borel models bounded above includes the cases treated by Blackwell in [4] and Strauch in [1].

All of the above-mentioned papers studied only homogeneous models. The nonhomogeneous case was considered by Furakawa in [1] and Hinderer in [1]. Hinderer also introduced generalizations of the classes of positive and negative models, analogous to our models bounded above or below. Summable models were considered by Hinderer [2].

The class of semicontinuous models with a finite interval of control was first studied by Dubins and Savage [1] (in a somewhat different form where the action a is identified with the distribution $p(\cdot|a)$); see Chapter 2, §16. Maitra [2,3] considered this class in the framework of the Blackwell theory. The measurable selection theorem in Chapter 2, §6 of this book is due to Kuratowski and Ryll-Nardzewski [1]. Our proof of this theorem and of its consequences is taken from Dynkin [5].

The theory of semicontinuous models is based on the fact that the operator T preserves invariant the class of upper semicontinuous functions. Blackwell, Freedman and Orkin [1] and Freedman [1] studied models (on a finite control interval) connected, in the same way, with some other functional classes.

In the first of these papers, this is the class of nonnegative semianalytic functions: a function f is semianalytic if the set $\{x: f(x) > c\}$ is analytic for every real c. The authors consider (in general, non-Borelian) models with analytic state and action spaces and nonnegative semianalytic reward functions, and prove the existence of simple uniformly ε-optimal strategies measurable with respect to the σ-algebra generated by the analytic sets. In distinction to our approach, they identify a with $p(\cdot|a)$, treat only nonrandomized strategies and include in the history h only the states but not the actions. The invariance property of the class of semianalytic functions was earlier used by Zvonkin [1] to investigate some Markov control processes with continuous time parameter.

Freedman [1] considered models with topological state and action spaces in which the operator T preserves either the class of nonnegative lower semicontinuous functions or the class of nonnegative functions f with the property: every set $\{x: f(x) > c\}$ is a countable union of closed sets (e.g., is of class F_σ).

The connections between models in the general sense and the models of Dubins-Savage type [1] were studied recently by Blackwell [7].

Maximization of the Average Reward per Unit Time
(General Theory of Chapter 7)

The strategy improvement procedure in the finite models described in Chapter 7, §5, is due to Howard [1]. He showed that after a finite number of steps this procedure will lead to a strategy which is the best in the class of all stationary strategies. Here the question remained open, as to whether there existed a still better nonstationary strategy. A negative answer to this question was obtained independently by Derman [1], and Viskov and Shiryaev [1]. (A paper [1] of Romanovskii was also devoted to this question.) The asymptotic formula of Chapter 7, §6 is due to Blackwell [2].

Countable models with constant asymptotic value were investigated by Derman [3] and Ross [1]. General models with constant v were investigated by Taylor [1], Ross [2], Gubenko and Shtatland [1], and with arbitrary v by Gubenko [1]; this was the first time that a canonical system of equations, as in Chapter 7, §2, had been written down in the general case. The concept of a canonical strategy was proposed by Yushkevich [1] (for the case of finite models, Denardo and Fox [1] were very close to this concept). Models with minorants, and some generalizations of these models, were considered by Gubenko and Shtatland [1], and special cases have been in the cited papers of Taylor, Derman, and Ross. See also Hordijk [1]. There one will find still further conditions guaranteeing the existence of asymptotically optimal strategies.

Example 1 of Chapter 7, §8, in which the state space is finite, the action spaces compact, and there is no asymptotically optimal strategy, is contained in the survey of Bather [1]. Martin-Löf proved in [1] for such models the existence of a stationary asymptotically optimal strategy under the assumption that for any choice of actions all the states form one ergodic class (and some further assumptions of a general character). Fainberg in [1] using other methods, extended this result to the case when some of the states are transient. In §13 of Chapter 7 we present a result of Chitashvili [1] and Fainberg [2]. Fainberg's proof is valid also for upper values.

A large number of papers are devoted to more detailed investigations of finite models, for instance the study of the more delicate behavior of the reward over the interval $[0,n]$ as $n \to \infty$ (which is connected with the asymptotics of the total reward over an infinite time as $\beta \uparrow 1$), algorithms for calculating the asymptotic value and finding a stationary optimal strategy, using in particular linear programming methods, and other questions. For further information on this topic we refer the reader to the book [4] of Derman, which deals specially with finite models.

Models with Incomplete Information
(General Theory of Chapter 8)

The general schema of control with incomplete information was proposed by Shiryaev [1,2] and Dynkin [2]. The reduction to models with complete information was carried out for models with a countable state space and Borelian action spaces by Savarigi and Yoshikawa [1], and for general Borel models by Rhenius [1] and Yushkevich [2], independently.

Concave Models and Models of Economic Development
(Introduction, and Chapter 9)

The model of Gale [1] is a natural generalization of the model of economic development proposed by von Neumann [1]. In the von Neumann model there are finitely many basic production processes, which can be carried out with arbitrary intensity. (A special case of this is the Leontiev model, where the output of each production process consists of only one product, and each product can be obtained from a unique production process.) Optimal plans in the deterministic case were investigated by Gale [3]. The proofs of Gale were improved by Brock [1].

A stochastic version of the Gale model and a concave model over a finite interval were studied by Dynkin [4,5,6]. The content of Chapter 9 consists in the exposition of these papers. The case of an infinite interval was investigated in the papers [1,2] of Evstigneev, [1] of Kuznetsov, and [1] of Taksar. Another variant of models of economic production under uncertainty was worked out by Radner and his colleagues; see Radner [2].

The importance of prices for optimization problems was noted as far back as 1940 by Kantorovich [2]. The further development of these ideas is represented by the well-known Kuhn-Tucker theorem [1].

For further information on (deterministic) multi-sector models of economic development, we refer to a survey by Gale [4] and the books of Gale [2] and Nikaido [1].

The Problem for Allocation of a Resource Between Production and Consumption (Introduction, Chapter 2, §7, and Chapter 6, §9)

A one-sector model of production and consumption, taking account of random risks, was first studied by Phelps in [1], starting from an idea of Ramsey [1] and working out the cases of power and logarithmic utility. More refined questions were investigated in recent papers of Brock and Mirman [1,2], Mirman [1,2], Beckmann [2], where more detailed bibliographical information can be found.

The Water Regulation Problem (Introduction, and Chapter 2, §8)

In a rather more developed form, this problem has great practical importance; see, for example, Moran [1].

The Problem of Allocation of Stakes (Introduction, Chapter 2, §9 and Chapter 6, §10)

The optimality of the bold strategy for a gambler who needs to accumulate a fixed sum in an unfavorable game was established by Dubins and Savage [1]. That monograph has played an important rôle in the development of stochastic control processes. We have already noted above the results on semicontinuous models contained in it; and our proof in Chapter 2, §9 was taken from Chapter 5 in it.

The case of a favorable game was investigated by Breiman [1]. Berry, Heath and Sudderth [1] have found a strategy which is ε-optimal for all values of the probability p of the gain simultanuously.

The Problem for Allocation of a Resource Among Consumption and Several Productive Sectors (Introduction, Chapter 2, §10, and Chapter 6, §11)

This model was investigated by Samuelson [1], on the assumption that the output in one of the sectors was not random.

The Replacement Problem (Introduction, Chapter 1, §11, Chapter 6, §5, and Chapter 7, §11)

A great deal of literature in the journals has been devoted to various variants of this problem. Numerical examples are discussed in the book [1] of Bellman and Dreyfus. We note also the monograph [1] of Jorgenson, McCall and Radner, containing, in particular, a detailed bibliography. Our exposition does not make use of these sources.

The Stabilization Problem (Introduction, Chapter 2, §11, Chapter 6, §12, Chapter 7, §12, and Chapter 8, §5)

The problem of control of a one-dimensional linear stochastic system with a quadratic criterion (and full information) was first considered by Simon in [1], where he proved that for optimal control it suffices to know only the means and the variances of the random oscillations. Theil [1] extended that result to the multidimensional case, and Kalman and Köpcke [1] worked out a corresponding algorithm.

The problem of control of a Gaussian system with incomplete information was considered by Joseph and Tou in [1]. They established one of the variants of the separation theorem. (An algorithm for the best estimate of the unobservable state was proposed earlier by Kalman [1]). The separation theorem presented here is given also in DeGroot's book [1], in Chapter 14, §11.

The problem described above and its generalizations have numerous important applications, and an enormous literature is devoted to it. We mention only the books of Feldbaum [1], Aoki [1], Sawaragi, Sunahara, and Nakamizo [1], and Åström [1].

The Bus, Streetcar, or Walk Problem (Chapter 1, §10 and Chapter 6, §4)

Generalizations of this amusing problem may be found in the journal literature, e.g. Ambarjan [1].

The Two-armed Bandit Problem (Chapter 8, §3)

Our exposition is a simplified variant of the paper [1] of Feldman. The same paper was used by DeGroot in [1], Chapter 14, §§5–7, where the reader will find historical notes and references to the literature.

We note the paper [1] of Cover and Hellman, in which they found the best strategy realizable with a fixed finite memory. They also used an analogous approach to other optimal control problems.

Borel spaces (Appendix 1)

The isomorphism of Borel spaces, sometimes called standard Borel spaces, was proved in the monograph [1] of Kuratowski. Our exposition makes use of Kuratowski's proof and of Chapter I of Parthasarathi's book [1].

Analytic sets (Appendix 2)

The $\mathscr{A}$-operation was introduced by P. S. Aleksandrov in 1916. Then Luzin formulated the problem: is every set obtained from intervals using the $\mathscr{A}$-operation a Borel set? Suslin gave a negative answer to that question, and thus opened up a new class of sets, which now bear the name analytic sets (a term introduced by Luzin). The theorems on the separation of analytic sets, and on their measurability relative to Lebesgue measure, were proved by Luzin. The theory of analytic sets was treated by Luzin in the monograph [1], from which we have taken the examples presented in §5 of Appendix 2, and in §3 of Appendix 3. The universal measurability of analytic sets is proved in Appendix 2 in the same way as Choquet proved capacibility of analytic sets. For the $\mathscr{A}$-operation, and analytic sets, see also Saks [1]. More modern expositions can be found in Chapter 3 of Meyer's book [1] or in Dellacherie [1], Chapter 1.

The Measurable Selection Theorems (Appendix 3)

The measurable selection theorems of §1 of Appendix 3 was first proved by Yankov [1] (for the projection of a plane analytic set onto a line and Lebesgue measure). Nine years later it was reproved by von Neumann, as Lemma 5 in [2], apparently without knowledge of the work of Yankov. Von Neumann considered a more general case of a continuous mapping of an arbitrary analytic set into the line. However, in view of the isomorphism of Borel spaces, this case, as well as the still more general case taken up in §1 of Appendix 3, can be reduced to the situation considered by Yankov. It seems that many other authors are unacquainted with the Yankov paper, and refer to that lemma as the von Neumann lemma. The proof we give differs from the earlier ones.

The more general measurable selection theorem in which an arbitrary measure space maps measurably into a Borel space, was proved by Aumann, as Theorem 2 in [1]. Another version of the proof was presented in the monograph [1] of Hildenbrand. In that monograph one will find further references to the literature.

The theorem of §2 of Appendix 3 belongs to Blackwell and Ryll-Nardzewski [1]. Its proof has been simplified thanks to the universal measurability of the analytic sets, and is carried out according to the same plan as the proof of the theorem of §1 of Appendix 3.

Conditional Distributions (Appendix 4)

Actually the theorem on the existence of conditional distributions is contained already in Chapter 1, §9 of Doob's book [1]. In the form we need it is proved by Parthasarathi in Chapter 5, §8, of [1]. Our proof is somewhat different. The concept of a support system was introduced by Dynkin in [3].

The existence of conditional distributions, as well as a number of nice properties of Borel spaces, hold also for a wider class of Luzin spaces, introduced by Blackwell in [1].

Bibliography*

Ambaryan, S. L.

(1,1967)   A choice of optimal routes of passenger movement at a given urban transportation network. *Econ. and Mat. Methods* **3**, 862–871 (in Russian).

Aoki, M.

(1,1967)   *Optimization of Stochastic Systems.* Academic Press.

Aris, R.

(1,1964)   *Discrete Dynamic Programming.* Blaisdell.

Arrow, K. J.

(1,1971)   *Essays in the Theory of Risk-bearing.* Markham Publ. Co., Chicago.

Arrow, K. J., Harris, T., Marschak, J.

(1,1951)   Optimal inventory policy. *Econometrica* **19**, 250–272.

Arrow, K. J., Karlin, S., Scarf, H.

(1,1958)   *Studies in the Mathematical Theory of Inventory and Production.* Stanford University Press.

(2,1962)   *Studies in Applied Probability and Management Science.* Stanford University Press.

* Books and articles published originally in Russian are listed as follows: first the date of the first Russian edition is given in parentheses, then the title of the English translation and the date of the most recent English edition. Most articles have appeared in cover-to-cover translations of the corresponding Russian journals:

Theory of Prob. (published by SIAM)—Teoriya veroyatnostei i ee primeneniya,
  Soviet Math., Doklady (published by AMS)—Doklady Akademii Nauk SSSR
  Math. of the USSR—Sbornik (published by AMS)—Matematicheskii sbornik
  Russian Math. Surveys (published by the London Math. Society)—Uspekhi matematicheskikh
  Cybernetics-Kibernetika (Kiev)

Åström, K. J.

(1,1970)   *Introduction to Stochastic Control Theory.* Academic Press.

Aumann, R. J.

(1,1969)   Measurable utility and the measurable choice theorems. *La Decision*, **2** (Actes
            Coll. du Internat., Aix-en-Province 1967), 15–26. Edition de CRNS.

Bather, J. A.

(1,1973)   Optimal decision procedures for finite Markov chains. I: Examples. *Adv. Appl.
            Prob.* **5**, 328–339, II: Communicating systems, ibid., 521–540, III: General
            convex systems, ibid., 541–553.

Beckmann, M. J.

(1,1968)   *Dynamic Programming of Economic Decisions.* Springer-Verlag.

(2,1974)   Resource allocation over time. Some dynamic programming models. *Mathe-
            matical Models in Economics.* North-Holland, 171–178.

Bellman, R.

(1,1957)   *Dynamic Programming.* Princeton Univ. Press.

(2,1957)   A Markovian decision process. *J. Math. Mech.* **6**, 679–684.

(3,1961)   *Adaptive Control Processes.* Princeton Univ. Press.

Bellman, R., Blackwell, D.

(1,1949)   *On a Particular Non-Zero Sum Game.* Rand McNally.

Bellman, R., Dreyfus, S.

(1,1962)   *Applied Dynamic Programming.* Princeton Univ. Press.

Berry, D., Heath, D., Sudderth, W.

(1,1974)   Red and black with unknown win probability. Ann. Stat. **2**, 603–608.

Bertsekas, D. P.

(1,1976)   Dynamic Programming and Stochastic Control. Academic Press.

Blackwell, D.

(1,1956)   On a class of probability spaces–"Proc. 3rd Berkeley Sympos. on Math. Stat.
            and Prob. 1954–1955," v. 2, 1–6.

(2,1962)   Discrete dynamic programming. *Ann. Math. Stat.* **33**, 719–726.

(3,1964)   Memoryless strategies in finite-stage programming. *Ann. Math. Stat.* **35**, 863–865.

(4,1965)   Discounted dynamic programming. *Ann. Math. Stat.* **36**, 226–235.

(5,1967)   Positive dynamic programming. Proc. of the 5th Berkeley Symposium, Vol. 1, 415–418.

(6,1970)   On stationary strategies. *Journal of the Royal Stat. Society, Ser* A, **133** Part 1, 33–37.

(7,1976)   The stochastic process of Borel gambling and dynamic programming. *Ann. Stat.* 4, 370–374.

Blackwell, D., Freedman, D., Orkin, M.

(1,1974)   The optimal reward operator in dynamic programming. *Ann. Prob.* **2**, 926–941.

Blackwell, D., Ryll-Nardzewski, C.

(1,1963)   Non-existence of everywhere proper conditional distributions. *Ann. Math. Stat.* **34**, 223–225.

Boltyanskii, V. G.

(1,1973)   *Optimal Control of Discrete Systems.* Nauka (in Russian).

Bourbaki, N.

(1,1953)   *Espaces Vectoriels Topologiques.* Hermann.

Breiman, L.

(1,1961)   Optimal gambling systems for favorable games.–"Proc. 4th Berkeley Sympos. on Math. Stat. and Prob." v. 1, 67–78.

Brock, W. A.

(1,1970)   On existence of weakly maximal programs in a multisector economy. *Rev. Econ. Studies* **37**, 275–280.

Brock, W. A., Mirman, L. J.

(1,1972)   Optimal economic growth and uncertainty: the discounted case. *J. Econ. Theory* **4**, 479–513.

(2,1973)   Optimal economic growth and uncertainty: the no discounting case. *International Economic Review* **14**, 560–573.

Chitashvili, R. Ya.

(1,1975)   A controlled finite Markov chain with an arbitrary set of decisions. Theory of Probability **20**, 839–846.

Cover, T. M., Hellman, M. E.

(1,1970)   The two-armed bandit problem with time-invariant finite memory. *IEEE Trans. Inform. Theory* **IT-16**, 185–195.

Cramer, H.

(1,1946)   *Mathematical Methods of Statistics*. Princeton Univ. Press.

De Groot, M.

(1,1970)   *Optimal Statistical Decisions*. McGraw-Hill.

De Leve, G.

(1,1964,   *Generalized Markovian Decision Processes*. I: *Model and Method*. 1964, II:
1970)   *Probabilistic background*. 1964, III: *Applications* (together with H. G. Tijms
            and P. J. Weeda). 1970. Mathematisch Centrum.

Dellacherie, C.

(1,1972)   Capacités et Processus Stochastiques. Springer-Verlag.

Denardo, E. V., Fox, B. L.

(1,1968)   Multichain Markov renewal programs. *SIAM J. Appl. Math.* **16**, 468–487.

Derman, C.

(1,1962)   On sequential decisions and Markov chains. *Management Sci.* **9**, 16–24.

(2,1965)   Markovian sequential control processes—denumerable state space. *J. Math.
            Anal. Appl.* **10**, 295–302.

(3,1966)   Denumerable state Markovian decision processes—average cost criterion. *Ann.
            Math. Stat.* **37**, 1545–1553.

(4,1970)   *Finite State Markovian Decision Processes*. Academic Press.

Doob, J. L.

(1,1953)   *Stochastic Processes*. Wiley.

Dubins, L. E., Savage, L. J.

(1,1965)   *How to Gamble if You Must*. McGraw–Hill.

Dunford, N., Schwartz J. T.

(1,1958)   *Linear Operators* Part I. Wiley-Interscience.

Dvoretzky, A., Keifer, J., Wolfowitz, J.

(1,1952)   The inventory problem. *Econometrica* **20**, 187–222, 450–466.

Dynkin, E. B.

(1,1959)   *Die Grundlagen der Theorie der Markoffschen Prozesse*. Springer-Verlag. 1961.

(2,1965)   Controlled random sequences. *Theory of Probability* **10**, 1–14.

(3,1969)   Exit spaces of Markov processes. *Russian Math Surveys.* **24**, 4, 89–157.

(4,1971)   Some probabilistic models of developing economies. *Soviet Math., Doklady* **12**, 1422–1425.

(5,1972)   Stochastic concave dynamic programming. *Math. of the USSR, Sbornik* **16**, 501–515.

(6,1974)   Optimal programs and stimulating prices in probabilistic models of economic development *Mathematical Models in Economics*, Editors J.Łoś and M. W.Łoś, 207–218. North-Holland.

(7,1976)   Economic equilibrium under uncertainty, *Computing Equilibria: How and Why.* Editors J.Łoś and M. W.Łoś. 41–60, North-Holland.

Dynkin, E. B., Ovseevich, A.

(1,1975)   On preference relations under uncertainty *Econ. and Mat. Methods* (in Russian), **11**, 393–395.

Evstigneev, I. V.

(1,1972)   Optimal economic planning with due consideration of stationary random factors. *Soviet Math. Doklady* 13, 1357–1359.

(2,1974)   Optimal stochastic programs and their stimulating prices. *Mathematical Models in Economics.* 219–252, North-Holland.

Fainberg, E. A.

(1,1975)   On controlled finite state Markov processes with compact control sets. *Theory of Probability* **20**, 856–862.

(2,1978)   The existence of a stationary ε-optimal policy for a finite Markov chain. *Theory of Probability* **23**, 297–313.

Feldbaum, A. A.

(1,1965)   *Optimal Control Systems.* Academic Press.

Feldman, D.

(1,1962)   Contributions to the "two-armed bandit" problem. *Ann. Math. Stat.* **33**, 847–856.

Feller, W.

(1,1950)   *Probability Theory and its Applications*, vol. 1. Wiley. (reprinted 1968).

Fleming, W. H., Rishel, R. W.

(1,1975)   *Deterministic and Stochastic Optimal Control.* Springer-Verlag.

Freedman, D.

(1,1974)   The optimal reward operator in special cases of dynamic programming problems. *Ann. Prob.* **2**, 942–949.

Frid, E. B.

(1,1970)   On a problem of D. Blackwell from the theory of dynamic programming. *Theory of Probability* **15**, 719–722.

Furukawa, N.

(1,1968)   A Markov decision process with non-stationary laws. *Bull. Math. Stat.* **13**, 41–52.

Gale, D.

(1,1956)   The closed linear model of production. *Linear Inequalities and Related Systems* Editors H. W. Kuhn and A. W. Tucker Princeton Univ. Press.

(2,1960)   *Theory of Linear Economic Models.* McGraw–Hill.

(3,1967)   On optimal development in a multisector economy. *Rev. Econ. Studies* **34**, 1–18.

(4,1968)   A mathematical theory of optimal economic development. *Bull. Amer. Math. Soc.* **74**, 207–223.

Girlich, H. J.

(1,1973)   *Diskrete Stochastische Entscheidungsprozesse.* Teubner.

Gnedenko, B. V.

(1,1950)   *The Theory of Probability.* Chelsea (reprinted 1963).

Gödel, K.

(1,1938)   The consistency of the axiom of choice and the generalized continuum hypothesis. *Proc. Nat. Acad. Sci. USA* **24**, 556.

Gubenko, L. G.

(1,1972)   Controlled Markov and semi-Markov models and some concrete optimization problems for stochastic systems. Ph.D. dissertation, Kiev.

Gubenko, L. G., and Shtatland, E. S.

(1,1972)   On Markov discrete time processes decision. *Theory Prob. and Math. Statistics* **7**, 51–64 (in Russian).

(2,1972)   Controllable semi-Markov processes. *Kibernetica* **2**, 26–29.

Halmos, P. R.

(1,1950)   *Measure Theory.* Van-Nostrand (reprinted by Springer-Verlag, 1974).

Hardy, G. H., Littlewood, J. E., Polya, G.

(1,1934)   *Inequalities.* Cambridge Univ. Press.

Hildenbrand, W.

(1,1974)   *Core and Equilibria of a Large Economy*. Princeton Univ. Press.

Hinderer, K.

(1,1970)   *Foundations of Non-stationary Dynamic Programming with Discrete Time Parameter*. Springer-Verlag.

(2,1971)   Instationäre dynamische Optimierung bei schwachen Voraussetzungen uber die Gewinnfunktionen. *Abh. Math. Sem. Univ. Hamburg* **36**, 208–223.

Hordijk, A.

(1,1974)   *Dynamic Programming and Markov Potential Theory*. Mathematisch Centrum.

Howard, R. A.

(1,1960)   *Dynamic Programming and Markov Processes*. Wiley.

(2,1971)   *Dynamic Probabilistic Systems*, v. 2: *Semi-Markov and Decision Processes*. Wiley.

Jorgenson, D. W., McCall, J. J., Radner, R.

(1,1967)   *Optimal replacement policy*. North-Holland.

Joseph, P. D., Tou, J. T.

(1,1961)   On linear control theory. *AIEE Trans.* **80**, 2: *Applications and Industry*. 193–196.

Kalman, R. E.

(1,1960)   A new approach to linear filtering and prediction problems. *J. of Basic Engineering*, **82D**, 35–44.

Kalman, R. E., Koepke, R. W.

(1,1958)   Optimal synthesis of linear sampling control systems using generalized performance indexes. *ASME Trans.* **80**, 1820–1826.

Kantorovich, L. V.

(1,1939)   Mathematical methods in the organization and planning of production. *Management Science* 6, 4, 366–422.

(2,1940)   A new method of solving some classes of extremal problems. *Dokl. Akad. Nauk. SSSR* **28**, 211–215 (in Russian).

Karlin, S.

(1,1955)   The structure of dynamic programming models. *Naval Res. Logistic Quart.* **2**, 285–294.

Khinchin, A.

(1,1935)   *Continued fractions*. Univ. of Chicago Press (reprinted 1964).

Kolmogorov, A. N., Fomin, S. V.

(1,1954)   *Elements of the Theory of Functions and Functional Analysis.* Graylock Press,
            1957.

Krylov, N. V.

(1,1964)   On the existence of ε-optimal homogeneous Markov strategies for a directed
            chain. *Soviet Math., Doklady* **5**, 497–502.

(2,1965)   Construction of an optimal strategy for a finite controlled chain. *Theory of
            Probability* **10**, 45–54.

(3,1977)   *Controlled Processes of Diffusion Type.* Nauka (in Russian).

Kuhn, H. W., Tucker, A. W.

(1,1951)   Nonlinear programming. *Proc. 2nd Berkeley Sympos. on Math. Stat. and Prob.*
            1950, 481–492.

Kuratowski, K.

(1,1966)   *Topology.* Academic Press.

Kuratowski, K., Ryll-Nardzewski, C.

(1,1965)   A general theorem on selectors. *Bull. Acad. Polon. Sci. Ser. Sci. Math. Astr.
            Phys.* **13**, 397–403. (in Polish).

Kushner, H.

(1,1971)   *Introduction to Stochastic Control.* Holt.

Kuznetsov, S. E.

(1,1974)   Weakly optimal programs in models with changing technology. *Mathematical
            Models in Economics.* 259–270. North-Holland

Luzin, N. N.

(1,1930)   *Leçons sur les ensembles analytiques et leur applications.* Gauthier-Villars.

Maitra, A.

(1,1965)   Dynamic programming for countable state systems. *Sankhya*, Ser. A **27**, 241–248.

(2,1968)   Discounted dynamic programming on compact metric spaces. *Sankhya*, Ser. A
            30, 211–216.

(3,1969)   A note on positive dynamic programming. *Ann. Math. Stat.* **40**, 316–318.

Martin, J. J.

(1,1967)   *Bayesian Decision Problems and Markov Chains.* Wiley.

Martin-Löf, A.

(1,1967)   Existence of a stationary control for a Markov chain maximizing the average
            reward. *Operations Res.* **15**, 866–871.

Meyer, P. A.

(1,1966)   *Probability and Potentials.* Blaisdell.

Mine, H., Osaki, S.

(1,1970)   *Markovian Decision Processes.* American-Elsevier.

Mirman, L. J.

(1,1971)   Uncertainty and optimal consumption decisions. *Econometrica* **39**, 179–185.

(2,1973)   The steady state behavior of a class of one-sector growth models with uncertain
            technology. *J. Econ. Theory* **6**. 219–242.

Moran, P. A. P.

(1,1959)   *The Theory of Storage.* Wiley.

Natanson, I. P.

(1,1950)   *Theory of Functions of a Real Variable.* Ungar (reprinted 1960).

Neveu, J.

(1,1965)   *Mathematical Foundations of the Calculus of Probability.* Holden-Day.

Nikaido, H.

(1,1968)   *Convex Structures and Economic Theory.* Academic Press.

Novikov, P. S.

(1,1951)   On the consistency of some propositions of the descriptive theory of sets. *Trudy
            Mat. Inst. Steklov* **38**, 279–316. (English translation: *Am. Math. Soc. Transla-
            tions, Ser.* 2, **29** (1963) 51–90).

Orkin, M.

(1,1974)   On stationary policies—the general case. *Ann. Stat.* **2**, 219–222.

Parthasarathy, K. R.

(1,1967)   *Probability Measures on Metric Spaces.* Academic Press.

Phelps, E. S.

(1,1962)   The accumulation of risky capital: a sequential utility analysis. *Econometrica*
            30, 729–743.

284

Radner, R.

(1,1967)   Dynamic programming of economic growth. *Activity in the Theory of Growth and Planning*, 111–141, MacMillan.

(2,1973)   Optimal stationary consumption with stochastic production and resources. *J. Econom. Theory* **6**, 68–90.

Ramsey, F.

(1,1928)   A mathematical theory of savings. *Econ. J.* **38**, 543–559.

Rhenius, D.

(1,1974)   Incomplete information in Markov decision models. *Ann. Stat.* **2**, 1327–1334.

Rockefellar, R. T.

(1,1970)   *Convex Analysis*. Princeton Univ. Press.

Romanovskii, I. V.

(1,1965)   Existence of an optimal stationary policy in a Markov decision process. *Theory of Probability* **10**, 120–122.

(2,1967)   Optimal stationary control of discrete deterministic processes. *Kibernetica* **2**, 66–78.

(3,1971)   Deterministic processes of dynamic programming with supplementary constraints. *Kibernetica* **5**, 68–71.

Ross, S. M.

(1,1968)   Non-discounted denumerable Markovian decision models. *Ann. Math. Stat.* **39**, 412–423.

(2,1968)   Arbitrary state Markovian decision process. *Ann. Math. Stat.* **39**, 2118–2122.

(3,1970)   *Applied Probability Models with Optimization Applications*. Holden-Day.

Saks, S.

(1,1937)   *Theory of the Integral*. Hafner.

Samuelson, P. A.

(1,1969)   Lifetime portfolio selection by dynamic stochastic programming. *Rev. Econ. and Stat.* Harvard Univ. **51**, 239–246.

Sawaragi, Y., Sunahara, Y., Nakamizo, T.

(1,1967)   *Statistical decision theory in adaptive control systems*. Academic Press.

Sawaragi, Y., Yoshikazu, T.

(1,1970)   Discrete time Markovian decision process with incomplete state observation. *Ann. Math. Stat.* **41**, 78–86.

Shapley, L. S.

(1,1953)   Stochastic games. *Proc. Nat. Acad. Sci. USA* **39**, 1095–1100.

Shiryaev, A. N.

(1,1964)   On the theory of decision functions and control by an observation process with
             incomplete data. *Trans. 3rd Prague Conf. on Inform. Theory etc.* 1962, 657–681.
             Publ. Czech. Acad. Sci., Prague.

(2,1967)   Some new results in the theory of controlled random sequences. *Trans. 4th
             Prague Conf. on Inform. Theory etc.* 1965, 131–203. Prague (in Russian).

Simon, H. A.

(1,1956)   Dynamic programming under uncertainty with a quadratic criterion function.
             Econometrica **24**, 74–81.

Solovay, R.

(1,1970)   A model of set theory in which every set of reals is Lebesgue measurable. *Ann.
             of Math.* (2) **92**, 1–56.

Strauch, R. E.

(1,1966)   Negative dynamic programming. *Ann. Math. Stat.* **37**, 871–890.

Striebel, Ch.

(1,1975)   *Optimal Control of Discrete Time Stochastic Systems.* Springer-Verlag.

Taksar, M. I.

(1,1974)   Optimal planning over infinite time interval under random factors. *Mathematical
             Models in Economics.* 289–298, North-Holland.

Taylor, H. M. III

(1,1965)   Markovian sequential replacement processes. *Ann. Math. Stat.* **36**, 1677–1694.

Theil, H.

(1,1957)   A note on certainty equivalence in dynamic planning. *Econometrica* **25**, 346–349.

Viskov, O. V., Shiryaev, A. N.

(1,1964)   On controls leading to optimal stationary models. *Trudy Mat. Inst. Steklov
             Akad. Nauk SSSR*, **71**, 35–45 (in Russian).

von Neumann, J.

(1,1937)   Ueber ein oekonomisches Gleichungssystem und eine Verallgemeinerung des
             Brouwerschen Fixpunktsatzes.–"Ergebnisse eines mathematischen Kollo-
             quiums" 1935–1936, 8, Leipzig, Wien.

(2,1949)   On rings of operators. Reduction theory. *Ann. Math.* (2) **50**, 401–485.

Wagner, H. M.

(1,1969)   *Principles of Management Science with Applications to Executive Decisions.* Prentice-Hall.

Wald, A.

(1,1947)   *Sequential Analysis.* Wiley.

(2,1950)   *Statistical Decision Functions.* Wiley.

Widder, D.

(1,1946)   *Laplace Transform.* Princeton Univ.

Yankov, V.

(1,1941)   Sur l'uniformization des ensembles A. *C. R. Acad. Sci.* URSS **30**, 597–598.

Yushkevich, A. A.

(1,1972)   On a class of strategies in general Markov decision models. *Theory of Probability* **18**, 777–779.

(2,1976)   Reduction of a controlled Markov model with incomplete data to a problem with complete information in the case of Borel state and control spaces. *Theory of Probability* **21**, 153–158.

(3,1977)   Controlled Markov models with countable state space and continuous time. *Theory of Probability* **22**, 215–235.

(4,1977)   Controlled jump-type Markov processes. *Soviet Math., Doklady* **18**, 351–355.

(5,1978)   Analytically measurable strategies in controlled jump Markov processes (in Russian). "Trans. 8th Prague Conf. on Inform. Theory etc." Vol. A, 335–341.

Yushkevich. A. A., Fainberg E. A.

(1,1979)   On homogeneous Markov models with continuous time and finite or countable state space. *Teoriya Veroyatnostei i ee Primeneniya* 24, 155–160 (in Russian).

Zvonkin, A. K.

(1,1971)   On sequentually controlled Markov Processes. *Math of the USSR, Sbornik* **86(128)**, 607–617.

Index

Grundlehren der mathematischen Wissenschaften

A Series of Comprehensive Studies in Mathematics

A Selection